Synthesis, Processing and Application of Micro and Nanostructured Materials

Synthesis, Processing and Application of Micro and Nanostructured Materials

Special Issue Editor

Bogdan Stefan Vasile

MDPI • Basel • Beijing • Wuhan • Barcelona • Belgrade • Manchester • Tokyo • Cluj • Tianjin

Special Issue Editor
Bogdan Stefan Vasile
POLITEHNICA University of
Bucharest
Romania

Editorial Office
MDPI
St. Alban-Anlage 66
4052 Basel, Switzerland

This is a reprint of articles from the Special Issue published online in the open access journal *Nanomaterials* (ISSN 2079-4991) (available at: https://www.mdpi.com/journal/nanomaterials/special_issues/Micro_Nanostructured).

For citation purposes, cite each article independently as indicated on the article page online and as indicated below:

LastName, A.A.; LastName, B.B.; LastName, C.C. Article Title. *Journal Name* **Year**, *Article Number*, Page Range.

ISBN 978-3-03928-967-7 (Hbk)
ISBN 978-3-03928-968-4 (PDF)

Cover image courtesy of Adrian Vasile Surdu.

Contents

About the Special Issue Editor

Bogdan Stefan Vasile received his M.S. degree in composite materials in 2008 and his Ph.D. in chemical engineering in 2011 from the University Politehnica of Bucharest, Romania. His Ph.D. research activities have focused on the synthesis and characterization of zirconia and alumina nanopowders and nanocomposites. Since 2018, he has been a researcher at University Politehnica in Bucharest, Faculty of Applied Chemistry and Materials Science, National Research Center for Micro and Nanomaterials. His current research interests involve the unconventional chemical synthesis, characterization, and multipurpose applications of advanced micro and nanomaterials. His main specialization is in the field of transmission electron microscopy (TEM), scanning transmission electron microscopy (STEM), energy dispersive X-Ray analysis (EDX), energy filtered transmission electron microscopy (EFTEM), and energy electron loss spectroscopy (EELS). His specializations in the field of microscopy and nanomaterials were achieved with the FEI Application Laboratory, Anadolu University and the Institute for Electron Microscopy of Graz University of Technology. His overall research activities have allowed him to publish over 165 ISI papers with an h-index of 27. His passion for microscopy led to the foundation of the Electron Microscopy Society in Romania as one of the founding members; he was also the first President. From 2008, he has been the Leader of the Structural Analysis and Electron Microscopy laboratory at UPB and Technical Director at both the National Research Center for Food Safety (CNpSA) and the National Research Center for Micro and Nanomaterials (CNMN).

Preface to "Synthesis, Processing and Application of Micro and Nanostructured Materials"

Modern technology is a fundamental part of various scientific fields based on devices or structures that use materials with special properties as essential components. Specific and attractive applications have advanced with the development of micro and nanostructured materials. The progress in the synthesis and processing of micro/nanomaterials has opened various possibilities. Materials, at the micro scale, possess various properties that are of interest for many applications. However, these materials have multiple specific characteristics of nanostructured materials. Nanomaterials have received considerable interest from scientists from different fields, such as chemistry, biochemistry, biology, pharmacy, physics, material science, and engineering. New advances in both the synthesis and processing of micro and nanostructured materials have a positive interdependence regarding the applications in which these materials are involved. This is constantly demonstrated by the action and properties of nanoparticles in applications such as biomedical science, the food and space industries, mechanics, energy science, drug delivery systems, and optoelectronics. These types of materials also have economic and environmental implications as their emphasizes the possibility of overcoming issues related to weight, size, biocompatibility, energy consumption, and complicated strategies. They also have less of an environmental impact, which is an aspect considered throughout the development of a material for new applications. The papers published in this Special Issue outline various advances enabled by the use of micro and nanostructured materials, highlighting the properties and applications that are achievable by altering the composition and morphology of nanoparticles. In this book, uses of these materials with various applications in different fields are outlined. For readers with an interest in the synthesis, processing, and application of micro and nanostructured materials, this book provides a comprehensive insight into the recent contributions of scientific knowledge.

Bogdan Stefan Vasile
Special Issue Editor

Article

Defect Structure Determination of GaN Films in GaN/AlN/Si Heterostructures by HR-TEM, XRD, and Slow Positrons Experiments

Vladimir Lucian Ene [1,2], Doru Dinescu [2,3,*], Nikolay Djourelov [2], Iulia Zai [2,4],
Bogdan Stefan Vasile [1], Andreea Bianca Serban [2,3], Victor Leca [2] and Ecaterina Andronescu [1]

1   Department of Science and Engineering of Oxide Materials and Nanomaterials, Faculty of Applied
Chemistry and Materials Science, University Politehnica of Bucharest, 060042 Bucharest, Romania;
vladimir.ene@upb.ro (V.L.E.); bogdan.vasile@upb.ro (B.S.V.); ecaterina.andronescu@upb.ro (E.A.)

2   Extreme Light Infrastructure-Nuclear Physics (ELI-NP), 'Horia Hulubei' National R&D Institute for Physics
and Nuclear Engineering (IFIN-HH), 30 Reactorului Street, 077125 Măgurele, Romania;
nikolay.djourelov@eli-np.ro (N.D.); iulia.zai@eli-np.ro (I.Z.); andreea.serban@eli-np.ro (A.B.S.);
victor.leca@eli-np.ro (V.L.)

3   Doctoral School in Engineering and Applications of Lasers and Accelerators,
University Politehnica of Bucharest, 060042 Bucharest, Romania

4   Faculty of Physics, University of Bucharest, 077125 Măgurele, Romania

*   Correspondence: doru.dinescu@eli-np.ro

Received: 24 December 2019; Accepted: 19 January 2020; Published: 23 January 2020

Abstract: The present article evaluates, in qualitative and quantitative manners, the characteristics
(i.e., thickness of layers, crystal structures, growth orientation, elemental diffusion depths, edge, and
screw dislocation densities), within two GaN/AlN/Si heterostructures, that alter their efficiencies as
positron moderators. The structure of the GaN film, AlN buffer layer, substrate, and their growth
relationships were determined through high-resolution transmission electron microscopy (HR-TEM).
Data resulting from high-resolution X-ray diffraction (HR-XRD) was mathematically modeled to
extract dislocation densities and correlation lengths in the GaN film. Positron depth profiling was
evaluated through an experimental Doppler broadening spectroscopy (DBS) study, in order to quantify
the effective positron diffusion length. The differences in values for both edge $\left(\rho_d^e\right)$ and screw $\left(\rho_d^s\right)$
dislocation densities, and correlation lengths (L^e, L^s) found in the 690 nm GaN film, were associated
with the better effective positron diffusion length (L_{eff}) of $L_{eff}^{GaN2} = 43 \pm 6$ nm.

Keywords: gallium nitride; epitaxial thin films; dislocations; positron diffusion length

---

1. Introduction

Binary semiconductors, such as InN, AlN, GaAs, InAs, InP, GaN, AlSb, etc., and their alloys,
cover an extended range of structures useful in high-end device technology [1,2]. Due to the direct
bandgap that most of these materials possess, efficient emission and absorption of light is allowed.
Many binary compounds also exhibit a very low electron effective mass, thus a high mobility, which
makes them ideal candidates for developing high-speed devices [3]. Among these compounds, GaN has
shown impressive advantages. Because of its geometric and electronic structure made up of covalent
bonds between Ga and N, the wide energy band gap allows it to reach operating temperatures higher
than 350 °C [2]. A second advantage is the high mobility (>1200 cm^2 V^{-1} s^{-1}) of the two-dimensional
electron gas (formed at interfaces with e.g., AlN) that leads to low channel resistance and high current
density (>1 A mm^{-1}), and a breakdown field of 3.3 MV cm^{-1} that is 11 times higher than that of
silicon (0.3 MV cm^{-1}) [4,5]. GaN is widely used in applications that require either n-type or p-type
doped semiconductors for charge carrier injection in different devices [6]. New methods of obtaining

Ga based films using liquid Ga [7,8] for reactive depositions have emerged in recent years and the fundamentals behind liquid metal enabled synthesis, along with the related surface functionalization aspects [9] showed promising possibilities concerning the growth of GaN thin films. Despite this, the fabrication of defect-free GaN films still possesses interest in some fields, such as field assisted positron moderation [10].

Positron annihilation lifetime spectroscopy and Doppler broadening spectroscopy (DBS) have become the most used positron annihilation derived spectroscopy techniques suitable for non-destructive determinations of near surface crystallographic vacancies and dislocations in lattices, as well as optical and electronic properties of materials due to the high affinity of positrons to defects. Irrespective to the method used to obtain them, positrons manifest a broad energy distribution of about several hundreds of keV. In order to use the above-mentioned spectroscopy techniques for thin-film studies, positrons need to be moderated. The way to achieve this is to convert the fast positrons to slow positrons (with a low kinetic energy of few eV and a narrow bandwidth) by using a moderator material with negative work function for positrons (e.g., W or solid Ne) [11,12]. By varying the kinetic energy of the slow positrons, the depth at which they are implanted can be controlled [11]. The negative positron work function and the adequate branching ratio makes GaN a very promising candidate for field assisted positron moderation. A long positron diffusion length is expected due to the wide 3.4 eV bandgap. GaN studies have been undertaken and measurements have yielded values for the diffusion length of 19.3 ± 1.4 nm, surface branching ratio to free positrons of 0.48 ± 0.02 and positron work function of −2.4 ± 0.3 eV, respectively [13]. The moderator efficiency, usually smaller than 10^{-2}, is greatly reduced by atomic scale defects which can trap positrons.

GaN-based devices still encounter several obstructing issues, including high defect density and strain-induced polarization. In order to reduce the effects of these issues, a series of approaches were proposed in the last decade [1]. In the early stages, the main efforts were focused on improving both the qualities of the materials and the structuring of the device. The advanced growth techniques enabled management of the nanostructured layer interfaces, further enhancing the quantum efficiencies of devices. Substrates have a big influence on the growth mode and the final physical and chemical properties, determining the surface morphology, polarity, crystal orientation, composition, and elastic strains. When choosing a substrate, one of the most important criteria used is the mismatch parameter between the substrate and the deposited film. Lateral mismatch of lattices leads to a decrease of the thermal conductivity and accelerated diffusion of impurities. Vertical asymmetry causes a counter-phase interface. Thermal strain is induced in the film by the discrepancy between the thermal conductivities coefficients of substrates with respect to the epitaxial film. Chemical composition differences cause a contamination of the film which forms unstable electronic bonds and a mixed polarity that appears in the epitaxial film when the surface of the substrate is nonpolar [14]. Current reports of producing GaN films indicate that heteroepitaxial GaN films can be grown on different substrates such as Si, Al_2O_3, ZnO, TiO_2, SiC, with different orientations [15]. The stable phase of gallium nitride is the α-phase wurtzite structure. However, epitaxial layers can be achieved with the coexistence of wurtzite and zinc-blende (β-phase) phases due to the stacking sequence of nitrogen and gallium atoms. Both structures have polar axes and they do not have an inversion symmetry [16].

The aim of this study is to assess the quality of commercially available GaN epitaxial thin films, grown on Si, for their potential use as positron moderators. High-resolution transmission electron microscopy (HR-TEM) and high-resolution X-ray diffraction (HR-XRD) were performed in order to determine the GaN films' defect structures. The features of the heterostructures, such as layer thicknesses, interfaces, elemental diffusion, and dislocations were correlated with the effective positron diffusion lengths, evaluated by slow-positron DBS studies.

2. Materials and Methods

2.1. Materials

Two gallium nitride, GaN, thin films grown using an epitaxial growth technique on Si substrates were used in this study. The wafers were acquired from NTT Advanced Technology Corporation (Kanagawa, Japan) and are defined by high uniformity, high breakdown voltage, a sheet carrier density of approximatively 10^{13} cm^{-2}, and an electron mobility of over 2000 cm^2 V^{-1} s^{-1}. The two wafers, were further labeled as GaN300/Si and GaN700/Si, where the number stands for the claimed thickness of the GaN film, expressed in nm. No further details on structure, defects, and interfaces were made available by the producer.

2.2. Structural Analysis

2.2.1. Microstructural Characterization

The microstructure of the wafers was studied with the help of a Titan Themis 200 image corrected transmission electron microscope (FEI, Hillsboro, OR, USA), equipped with a high-brightness field emission gun (X-FEG) electron source and a Super-X detector for energy dispersive spectroscopy (EDS). The heterostructures were investigated at 200 kV by HR-TEM, coupled with selected area electron diffraction (SAED) and scanning transmission electron microscopy (STEM) for elemental line profiling. Prior to analysis, the wafers were mechanically polished and then ion beam milled at a voltage of 3 kV and current of 5 mA until perforation. Ion-beam milling was continued with decrements of voltage and current, in order to remove debris produced by the high voltage ion beam thinning.

For processing the elemental line profiles from EDS data, ImageJ software was used [17]. The visualization and analysis of crystal structures were made with SingleCrystal® (Oxford, England), and images of simulated crystals were generated using CrystalMaker®, a software by CrystalMaker Software Ltd., Oxford, England [18].

2.2.2. Defect Structure Determination

HR-XRD analysis was performed using a 9 kW Rigaku SmartLab diffractometer (Neu-Isenburg, Germany), with a rotating Cu anode (K_α = 1.5418 Å) and a HyPix-3000 high-resolution detector (Rigaku, Neu-Isenburg, Germany), in 0D mode. The data (ω—rocking curves of selected symmetrical and asymmetrical reflections) were recorded in double-axis configuration, in the parallel beam mode, using a parabolic mirror (cross beam optics module) and a four bounce Ge-220 monochromator (Rigaku, Neu-Isenburg, Germany), resulting in an axial divergence of the beam of 0.003° in the vertical diffraction plane of the goniometer. A narrow incidence slit of 1 mm was used to avoid the effect of sample curvature on the measurements. On the detector side, receiving slits (RS) of RS1 = 4 mm, and RS2 = 38.5 mm were used (open detector configuration), so that all diffuse scattering from the sample was accounted for. The wafers were first aligned with respect to the Si substrate, in order to avoid any measurement errors due to sample misalignment, then the rocking curve measurement of the selected GaN planes was performed.

The recorded data was processed using the theoretical model developed by Kaganer et al. [19], using an integral of the form:

$$I(\omega) = \frac{I_i}{\pi} \int_0^\infty \exp\left(-Ax^2 \ln\left(\frac{B+x}{x}\right)\right)\cos(\omega x)dx + I_{backgr} \tag{1}$$

where I_i is the integrated peak intensity and I_{backgr} is the background intensity. The A and B parameters were obtained by integral fitting on the experimental data. A and B describe the dislocation density and the dislocation correlation range, respectively, and can be expressed as:

$$A = f\rho_d b^2; \; B = \frac{gL}{b} \tag{2}$$

where b is the Burgers vector, ρ_d is the dislocation density, L is the dislocation correlation length, f and g are two dimensionless parameters which depend on the skew geometry of the diffraction setup:

$$f^e = \frac{0.7 \cos^2 \psi \cos^2 \phi}{4\pi \cos^2 \theta_B}; \; f^s = \frac{0.5 \sin^2 \psi \cos^2 \phi}{4\pi \cos^2 \theta_B}; \; g^e = \frac{2\pi \cos \theta_B}{\cos \phi \, \cos \psi}; \; g^s = \frac{2\pi \cos \theta_B}{\cos \phi \, \sin \psi} \tag{3}$$

where ψ is the angle between the sample surface and the scattering vector, ϕ is the angle between either incident or diffracted vector and the sample surface, and θ_B is the Bragg angle at which the diffraction interference takes place, according to the geometry described in Ref. [19]. Both f and g can be computed so that the density of dislocations, as well as the characteristic dislocation correlation length, can be obtained for either edge or screw defects, marked by the superscripts "e" and "s" in Equation (3). For edge dislocations, an asymmetrical lattice plane of the GaN network was considered, while for screw dislocations, a symmetrical plane of the same sample was used. For symmetric Bragg reflections (so, for screw dislocations), the setup implies that $\psi = \pi/2$ and $\phi = \theta_B$, resulting in $f = 1/8\pi$ and $g = 2\pi$, respectively [19].

2.3. Doppler Broadening Spectroscopy

With a great probability, the annihilation of a positron with an electron in condensed matter is followed by the emission of two gamma rays of energy $E_\gamma \approx 511$ keV. The longitudinal component of the annihilation pair momentum, p_L, determines the energy shift due to Doppler broadening, $\Delta E_\gamma = 511 - E_\gamma = p_L c/2$, where c is the speed of light. The Doppler broadening spectra of the annihilation radiation are sensitive to the electron momentum distribution of the site where the positron annihilated, since, the momentum distribution of the electrons in defects differs from that of electrons in the bulk material [20].

The DBS experiments were performed at the slow positron beam line of the Institute of High Energy Physics in Beijing, China. The gamma energy spectra were recorded by a HPGe detector (ORTEC, Zoetermeer, Netherlands), with a resolution of FWHM (full width at half maximum) = 0.97 keV estimated for 511 keV line. The detector was placed perpendicularly in respect to the positron beam axis, at a distance of 20 cm from the sample. The incident positron energy was controlled from $E_+ = 0.5$ to 25 keV. Each of the experimental spectra was collected over a period of 8 min for a fixed E_+, resulting in statistics of ~5×10^5 counts in the 511 keV region. The shape of the annihilation peak was analyzed by the sharpness parameter, S, defined as the sum of counts, in the central region of the peak ($|\Delta E_\gamma| < 0.78$ keV), relative to the total peak counts (N_{tot}), determined in the range between 500 and 522 keV. The triplet state of positronium (Ps) decays by emitting 3-gamma rays when it does not interact with the electrons of the material. The ratio, F_{Ps}, between the counts in the valley region (from 450 to 500 keV) in the energy spectrum to N_{tot} can give a relative estimate of the Ps emitted from the surface.

The implantation profile of positrons in a material with density ρ in g cm^{-3} can be described, according to Ref. [21], by:

$$P(z, E_+) = \frac{2z}{z_0} \exp\left(-\left(\frac{z}{z_0}\right)^2\right) \tag{4}$$

where z is the depth at which the positron is located, expressed in nm, $z_0 = 1.13 \, z_m$, and the mean penetration depth is

$$z_m = (36/\rho)E_+^{1.62} \text{ nm} \tag{5}$$

Different layer densities are taken into account by using the modified positron implantation profile described by:

$$P_\rho(z_\rho, E_+) = \rho(z_\rho)/\rho_0 P(z, E_+) \tag{6}$$

with $= \int_0^{z_\rho} \rho(\zeta)/\rho_0 d\zeta$, where ρ_0 is the density of the substrate. In the analysis of the experimental data, densities of 2.33, 3.26 and 6.15 g cm^{-3} were used for the Si substrate, AlN buffer layer, and GaN film, correspondingly.

Due to the correlation between the mean positron implantation depth, z_m, and E_+, the experimental data $S(E_+)$ and $F_{Ps}(E_+)$ represents depth profiles. The VEPFITsoftware (Delft University of Technology, Delft, Netherlands) was used to fit the experimental data [22]. In addition to the implantation, the processes that have to be taken into account to solve the positron transport problem are diffusion, drift (in case of electric field), and trapping or annihilation of free positrons. Surface related processes, such as Ps emission and positron surface trapping, are incorporated within the model. The influence of epithermal positrons, and that of thermal positrons which diffuse back to the surface, is also taken into account in the VEPFIT software.

The $S(E_+)$ is fitted using a model described by:

$$S(E_+) = S_e F_e(E_+) + S_s F_s(E_+) + \sum S_i \, F_i(E_+) \tag{7}$$

with $F_e(E_+) + F_s(E_+) + \sum F_i(E_+) = 1$, where $F_e(E_+)$ is the fraction of epithermal positrons annihilated at the surface, and $F_s(E_+)$ and $F_i(E_+)$ are the fractions of thermalized positrons annihilated at the surface and in the i-th layer. S_e, S_s, and S_i are characteristic parameters, respectively, corresponding to the annihilation of epithermal positrons and of thermalized positrons at the surface and in the bulk of i-th virtually uniform layer. VEPFIT uses discretization as a fast method of solving numerically the positron transport problem to obtain the fractions of annihilated positrons from the above described states. One of the parameters which is derived from the fit is the effective positron diffusion length (L_{eff}) for each layer. L_{eff} is limited by the layer defects and is described by:

$$L_{eff} = \left[D^+ / (k_t n_t + \lambda_b) \right] \tag{8}$$

where D^+ is the positron diffusion coefficient, λ_b is the annihilation rate of positrons in a defect-free material, and the product between the defect density, n_t, and the positron trapping rate, k_t, for vacancies, usually holds the value of 10^{15} s^{-1}.

Often, the information of the Ps emission from the surface, as derived from $F_{Ps}(E_+)$, is useful in the interpretation of the experimental results. Both depth profiles $S(E_+)$ and $F_{Ps}(E_+)$ can be fitted simultaneously by one and the same VEPFIT model.

3. Results and Discussion

3.1. Microstructural Characterization

3.1.1. TEM

Upon analyzing the structure of the two wafers, the existence of an AlN buffer layer was acknowledged. Such a buffer layer has the purpose of accommodating the GaN network to that of the Si substrate, thus decreasing the film strain and the amount of defects that would be generated during film growth due to lattice mismatch [23]. Although the lattice mismatch between GaN and Si (11$\bar{2}$1) is lower (16.9%) than in the case of AlN and Si (11$\bar{2}$1) (18.9%), the use of an AlN buffer layer is still recommended due to the low mismatch between AlN and GaN (2.4%) that can ultimately lead to a lower amount of defects in the final GaN film [24].

The TEM and SAED images in Figure 1 show the display of planes near the Si/AlN interface and near the AlN/GaN interface in the GaN300/Si and GaN700/Si samples.

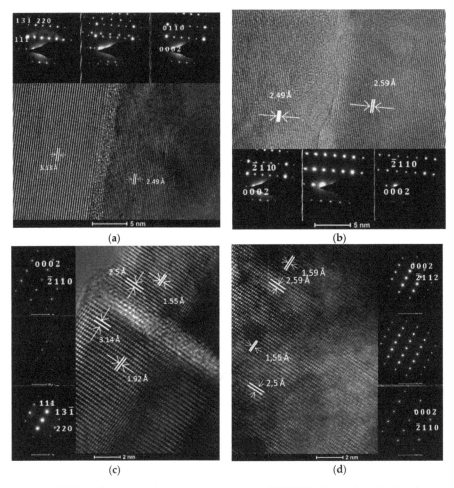

Figure 1. High-resolution transmission electron microscopy (HR-TEM) micrographs and selected area electron diffraction (SAED) patterns showing the display of atom planes in respect to their respective interfaces for: (**a**) GaN300/Si–Si/AlN interface, (**b**) GaN300/Si–AlN/GaN interface, (**c**) GaN700/Si–Si/AlN interface, (**d**) GaN700/Si–AlN/GaN interface.

Regarding the substrates from both wafers, the interplanary distance of 3.13 Å, corresponding to (11$\bar{2}$1) planes, confirm the $Fd\bar{3}m$ diamond-like cubic structure of Si (International Centre for Diffraction Data [ICDD] 00-005-0565), whereas the interplanary distance of 2.49 Å, corresponding to (0 0 0 2) planes, confirmed the $P63mc$ hexagonal structure of the AlN (ICDD 00-025-1133) buffer layer. Literature studies revealed that GaN has a better affinity to grow on Si (11$\bar{2}$1) rather than Si (0 0 0 1) because of the threefold symmetry of the Si (11$\bar{2}$1) and the six-fold arrangement for Si atoms that are present in the case of growing AlN/GaN along the (0 0 0 2) direction [25]. From the SAED patterns in Figure 1a,c, it can be deduced that through a semi-coherent interface of about 1 nm, containing point defects and a low degree of crystallinity, hexagonal AlN grew over cubic Si, with the relationship $Fd\bar{3}m$ Si (11$\bar{2}$1) ∥ (0 0 0 2) AlN $P63mc$. Interplanary distances of 2.49 Å, corresponding to (0 0 0 2) planes, highlight once more the $P63mc$ hexagonal structure of AlN. Regarding the film, interplanary distances of 2.59 Å, corresponding to (0 0 0 2) planes, confirm the $P63mc$ hexagonal structure of GaN (ICDD 00-050-0792). From the SAED patterns in Figure 1b,d, it can be deduced that through an interface containing linear dislocations, hexagonal GaN grew over hexagonal AlN with the relationship *P63mc* AlN (0 0 0 2) ‖ (0 0 0 2) GaN *P63mc*. In comparison with the thinner GaN film (GaN300/Si), the thicker one (GaN700/Si), although it possesses the same crystallographic relationship relative to the buffer layer, shows fewer point defects and linear dislocations at the interface, most likely due to the higher amount of time needed to deposit a thicker film, during which, the sample is kept, in the manufacturing process, at a temperature that favors the dislocation movement under thermal stress [26].

Simulated crystal models, based on the SAED patterns are presented in Figure 2, as overlays on the HR-TEM micrographs of the interfaces.

Figure 2. HR-TEM micrographs with display of simulated crystal lattices near the interface between Si substrate and AlN buffer layer in (**a**) GaN300/Si, (**c**) GaN700/Si and between AlN buffer layer and GaN film for (**b**) GaN300/Si, (**d**) GaN700/Si.

In order to assess layer thicknesses and elemental diffusion length, TEM, STEM and EDS were performed, which are highlighted in Figure 3.

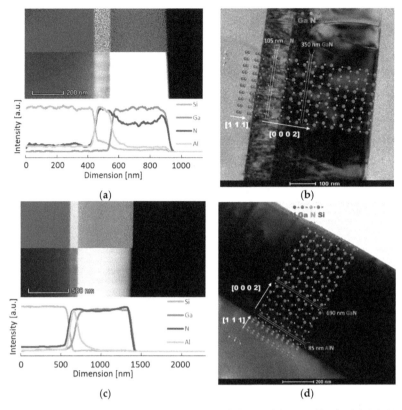

Figure 3. STEM micrographs with EDS mapping and elemental line profiles for (**a**) GaN300/Si, (**c**) GaN700/Si, and TEM micrographs showing the overview of the two wafers, (**b**) and (**d**), respectively.

The STEM study allowed assessing layer thicknesses for both samples, the thinner one having a 350 nm GaN film and a 105 nm AlN buffer layer, whereas for the thicker sample, a 690 nm GaN film and an 85 nm AlN buffer layer were found. For both samples, the epitaxial growth relationship can be described by the relationship: GaN *P63mc* (0 0 0 2) ∥ *P63mc* AlN (0 0 0 2) ∥ (11$\bar{2}$1) *Fd$\bar{3}$m* Si.

From the EDS maps and elemental line profiles presented in Figure 3a,c, it can be seen that both samples manifested Al diffusion at the edge of the GaN layer. Because the film remained highly crystalline and the interface between GaN/AlN is highly coherent, the diffusion of Al is most likely due to the native defects mediated by Al displacements either during high-temperature annealing of the film [27] or a film growth process that involves high temperatures.

3.1.2. XRD

In order to assess the threading dislocation density and correlation length of the GaN films, two pairs of rocking curves (ω scans) were measured: one for the (0 0 0 4) plane, to assess the screw values ρ_d^s and L^s, and another one for the (10$\bar{1}$5) plane, to determine the edge characteristics ρ_d^e and L^e. The collected and simulated omega scans, along with their respective full width at half maximum (FWHM), are shown in Figure 4.

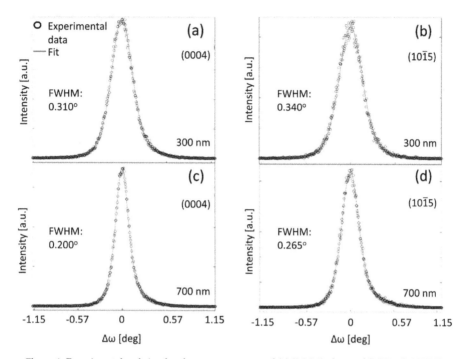

Figure 4. Experimental and simulated omega scans around (**a**) (0 0 0 4) planes of GaN in GaN300/Si, (**b**) (10$\bar{1}$5) planes of GaN in GaN300/Si, (**c**) (0 0 0 4) planes of GaN in GaN700/Si, (**d**) (10$\bar{1}$5) planes of GaN in GaN700/Si. Abbreviations: FWHM, full width at half maximum.

The experimental advantage of the open detector consists in obtaining high intensities, while the mathematical interpretation implies a simpler intensity distribution that can be described using a one-dimensional integral. While the model usually applies for any diffraction maximum, separation between two peaks that are in close proximity of each other requires a triple-axis configuration of the diffractometer, with analyzer at the detector side, making the mathematical processing in this case more complex, requiring a fit of a two-dimensional integral [19]. The (0 0 0 4) and (10$\bar{1}$5) planes are chosen because of lack of overlapping diffraction peaks near the respective ω (coming from either the buffer layer or substrate).

The length of the Burgers vector of edge dislocations was b^e = 0.32 nm and for screw dislocations - b^s = 0.52 nm. Parameters f^e and g^e, f^s, and g^s are calculated with Equation (3), and with the help of the extracted A and B, a series of threading dislocation densities and correlation lengths were calculated, the results being summarized in Table 1. The total threading dislocation density, ρ_d^t, is calculated as the sum of the two component densities (screw and edge), while the mean distance between two dislocations is given by $r_d = 1/(\rho_d^t)^{1/2}$ [28]. As shown in Table 1, the thicker GaN film manifests defect densities ρ_d^e = 2.24 × 10^{11} cm^{-2} and ρ_d^s = 1.35 × 10^{10} cm^{-2}, both lower than those of the thinner one, ρ_d^e = 4.19 × 10^{11} cm^{-2} and ρ_d^s = 1.85 × 10^{10} cm^{-2}. In the GaN700/Si wafer, the values for the dislocations correlation lengths, L^e = 41 nm and L^s = 220 nm, are higher compared to the corresponding values of the GaN300/Si wafer, L^e = 27 nm and L^s = 107 nm. The dislocation correlation length, also known as screening range, corresponds to the average size of cells in which the total Burger vector is equal to zero. The correlation lengths values suggest a reduced scattering of X-rays for the GaN film from the GaN700/Si wafer, also indicated by the smaller values of the FWHM, depicting a better quality of the film.

Table 1. Dislocation densities and correlation lengths for GaN in the GaN300/Si and GaN700/Si samples. The uncertainty of the presented values is within the least significant digit.

Sample	ρ_d^e [cm^{-2}]	ρ_d^s [cm^{-2}]	ρ_d^t [cm^{-2}]	r_d [nm]	L^e [nm]	L^s [nm]
GaN300/Si	4.19×10^{11}	1.85×10^{10}	4.37×10^{11}	15	27	107
GaN700/Si	2.24×10^{11}	1.35×10^{10}	2.35×10^{11}	21	41	220

3.2. Positron Implantation Profile

The depth profiles $S(E_+)$ for the GaN300/Si and GaN700/Si are shown in Figure 5. The sharp initial decrease of S for $E_+ \lesssim 1$ keV was due to annihilated epithermal positrons. Because of their high kinetic energy in the moment of annihilation, S_e did not reflect the material structure. At $E_+ \gtrsim 1$ keV, it can be seen that S slowly increased with E_+ in the GaN film range, while approaching the AlN buffer layer, a stronger increase starts (at $E_+ \gtrsim 11$ keV) and tends to reach a saturation level in the Si substrate (better seen in Figure 5b). Full saturation can be expected at high enough energies ($E_+ > 25$ keV) to have all implanted positrons annihilated entirely in the Si substrate.

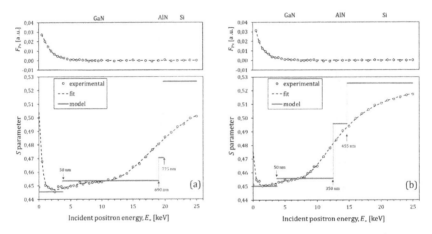

Figure 5. Plotted depth profiles $S(E_+)$ of (**a**) GaN700/Si and (**b**) GaN300/Si. The experimental errors are in the order of the experimental point size. The stairs represent the best parameters obtained by the fit of a 4-layer model to the experimental data by the VEPFIT software. The upper part of figure is the experimental data and the best fit of the relative Ps fraction, $F_{Ps}(E_+)$.

Based on the TEM information for the wafer layers, a three-layer (GaN, AlN, and Si) model was applied to fit the experimental S parameter by VEPFIT. The effective positron diffusion length in the Si substrates was fixed to 245 nm in accordance to available literature data [29,30]. The thicknesses of layers in the model were fixed to the values determined by the TEM analysis (see Section 3.1.1). The boundary depths of the layers were calculated by Equation (5) and indicated in Figure 5. These preliminary fits, for both samples, resulted in normalized chi squares (χ^2) of 1.47 and 1.40 for GaN700/Si and GaN300/Si, respectively, and also, into long $L_{eff} \sim 100$ nm for the GaN film in both cases. The curves of the preliminary fits were very close to the fits showed in Figure 5. However, the best fit parameters revealed that the $S_s \sim 0.445$ was found to be lower than $S_{GaN} \sim 0.455$ (specific to positrons annihilated in GaN film). No Ps was formed in the bulk of GaN, however, at the surface, the branching ratio showed that 12% of the positrons formed Ps [13]. The triplet state of Ps (*o*-Ps) annihilates in vacuum into three gamma rays that do not contribute to the 511-keV peak. Statistically, 25% of Ps is singlet form (*p*-Ps). The *p*-Ps annihilation in vacuum was characterized by a narrow Doppler shift distribution curve, thus, with a high S [31]. For sample-detector longitudinal geometry, the emission of Ps at low incident positron energy may cause asymmetry in the Doppler broadened peak [32]. It is caused by a shift of the centroid of the *p*-Ps contribution and may lead to an increase of the annihilation peak width. However, for the geometry described in Section 2.3, if Ps is emitted from the surface, the centroid of the *p*-Ps contribution will not be shifted. Therefore, S_s should have a larger value than S_{GaN}. As this relationship is not fulfilled for the preliminary fit results, it can be concluded that this type of fit is physically incorrect. Our attempts to force S_s to be greater (or at least equal) than S_{GaN}, without any increase in the number of layers of the model, led to bad fits with $\chi^2 > 7$. The latter indicates depth inhomogeneity of the GaN film.

In a defect-free GaN, produced by metalorganic vapor phase epitaxy, *S* decreased smoothly with the increase of E_+, which is typical for materials with long effective positron diffusion length (reported as $L_{eff}^{DF} = 135$ nm) [33]. For Mg-doped p-type GaN, a sharp decrease in *S* was observed for $E_+ \lesssim 1$ keV, similarly to what can be seen in Figure 5. Uedono et al. explain this behavior by the created local electric field due to band bending near the surface, which suppresses the back diffusion of the thermalized positrons to the surface [34]. As a result, less Ps is formed at the surface by the thermalized positrons, and the positron diffusion length is shortened in a near surface layer. Based on the above reasoning, the number of the layers of the fitting model is changed by splitting the GaN film into sublayers (GaN1 and GaN2). In order to have a better understanding of the near surface positron annihilation, simultaneous fit of $S(E_+)$ and $F_{Ps}(E_+)$ were performed by VEPFIT. Reasonable fits (see Figure 5) were obtained with two sublayers of the GaN film (4-layer model). The best fit parameters are summarized in Table 2.

Table 2. Best fit parameters obtained by VEPFIT from the $S(E_+)$ and $F_{Ps}(E_+)$ depth profiles. The values without error margins are fixed parameters.

Sample		GaN300/Si $\chi^2 = 1.15$			GaN700/Si $\chi^2 = 1.73$		
Layer		L_{eff} [nm]	S	d [nm]	L_{eff} [nm]	S	d [nm]
	Sublayer						
GaN	GaN1	14.3 ± 0.5	0.4501 ± 0.0006	50	13.1 ± 0.4	0.4456 ± 0.0004	50
	GaN2	22 ± 6	0.4558 ± 0.0004	300	43 ± 6	0.4536 ± 0.0003	640
AlN		26 ± 10	0.4957 ± 0.0019	105	4 ± 33	0.4707 ± 0.0032	85
Si		245	0.5254 ± 0.0005	-	245	0.5264 ± 0.0011	-

In a material in which Ps is not formed in the bulk, a higher *S* parameter means either more defects or bigger defects [34]. Also, more defects or more efficient positron trapping by defects (occurs for bigger defects) will result, according to Equation (4), in shorter L_{eff}. For GaN700/Si, the value $S_{GaN1} = 0.4456 \pm 0.0004$ is lower than $S_{GaN2} = 0.4456 \pm 0.0004$ and this relationship indicates lower quality of the GaN2 sublayer compared to GaN1. The fact that $L_{eff}^{GaN1} = 13.1 \pm 0.4$ nm is shorter than $L_{eff}^{GaN2} = 43 \pm 6$ nm seems to contradict the latter statement. However, the short L_{eff}^{GaN1} can be explained by the presence of local electric field directed inward the surface. Using the detector resolution and the *S* determination range given in Section 2.3, the characteristic parameter for *p*-Ps annihilation, $S_{p\text{-}Ps}$, was estimated to be 0.95. If the branching ratio of Ps formation by thermalized positrons on the surface is 12% [13], the *p*-Ps annihilation contribution will be 3%. This should lead to an 0.028 increase in S_s, compared to S_{GaN1}. As can be seen in Figure 5b, the S_s (see the parameter's stair at $E_+ = 0$) was very close to S_{GaN1}, indicating strong reduction in the Ps formation due to back-diffusion of thermalized positrons to the surface. The results for GaN300/Si in Figure 5a can be explained analogously.

For both samples, L_{eff}^{GaN2} (see Table 2) are shorter than the defect free value $L_{eff}^{DF} = 135$ nm. Saleh and Elhasi [35] suggested that the observed low $L_{eff} < 60$ nm values of positron diffusion length in GaN are due to positron interaction with dislocations. The dislocations can shorten L_{eff} by enhanced scattering of thermal positrons on them, and while vacancies tend to reside along them, they induce negative charge densities [36], trapping positrons more efficiently. In the case of the present study, the TEM analysis and the XRD defect assessment pointed out higher dislocation densities in the GaN film of the GaN300/Si wafer compared to GaN700/Si wafer (see Table 1). This is in agreement with the shorter effective positron diffusion length $L_{\text{eff}}^{\text{GaN2}} = 22 \pm 6$ nm (higher $S_{\text{GaN2}} = 0.4558 \pm 0.0004$) in the GaN300/Si wafer, compared to $L_{\text{eff}}^{\text{GaN2}} = 43 \pm 6$ nm ($S_{\text{GaN2}} = 0.4536 \pm 0.0003$) in GaN700/Si wafer. Another explanation for the last relationships could be that the highly defect GaN/AlN interface region in the GaN film, has stronger influence on the S_{GaN2} and $L_{\text{eff}}^{\text{GaN2}}$ for the thinner GaN film.

It is important to note that the above VEPFIT analysis, summarized in Table 2, was done without considering any electric field or interface layers (interpenetrating) neither between GaN and AlN nor between AlN and Si. The presence of such interpenetrating interface layers were derived from the elemental line profiles in Figure 3 (better seen in Figure 3a). The significant polarization, due to charge densities present at semiconductor heterojunction interfaces, creates an internal electric field which influences the positron transport trough the heterojunction interface. The mechanism of forming the potential barrier is due to the equalization of the Fermi levels of the two materials by charge transfer [37]. The positrons cannot diffuse equally well in both directions across such interface. For example, the diffusing positrons in the GaN layer are pushed back by the potential barrier at the interface, while these which diffuse in AlN will fall in a well at the barrier. The positrons tend to localize at the well barrier interface in a nitride heterostructure, as shown by theoretical calculations [38], and observed experimentally for GaN/SiC hetrojunction [39]. These effects can lead to a wrong estimation by VEPFIT analysis of both thicknesses of the layers and effective positron diffusion lengths. However, in our case, the $S(E_+)$ points increased rather smoothly with the increase of E_+ in the region of the AlN buffer layer (see Figure 5a,b) and the specific parameters for the AlN (see Table 2) are determined by large uncertainties even with no electric field. So, further complications of the model are not reasonable to be applied.

4. Conclusions

Two commercially available GaN/AlN/Si wafers were characterized by means of TEM and XRD in order to assess the relationship of the heterostructures characteristics (i.e., thickness of layers, crystal structures, preferred orientation growth, elemental diffusion, edge, and screw dislocation densities) with the positron diffusion depths, evaluated by DBS studies. Although epitaxial films show, in general, a high periodicity in the crystal lattice, there are inevitable defects that are bound to appear due to lattice mismatch between substrates, buffer layers, and films. Hence, within the epitaxial layers, defined by a [GaN $P6_3mc$ (0 0 0 2) || $P6_3mc$ AlN (0 0 0 2) || (11$\bar{2}$1) $Fd\bar{3}m$ Si] relationship, a correlation between elemental diffusion, dislocation densities, and positron depth profiles was assessed. XRD dislocation evaluations pointed out higher density of dislocations in the GaN300/Si wafer ($\rho_d^e = 4.19 \times 10^{11}$ cm^{-2}, $\rho_d^s = 1.85 \times 10^{10}$ cm^{-2}, $\rho_d^t = 4.37 \times 10^{11}$ cm^{-2}), implying a lower quality of the GaN film, compared to the one in the GaN700/Si wafer ($\rho_d^e = 2.24 \times 10^{11}$ cm^{-2}, $\rho_d^s = 1.35 \times 10^{10}$ cm^{-2}, $\rho_d^t = 2.35 \times 10^{11}$ cm^{-2}). This was also supported by the higher dislocation correlation lengths found in the GaN700/Si wafer ($L^e = 41$ nm and $L^s = 220$ nm) as well as the larger mean distance between two dislocations ($r_d = 21$ nm) which corresponded to larger average size of cells in which the total Burger vector is equal to zero, implying a higher crystallinity of the GaN film, compared to the one in the GaN300/Si ($L^e = 27$ nm, $L^s = 107$ nm, $r_d = 15$ nm). Elemental diffusion studies carried out by TEM have shown that outside each layer boundary, both Al and Ga cross their respective layer interface to a certain depth, justifying the need of using a model that includes two different GaN layers (for each wafer) to explain the results from the DBS studies. Because both wafers were grown using the same method, in similar conditions, the improvement in crystallinity of the top GaN film is associated with the decreased lengths for elemental interfusion, relative to the GaN film width. While atomic displacements intermediate defect formation and propagation, a shorter length of non-stoichiometry in the GaN film induces a better quality of the top film, lowering the amount of defects and thus improving the positron moderation capacity of the material. The studied materials, because of their high amounts of edge and screw dislocations, diffusion, and partial non-stoichiometry, still imply several limitations in their use in the field of positron moderation. The positron data revealed the lack of uniformity in defect depth distribution, a fact that could not be observed in HR-TEM, nor in XRD. Therefore, using a positron-based complementary technique holds significant value for structural characterization. The DBS experiment assessed the effective positron diffusion length for both wafers, with a larger value of $L_{eff}^{GaN2} = 43 \pm 6$ nm ($S_{GaN2} = 0.4536 \pm 0.0003$), corresponding to the GaN film found in the GaN700/Si wafer, compared with $L_{eff}^{GaN2} = 22 \pm 6$ nm ($S_{GaN2} = 0.4558 \pm 0.0004$) for the GaN film in the GaN300/Si wafer.

Author Contributions: The authors have participated to the paper as follows; Conceptualization, V.-L.E., D.D., and I.Z.; methodology, N.D., V.L.; validation, V.L., E.A., and N.D.; formal analysis, V.-L.E., B.-S.V., D.D., I.Z., and A.-B.S.; investigation in positron studies, D.D. and N.D.; investigation in TEM, V.-L.E., and B.-S.V.; investigation in XRD, I.Z. and V.L.; writing—original draft preparation, V.-L.E.; writing—review and editing, V.-L.E., N.D., and D.D.; supervision, V.L., N.D. and E.A.; project administration, N.D. All authors have read and agreed to the published version of the manuscript.

Funding: This research received no external funding.

Acknowledgments: The support of the EU-funding project POSCCE-A2-O2.2.1-2013-1/Priority Axe 2, Project No. 638/12.03.2014, ID 1970, SMIS-CSNR code 48652 is gratefully acknowledged for the equipment purchased for this project. The authors also wish to acknowledge the support from the ELI-NP—Phase II, a project financed through the European Regional Development Fund—the Competitiveness Operational Programme (1/07.07.2016, COP, ID 1334) and from ELI-RO program, funded by Institute of Atomic Physics (Magurele, Romania), contract no. 27-ELI/2016.

Conflicts of Interest: The authors declare no conflict of interest. The funders had no role in the design of the study; in the collection, analyses, or interpretation of data; in the writing of the manuscript, or in the decision to publish the results.

References

1. Tsai, Y.L.; Lai, K.Y.; Lee, M.J.; Liao, Y.K.; Ooi, B.S.; Kuo, H.C.; He, J.H. Photon management of GaN-based optoelectronic devices via nanoscaled phenomena. *Prog. Quantum Electron.* **2016**, *49*, 1–25. [CrossRef]
2. Pampili, P.; Parbrook, P.J. Doping of III-nitride materials. *Mater. Sci. Semicond. Process.* **2017**, *62*, 180–191. [CrossRef]
3. Kuech, T.F. III-V compound semiconductors: Growth and structures. *Prog. Cryst. Growth Charact. Mater.* **2016**, *62*, 352–370. [CrossRef]
4. Meneghini, M.; Tajalli, A.; Moens, P.; Banerjee, A.; Zanoni, E.; Meneghesso, G. Trapping phenomena and degradation mechanisms in GaN-based power HEMTs. *Mater. Sci. Semicond. Process.* **2018**, *78*, 118–126. [CrossRef]
5. Roccaforte, F.; Fiorenza, P.; Greco, G.; Nigro, R.L.; Giannazzo, F.; Iucolano, F.; Saggio, M. Emerging trends in wide band gap semiconductors (SiC and GaN) technology for power devices. *Microelectron. Eng.* **2018**, *78*, 118–126. [CrossRef]
6. Flack, T.J.; Pushpakaran, B.N.; Bayne, S.B. GaN Technology for Power Electronic Applications: A Review. *J. Electron. Mater.* **2016**, *45*, 2673–2682. [CrossRef]
7. Carey, B.J.; Ou, J.Z.; Clark, R.M.; Berean, K.J.; Zavabeti, A.; Chesman, A.S.; Russo, S.P.; Lau, D.W.; Xu, Z.Q.; Bao, Q.; et al. Wafer-scale two-dimensional semiconductors from printed oxide skin of liquid metals. *Nat. Commun.* **2017**, *8*, 14482. [CrossRef]
8. Syed, N.; Zavabeti, A.; Messalea, K.A.; Della Gaspera, E.; Elbourne, A.; Jannat, A.; Mohiuddin, M.; Zhang, B.Y.; Zheng, G.; Wang, L.; et al. Wafer-Sized Ultrathin Gallium and Indium Nitride Nanosheets through the Ammonolysis of Liquid Metal Derived Oxides. *J. Am. Chem. Soc.* **2019**, *141*, 104–108. [CrossRef]
9. Daeneke, T.; Khoshmanesh, K.; Mahmood, N.; De Castro, I.A.; Esrafilzadeh, D.; Barrow, S.J.; Dickey, M.D.; Kalantar-Zadeh, K. Liquid metals: Fundamentals and applications in chemistry. *Chem. Soc. Rev.* **2018**, *47*, 4073–4111. [CrossRef]
10. Merrison, J.P.; Charlton, M.; Deutch, B.I.; Jorgensen, L.V. Field assisted positron moderation by surface charging of rare gas solids. *J. Phys. Condens. Matter* **1992**, *4*, L207–L212. [CrossRef]
11. Hugenschmidt, C. Positrons in surface physics. *Surf. Sci. Rep.* **2016**, *71*, 547–594. [CrossRef]
12. Beling, C.D.; Fung, S.; Ming, L.; Gong, M.; Panda, B.K. Theoretical search for possible high efficiency semiconductor based field assisted positron moderators. *Appl. Surf. Sci.* **1999**, *149*, 253–259. [CrossRef]
13. Jørgensen, L.V.; Schut, H. GaN-a new material for positron moderation. *Appl. Surf. Sci.* **2008**, *255*, 231–233. [CrossRef]

14. Kukushkin, S.A.; Osipov, A.V.; Bessolov, V.N.; Medvedev, B.K.; Nevolin, V.K.; Tcarik, K.A. Substrates for epitaxy of Gallium Nitride:new materials and techniques. *Rev. Adv. Mater. Sci.* **2008**, *17*, 1–32.

15. Yam, F.K.; Low, L.L.; Oh, S.A.; Hassan, Z. Gallium nitride: An overview of structural defects. In *Optoelectronics—Materials and Techniques*; IntechOpen Limited: London, UK, 2011; pp. 99–136.

16. Ambacher, O. Growth and applications of Group III-nitrides. *J. Phys. D Appl. Phys.* **1998**, *31*, 2653–2710. [CrossRef]

17. Schneider, C.A.; Rasband, W.S.; Eliceiri, K.W. NIH Image to ImageJ: 25 years of image analysis. *Nat. Methods* **2012**, *9*, 671–675. [CrossRef]

18. Palmer, D.C. *CrystalMaker*; Begbroke: Oxfordshire, UK, 2014.

19. Kaganer, V.M.; Brandt, O.; Trampert, A.; Ploog, K.H. X-ray diffraction peak profiles from threading dislocations in GaN epitaxial films. *Phys. Rev. B Condens. Matter Mater. Phys.* **2005**, *72*, 045423. [CrossRef]

20. Tuomisto, F.; Makkonen, I. Defect identification in semiconductors with positron annihilation: Experiment and theory. *Rev. Mod. Phys.* **2013**, *85*, 1583–1631. [CrossRef]

21. Van Veen, A.; Schut, H.; de Vries, J.; Hakvoort, R.A.; Ijpma, M.R. Analysis of positron profiling data by means of "VEPFIT". *AIP Conf. Proc.* **1991**, *218*, 171–198.

22. Van Veen, A.; Schut, H.; Clement, M.; de Nijs, J.M.M.; Kruseman, A.; IJpma, M.R. VEPFIT applied to depth profiling problems. *Appl. Surf. Sci.* **1995**, *85*, 216–224. [CrossRef]

23. Yu, H.; Ozturk, M.K.; Ozcelik, S.; Ozbay, E. A study of semi-insulating GaN grown on AlN buffer/sapphire substrate by metalorganic chemical vapor deposition. *J. Cryst. Growth* **2006**, *293*, 273–277.

24. Lahreche, H.; Vennéguès, P.; Tottereau, O.; Laügt, M.; Lorenzini, P.; Leroux, M.; Beaumont, B.; Gibart, P. Optimisation of AlN and GaN growth by metalorganic vapour-phase epitaxy (MOVPE) on Si (1 1 1). *J. Cryst. Growth* **2000**, *217*, 13–25. [CrossRef]

25. Mánuel, J.M.; Morales, F.M.; García, R.; Aidam, R.; Kirste, L.; Ambacher, O. Threading dislocation propagation in AlGaN/GaN based HEMT structures grown on Si (111) by plasma assisted molecular beam epitaxy. *J. Cryst. Growth* **2012**, *357*, 35–41. [CrossRef]

26. Yamaguchi, M.; Yamamoto, A.; Tachikawa, M.; Itoh, Y.; Sugo, M. Defect reduction effects in GaAs on Si substrates by thermal annealing. *Appl. Phys. Lett.* **1988**, *53*, 2293–2295. [CrossRef]

27. Bogusławski, P.; Rapcewicz, K.; Bernholc, J.J. Surface segregation and interface stability of AlN/GaN, GaN/InN, and AlN/InN {0001} epitaxial systems. *Phys. Rev. B* **2000**, *61*, 10820–10826. [CrossRef]

28. Romanitan, C.; Gavrila, R.; Danila, M. Comparative study of threading dislocations in GaN epitaxial layers by nondestructive methods. *Mater. Sci. Semicond. Process.* **2017**, *57*, 32–38. [CrossRef]

29. Zubiaga, A.; García, J.A.; Plazaola, F.; Tuomisto, F.; Zúñiga-Pérez, J.; Muñoz-Sanjosé, V. Positron annihilation spectroscopy for the determination of thickness and defect profile in thin semiconductor layers. *Phys. Rev. B* **2007**, *75*, 205–305. [CrossRef]

30. Schultz, P.J.; Tandberg, E.; Lynn, K.G.; Nielsen, B.; Jackman, T.E.; Denhoff, M.W.; Aers, G.C. Defects and Impurities at the Si/Si(100) Interface Studied with Monoenergetic Positrons. *Phys. Rev. Lett.* **1988**, *61*, 187–190. [CrossRef]

31. Jean, Y.C.; Mallon, P.E.; Schrader, D.M. *Principles and Applications of Positron and Positronium Chemistry*; World Scientific Publishing Co.Pte.Ltd.: Singapore, 2003.

32. Van Petegem, S.; Dauwe, C.; Van Hoecke, T.; De Baerdemaeker, J.; Segers, D. Diffusion length of positrons and positronium investigated using a positron beam with longitudinal geometry. *Phys. Rev. B Condens. Matter Mater. Phys.* **2004**, *70*, 115410. [CrossRef]

33. Uedono, A.; Ishibashi, S.; Tenjinbayashi, K.; Tsutsui, T.; Nakahara, K.; Takamizu, D.; Chichibu, S.F. Defect characterization in Mg-doped GaN studied using a monoenergetic positron beam. *J. Appl. Phys.* **2012**, *111*, 014508. [CrossRef]

34. Krause-Rehberg, R.; Leipner, H.S. *Positron Annihilation in Semiconductors—Defect Studies*; Springer-Verlag: Berlin/Heidelberg, Germany, 1999.

35. Saleh, A.S.; Elhasi, A.M. Investigation of Positron Annihilation Diffusion Length in Gallium Nitride. *Am. J. Mod. Phys.* **2014**, *3*, 24–28. [CrossRef]

36. Pi, X.D.; Coleman, P.G.; Tseng, C.L.; Burrows, C.P.; Yavich, B.; Wang, W.N. Defects in GaN films studied by positron annihilation spectroscopy. *J. Phys. Condens. Matter* **2002**, *14*, L243–L248. [CrossRef]

37. Puska, M.J.; Lanki, P.; Nieminen, R.M. Positron affinities for elemental metals. *J. Phys. Condens. Matter* **1999**, *1*, 6081–6094. [CrossRef]

38. Makkonen, I.; Snicker, A.; Puska, M.J.; Mäki, J.M.; Tuomisto, F. Positrons as interface-sensitive probes of polar semiconductor heterostructures. *Phys. Rev. B Condens. Matter Mater. Phys.* **2010**, *82*, 041307. [CrossRef]
39. Hu, Y.F.; Shan, Y.Y.; Beling, C.D.; Fung, S.; Xie, M.H.; Cheung, S.H.; Tu, J.; Brauer, G.; Anwand, W.; Tong, D.S. GaN Thin Films on SiC Substrates Studied Using Variable Energy Positron Annihilation Spectroscopy. In *Materials Science Forum*; Trans Tech Publications Ltd.: Zurich-Uetikon, Switzerland, 2001; Volume 363, pp. 478–480.

MDPI

Article

Morphological, Optical, and Electrical Properties of p-Type Nickel Oxide Thin Films by Nonvacuum Deposition

Chien-Chen Diao [1], Chun-Yuan Huang [2], Cheng-Fu Yang [3,*] and Chia-Ching Wu [2,*]

[1]   Department of Electronic Engineering, Kao Yuan University, Kaohsiung 821, Taiwan; ccd@kyu.edu.com
[2]   Department of Applied Science, National Taitung University, Taitung 950, Taiwan; laputa@nttu.edu.tw
[3]   Department of Chemical and Materials Engineering, National University of Kaohsiung,
      Kaohsiung 811, Taiwan
*   Correspondence: cfyang@nku.edu.com (C.-F.Y.); ccwu@nttu.edu.tw (C.-C.W.)

Received: 9 March 2020; Accepted: 26 March 2020; Published: 29 March 2020

Abstract: In this study, a p-type 2 at% lithium-doped nickel oxide (abbreviation L2NiO) solution was prepared using $Ni(NO_3)_2 \cdot 6H_2O$, and $LiNO_3 \cdot L2NiO$ thin films were deposited using an atomizer by spraying the L2NiO solution onto a glass substrate. The sprayed specimen was heated at a low temperature (140 °C) and annealed at different high temperatures and times. This method can reduce the evaporation ratio of the L2NiO solution, affording high-order nucleating points on the substrate. The L2NiO thin films were characterized by X-ray diffraction, scanning electron microscopy, UV–visible spectroscopy, and electrical properties. The figure of merit (FOM) for L2NiO thin films was calculated by Haacke's formula, and the maximum value was found to be $5.3 \times 10^{-6} \ \Omega^{-1}$. FOM results revealed that the L2NiO thin films annealed at 600 °C for 3 h exhibited satisfactory optical and electrical characteristics for photoelectric device applications. Finally, a transparent heterojunction diode was successfully prepared using the L2NiO/indium tin oxide (ITO) structure. The current–voltage characteristics revealed that the transparent heterojunction diode exhibited rectifying properties, with a turn-on voltage of 1.04 V, a leakage current of $1.09 \times 10^{-4} \ A/cm^2$ (at 1.1 V), and an ideality factor of $n = 0.46$.

Keywords: lithium-doped nickel oxide; non-vacuum deposition; figure of merit; heterojunction diode

1. Introduction

At present, numerous applications, such as touch panels, light-emitting diodes, and solar cells, require transparent, conductive coatings [1–3]. Thus far, materials belonging to the transparent conducting oxide (TCO) family have been frequently used for this purpose. Most of the industry standard TCO are n-type wide bandgap oxides (Eg > 3.1 eV), such as In_2O_3, SnO_2, and ZnO, whose conductivity can be further tuned by aliovalent doping or the formation of oxygen vacancies [4]. In contrast, the development of p-type TOS remains a challenge. Recently, semi-transparent p-type conducting films of the nickel oxide (NiO) have attracted considerable attention because of their importance in several scientific applications, including (i)material for electrochromic display devices [5,6], (ii) functional sensor layers in chemical sensors [7], (iii) transparent electronic devices [8] and (iv) the magnetic properties of nanoparticles [9–12]. A stoichiometric NiO thin film is an insulator at room temperature (resistivity is $\sim 10^{13} \ \Omega \cdot cm$) [13]. Much effort has been made to explain the insulating behavior of NiO. NiO crystallizes in a rock-salt crystal structure, in which Ni cations have a nominal valence state of 2+ ($3d^8$) in octahedral coordination (see Figure 1). Due to a strong electron correlation in 3d orbitals, it has an optical bandgap of 3.4–4.0 eV [14]. In addition, according to the literature, at temperatures above the Néel temperature (523 K), the crystal structure of NiO is cubic, whereas below the Néel temperature, the crystals become slightly distorted and acquire a rhombohedral structure which accompanies the antiferromagnetic ordering [15].

NiO thin films can be grown by several chemical and physical methods, including magnetron sputtering [16,17], evaporation [18], the sol–gel method [19], laser ablation deposition [20], and spray pyrolysis (SP) [21,22]. NiO thin films with low resistivity (1.4×10^{-1} $\Omega \cdot$cm) can be deposited by sputtering [23]. Compared with vacuum deposition, SP is a relatively simple, cost-effective nonvacuum deposition method for fabricating TCO thin films for large-area coating. However, the resistivity of the doped NiO thin films fabricated by SP is ~10^4 $\Omega \cdot$cm [24]; this resistivity is several orders of magnitude greater than that observed for sputter-deposited NiO thin films. Conventional SP involves spraying a nickel nitrate solution onto a preheated glass substrate at a temperature greater than 300 °C, followed by evaporation, solute precipitation, and pyrolytic decomposition. With the increase in the substrate temperature, the evaporation ratio of the solution on the substrate is extremely swift, leading to the formation of inferior NiO thin films. To solve this problem, a modified spray method was used in this study. First, the substrate temperature is slightly greater than the boiling point of the spray solution. The evaporation ratio of the spray solution decreases, affording high-order nucleation points by the spraying of the solution onto the substrate. The thin films were then formed and further annealed at high temperatures to afford a crystalline structure. Finally, high-quality thin films were obtained and subsequently applied as a photoelectric device.

To improve the conductivity of the NiO thin film, three improved mechanisms were used: (i) holes generated from nickel vacancies, (ii) oxygen interstitial atoms, and (iii) monovalent atoms used as a dopant. Monovalent atoms can be used as the dopant to increase the electrical conductivity of the NiO thin films [25,26]. In this study, a modified spray method was employed for the deposition of 2 at% Lithium (Li)-doped NiO (L2NiO) thin films with a high electrical conductivity. The monovalent atoms of Li can be substitution Ni atoms. In addition, the effects of annealing temperatures and times on the physical, optical, and electrical properties of the L2NiO thin films were investigated. X-ray photoelectron spectroscopy (XPS) was used to investigate the variations in the characteristics of the L2NiO thin films. Finally, a transparent heterojunction diode device comprising an L2NiO thin film and an indium–tin oxide (ITO) thin film was fabricated for future applications.

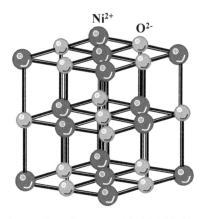

Figure 1. Crystal structure of the NiO thin film.

2. Experimental Methods

Lithium-doped nickel oxide (LNiO) thin films were deposited on a Corning glass substrate by the modified spray method. The spray solution was prepared by mixing nickel nitrate (Ni(NO$_3$)$_2 \cdot$6H$_2$O, Alfa Aesar, MA, USA) and lithium nitrate (LiNO$_3$, J.T. Baker, NJ, USA) in deionized (DI) water. A 1 M L2NiO spray solution was prepared by doping 2 at% Li in NiO. The modified spray method involved spraying a L2NiO solution at 140 °C, which then evaporated, affording high-quality L2NiO thin films on the Corning glass substrate. The L2NiO thin films were deposited under the following conditions: solution volume = 40 mL, deposition rate = 10 mL/min. The distance between the Corning glass substrate and the nozzle was approximately 20 cm, and compressed air was used as the carrier gas. Annealing temperatures and times were 400–600 °C and 1–3 h, respectively, for the crystallization of the L2NiO thin films. Finally, to fabricate the transparent heterojunction diodes, L2NiO thin films were deposited on an ITO glass substrate, and the top and bottom aluminum (Al) electrodes were deposited by electron-beam evaporation. The surface morphology of the L2NiO thin films were examined by high-resolution scanning electron microscopy (HR-SEM, Hitachi, Japan). The resulting interface layer morphology between the L2NiO and ITO thin film was characterized by high-resolution transmission electron microscopy (HR-TEM, JOEL, Japan). The phase and crystallinity of the L2NiO thin films were measured by X-ray diffraction (XRD, Bruker, MA, USA) using CuKα radiation in the 2θ range of 20°–80°. The bonding state and element content of the L2NiO thin films were investigated using X-ray photoemission spectroscopy (XPS, ULVAC·PHI, Japan). The XPS using a monochromatic Al Kα X-ray (hν = 1486.6 eV) source was carried out at normal emission with an electron energy analyzer. The resistivity, carrier concentration, and mobility were measured by Hall effect measurements using the Van der Pauw method. The optical transmittance of the L2NiO thin films was measured using a UV–vis system (Agilent, CA, USA), and the transmittance spectrum was recorded as a function of the wavelength in the range of 200 to 1100 nm. The current–voltage (I–V) properties of the transparent heterojunction diode was measured using an HP4156 semiconductor parameter analyzer (Agilent, CA, USA).

3. Results and Discussion

Figure 2 shows the HR-SEM images of the L2NiO thin film with different annealing temperatures and times. The HR-SEM image of the L2NiO thin film which had annealed at 400 °C for 1 h revealed a smooth surface and no grain growth (Figure 2a). With the further increase in the annealing time to 3 h at 400 °C, the surface morphology revealed small grain sizes (Figure 2b). The average grain size of the film annealed at 400 °C for 3 h was 38 nm. Surface SEM morphologies shown in Figure 2c,d were compared by the increase in the annealing temperature of the L2NiO thin films from 500 °C to 600 °C for 3 h, and the grain sizes slightly increased. At annealing temperatures of 500 °C and 600 °C for 3 h, the average grain sizes of the L2NiO thin films were 45 nm and 58 nm, respectively. As a result of annealing at higher temperatures, the surface atoms on the substrate acquire more energy, and these atoms can move to suitable nucleation sites. In addition, the low activation energy ions doped in the thin film can easily escape from trap sites and transfer to nucleation sites. Crystalline thin films can be obtained when an increased number of better nucleation sites are formed on the substrate. In this study, the low activation energy of the Li ions doped in the NiO thin film leads to the increase in grain size with the increase in the annealing temperatures and times [27]. Compared with previous reports, the crystalline grain structure of the L2NiO thin films deposited by the modified spray method is better than that obtained by SP [28,29], because SP involves the deposition of the solution onto a preheated (>300 °C) substrate, but the evaporation ratio of the solution is extremely swift, affording poor nucleating points. Therefore, the surface morphology of the thin film is not good. The thickness of the L2NiO thin film with different annealing temperatures and times is shown in cross-section SEM images (Figure S1). The thickness of the L2NiO thin film annealed at 400 °C for 1 h was 202 nm. The thickness of the L2NiO thin films increased slightly as the annealing temperatures and times increased.

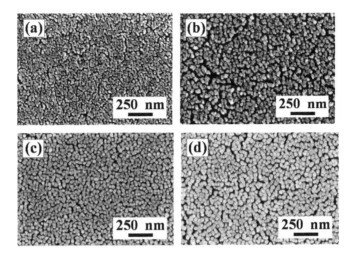

Figure 2. Surface SEM images of the L2NiO thin films as a function of annealing temperatures and times: (**a**) 400 °C for 1 h, (**b**) 400 °C for 3 h, (**c**) 500 °C for 3 h, and (**d**) 600 °C for 3 h.

The crystalline structure of the L2NiO thin films was examined by XRD using CuKα (λ = 0.1542 nm) radiation. Figure 3 shows the XRD patterns of the L2NiO thin films with different annealing temperatures and times. The observed XRD patterns of the L2NiO thin films were compared with the Joint Committee on Powder Diffraction Standards (JCPDS) data; they were in good agreement with the standard diffraction pattern of NiO (JCPDS card no. 47-1049). Diffraction peaks for the L2NiO thin films were observed at 2θ values of 37.3°, 43.2°, and 63.1°, which correspond to the (111), (200), and (220) planes, respectively. The L2NiO thin films were polycrystalline without any other detectable secondary phase. Diffraction results revealed that the L2NiO thin film annealed at 400 °C for 1 h exhibited an approximate amorphous structure due to its weak-intensity diffraction peaks (Figure 3a). However, with the increase in the annealing temperatures and times from 400 °C to 600 °C and 1 to 3 h, respectively, diffraction intensities for the (111), (200), and (220) planes slightly increased (Figure 3b–d). The increase in the diffraction intensity was related to the grain sizes of the L2NiO thin films. Figure 3 (right side) also shows the grazing incidence angle X-ray diffraction patterns (GIAXRD) of the L2NiO films in the 2θ range of 42° to 45°. The full-width half-maximum for the diffraction peak of the (200) plane of the L2NiO thin films decreased from 0.38 to 0.25. The crystallite size of the L2NiO thin films was then calculated using the Scherrer equation. With the increase in the annealing temperatures and times, the grain sizes increased from 39 nm to 60 nm. The results obtained for the various grain sizes were similar to those from SEM (Figure 2). In addition, with the increase in the annealing temperatures and times, the (200) plane was slightly shifted to high angles. According to Bragg's law ($n\lambda = 2d\sin\theta$) and $d = a/(h^2 + k^2 + l^2)^{1/2}$, the lattice constant (a) slightly decreased from 4.178 Å to 4.169 Å with the increase in the annealing temperatures and times, indicating that the larger radius of Ni^{2+} (0.69 Å) can be substituted by the smaller radius of Li^+ (0.68 Å); this subsequently leads to the decreased lattice constant of L2NiO thin films [30].

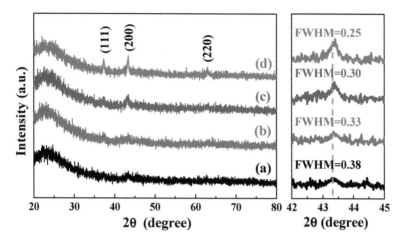

Figure 3. X-ray diffraction patterns of the L2NiO thin films as a function of the annealing temperatures and times: (**a**) 400°C for 1 h, (**b**) 400°C for 3 h, (**c**) 500°C for 3 h, and (**d**) 600°C for 3 h.

The crystal structure parameter of the L2NiO thin films produced with a 600 °C annealing temperature for 3h was fitted using the cubic structural model, with the atomic positions being described in the space group Fm3m. The fitted profiles of the L2NiO thin films for XRD data at 600 °C annealing temperature for 3h is shown in Figure 4. The final refinement convergence of the L2NiO thin films produced with a 600 °C annealing temperature for 3h was achieved with $\chi^2 = 1.38$, and the measured result agreed well with the simulation value. i.e., the lattice constant (a) of the L2NiO thin was 4.1686 Å, similar to the value calculated using Bragg's law. The refined values of all thin film are also tabulated in Table 1.

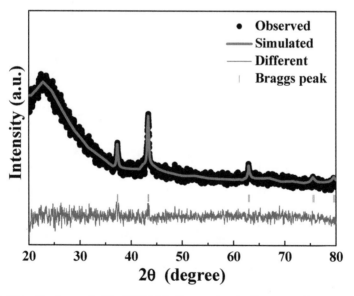

Figure 4. Rietveld refinement of the L2NiO thin films produced with an annealing temperature of 600°C for 3 h.

Table 1. Refined values of the L2NiO thin films with different annealing temperatures and times.

Parameter	Nondoped NiO (R [31])	L2NiO (400°C, 1 h)	L2NiO (400°C, 3 h)	L2NiO (500°C, 3 h)	L2NiO (600°C, 3 h)
$a = c = b$ (Å)	4.1801	4.1774	4.1738	4.1701	4.1686
$\alpha = \beta = \gamma$	90°	90°	90°	90°	90°
Volume (Å^3)	73.01	72.49	72.44	72.41	72.38

Figure 5a shows the optical transmittance spectra of the L2NiO thin films in the 250–1100 nm range. For the L2NiO thin films annealed at 400 °C for 1 h and those annealed at 400 °C, 500 °C, and 600 °C for 3 h, average transmittance values in the visible region (400 to 700 nm) were 46.8%, 72.3%, 84.6%, and 87.9%, respectively. The increase in the average transmittance of the L2NiO thin films was related to the increase in the grain size and decrease in the grain boundary, leading to the low scattering effect in L2NiO thin films. Surface SEM images revealed that the grain size of the L2NiO thin films increased with different annealing temperatures and times; this result was in agreement with the optical transmittance results. In the ultraviolet range, with the increase in the annealing temperature from 400 °C to 600 °C at an annealing time of 1 h to 3 h, the absorption edge was shifted toward a short wavelength region. The blue-shift can be explained by the Burstein–Moss shift effect [32–34].

(a)

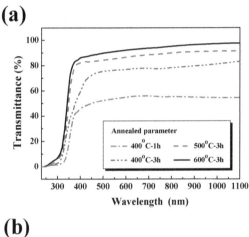

(b)

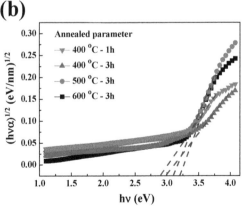

Figure 5. (a) Optical transmittance spectra and (b) optical bandgap of the L2NiO thin films as a function of annealing temperatures and times.

The optical energy band gap (E_g) is an important physical parameter that mainly determines the electrical and optical characteristics of materials. The energy band gap of the L2NiO thin films were determined by applying the Tauc and Davis-Mott models [35]:

$$(\alpha h v)^{1/2} = B\left(h v - E_g^{op}\right) \tag{1}$$

where α is the absorption coefficient, h is Planck's constant, v is the frequency of the incident photon, and B is the absorption edge width. The energy band gap was determined by the extrapolation of a straight linear region of the plots to $h v = 0$ [36]. Figure 5b shows $(\alpha h v)^{1/2}$ vs. $h v$ for L2NiO thin films with different annealing temperatures and times. With the increase in the annealing temperatures from 400 °C to 600 °C and the annealing times from 1 h to 3 h, the energy band gap of the L2NiO thin films increased from 2.89 to 3.21 eV (Figure 5b), suggesting that the Burstein–Moss effect may affect the band gap shift [33,34]. In the Burstein–Moss effect, the band-gap shift is mainly related to the high carrier concentration and/or low effective mass. According to the Burstein–Moss effect, the divergence of the band gap is expressed as

$$\Delta E_g^{BM} = \frac{h^2}{2m_{vc}^*}\left(3\pi^2 n\right)^{\frac{2}{3}} \tag{2}$$

where ΔE_g^{BM} is the shift value between the doped semiconductor and undoped semiconductor, m_{cv}^* is the reduced effective mass, and n is the carrier concentration. The absorption edge of the L2NiO thin films were observed in a shorter wavelength region because of the increase in the carrier concentration (n) (Figure 5a).

Figure 6 shows the optical energy band gap, carrier concentration (n), mobility (μ), and resistivity (ρ) of the L2NiO thin films with different annealing temperatures and times. All samples exhibited p-type properties. With the increase in the annealing temperatures and times, the mobility of the L2NiO thin films increased from 2.39 to 11.96 cm^2/Vs because of the increase in the grain size and decrease in the grain boundary, causing the carrier to encounter less hindering materials, and subsequently resulting in increased carrier mobility. Meanwhile, the increase in the annealing temperatures and times caused an increase in the carrier concentration. The number of Li atoms substituting the sites of Ni atoms increased, leading to the substitution of a large number of Ni^{2+} by Li ions in the normal crystal sites and creating holes at a high annealing temperature. Therefore, the carrier concentration of the L2NiO thin films increased; this result can be obtained by Equation (3).

$$\frac{1}{2}O_2^{(g)} + Li_2O \Leftrightarrow 2O_x^O + 2Li_{Ni}^{'} + 2\overset{\cdot}{h} \tag{3}$$

With the increase in the annealing temperatures and times, the Li concentration of the L2NiO thin films increased, as demonstrated by the XPS analysis shown in Table 2 and the XRD analysis shown in Figure 3 (right side). The resistivity of the film is known to be proportional to the reciprocal of the product of carrier concentration and mobility, as follows:

$$\rho = 1/(n \times \mu) \tag{4}$$

Table 2. Elements of L2NiO thin films as a function of annealing temperature and time.

	Ni (at%)	O (at%)	Li (at%)	O/Ni
L2NiO-400°C-1 h	45.23	52.96	1.81	1.171
L2NiO-400°C-3 h	45.1	53.07	1.83	1.176
L2NiO-500°C-4 h	44.9	53.23	1.87	1.185
L2NiO-600°C-3 h	44.62	53.50	1.88	1.199

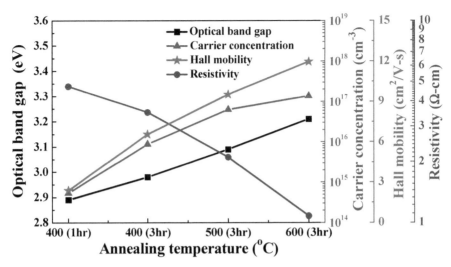

Figure 6. Optical band gap, carrier concentration, mobility, and resistivity of the L2NiO thin films as a function of annealing temperatures and times.

Therefore, with the increase in the carrier concentration and mobility, the resistivity of the L2NiO thin films decreased from 4.73 to 1.08 Ω·cm. Compared with previous reports, the resistivity of L2NiO thin films was slightly less than those of undoped NiO thin films [36,37].

Figure 7 shows the Ni 2p$_{3/2}$ XPS spectra of the L2NiO thin films with different annealing temperatures and times. During the Gaussian fitting process, binding energies for the NiO and Ni$_2$O$_3$ peaks were observed at 854.0 eV and 855.8 eV, respectively, in the Ni 2p$_{3/2}$ XPS spectra of the L2NiO thin films annealed at 400 °C for 1 h (right side of Figure 7) [38,39]. The Ni$_2$O$_3$ and NiO peaks corresponded to Ni^{3+} and Ni^{2+}, respectively. With the increase in annealing temperatures and times for the L2NiO thin films, the intensity of the Ni^{3+} bonding state slightly increased over that of the Ni^{2+} bonding state. This result is related to the insertion of an excess amount of oxygen ions in the interstitial sites of the L2NiO thin films due to annealing conducted in the atmosphere; this leads to the formation of Ni^{3+} ions and holes, which can be represented by Equation (5).

$$\frac{1}{2}O_2^{(g)} \Leftrightarrow O_i'' + 2\dot{h} \tag{5}$$

Meanwhile, the Ni 2p$_{3/2}$ results were also confirmed from the O1s XPS spectra of the L2NiO thin films with annealing temperatures and times, as shown in Figure 8. With the increase in the annealing temperature and times, the intensity of the O1s peak increased appreciably. In the O1s XPS spectra, the deconvolution of the electron binding energy of NiO (529.3 eV) and Ni$_2$O$_3$ (531.7 eV) was observed for the L2NiO thin films [38,40,41]. The increase in the magnitude of the Ni^{3+} bonding state was slightly greater than that of the Ni^{2+} bonding state, demonstrating that the hole carrier concentration of the L2NiO thin films increased with the annealing temperatures and times. The increased hole carrier concentration of the L2NiO thin films was in agreement with the Hall measurement (Figure 6).

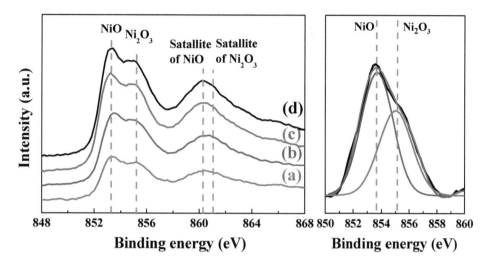

Figure 7. Ni 2p$_{3/2}$ XPS spectra of the L2NiO thin films as a function of annealing temperatures and times: (**a**) 400°C for 1 h, (**b**) 400°C for 3 h, (**c**) 500°C for 3 h, and (**d**) 600°C for 3 h.

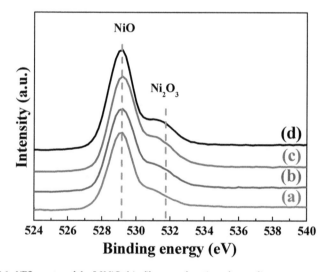

Figure 8. O 1s XPS spectra of the L2NiO thin films as a function of annealing temperatures and times. (**a**) 400°C for 1 h, (**b**) 400°C for 3 h, (**c**) 500°C for 3 h, and (**d**) 600°C for 3 h.

Figure 9 shows the figure-of-merit (FOM) of the L2NiO thin films with different annealing temperatures and times. To deposit L2NiO thin films with high transmission and low resistivity, the FOM values for the L2NiO thin films with different annealing temperatures and times were calculated by using Haacke's equation [42]:

$$\text{FOM} = \frac{T^{10}}{R_s} \tag{6}$$

where T is the average optical transmittance at 400–700 nm and R_s is the sheet resistance of the L2NiO thin films. With the increase in annealing temperatures and times, the FOM of the L2NiO thin films increased (Figure 9). The maximum FOM (5.3×10^{-6} Ω^{-1}) was obtained for the L2NiO thin films annealed at 600 °C for 3 h. FOM results revealed that L2NiO thin films exhibit satisfactory optical and electrical characteristics for photoelectric device applications.

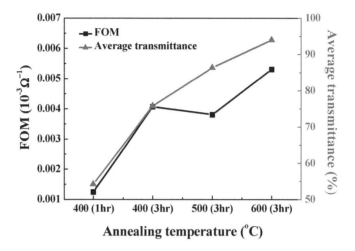

Figure 9. FOM values for L2NiO thin films as a function of annealing temperatures and time.

From the above results, it can be seen that the carrier concentration, mobility, and conductivity characteristics of the Li ions doped in NiO thin film improved compared to non-doped NiO thin film. For the possible fabrication of transparent heterojunction diodes, the L2NiO thin films were then deposited at an annealing temperature of 600 °C and an annealing time of 3 h onto an ITO glass substrate to form a p–n junction structure. The ITO thin film was deposited by the sputtering method. The thickness of the ITO thin film was 1000 Å with a 92% visible-light transmittance and a resistance of 12 Ω·cm, as shown in Figure S2a,b. Figure 10 shows the current–voltage (I–V) curve of the L2NiO/ITO transparent heterojunction diode. The I–V results confirmed that the fabricated L2NiO/ITO transparent heterojunction diode exhibited rectifying behavior with the use of aluminum (Al) as the electrodes. Before the measurement of the I–V properties of the L2NiO/ITO transparent heterojunction diode, ohmic contacts were confirmed to be present between the Al electrodes and the L2NiO and ITO thin films. Under forward bias, the turn-on voltage for the L2NiO/ITO transparent heterojunction diode was ~1.04 V (Figure 10a); this value is less than that (2.57 V) for a p-NiO/n-TZO diode [43], 2.5 V for a p-CuO/n-ZnO diode [44], and similar (1 V) to that for a p-NiO/n-ZnO diode [45]. Under a reverse voltage, the leakage current was 1.09×10^{-4} A/cm^2 at 1.1 V for the L2NiO/ITO transparent heterojunction diode (Figure 10b). The rectification ratio (R) for the L2NiO/ITO transparent heterojunction diodes was calculated using Equation (7) as follows to obtain a value of 17.3 (at 1.1 V):

$$R = \frac{\text{Forward current}}{\text{Reverse current}} \tag{7}$$

The ideality factor (n) can be calculated from the slope of the linear region of the forward-bias log(I)–V curve, which can be derived from Equation (8) and Figure 11:

$$n = \frac{q}{kT} \times \left[\frac{dV}{d\ln(I)} \right] \tag{8}$$

where k is the Boltzmann constant, T is the temperature in kelvin, and q is the electron charge. The ideality factor for the L2NiO/ITO transparent heterojunction diode was $n = 0.46$, which was less than the ideal value of $n = 1$ (Equation (8)). The high leakage current and low ideality factor result from imperfections between the heterojunction interfaces of the L2NiO and ITO thin films. Ajimsha et al. reported that in an oxide layer, these imperfections were caused by the presence of different crystal-type structures [46].

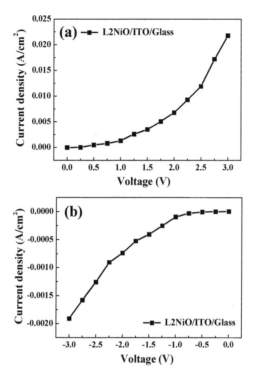

Figure 10. Current–voltage curve of the L2NiO/ITO transparent heterojunction diode: (**a**) forward current and (**b**) reverse current.

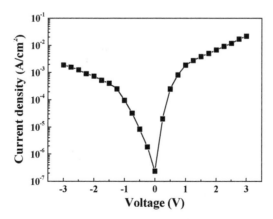

Figure 11. Log current–voltage curve of the L2NiO/ITO transparent heterojunction diode.

To further investigate the interface between the L2NiO and ITO thin films, TEM images were recorded. Figure 12a shows a magnified TEM image of the interface between the L2NiO thin film annealed at 600 °C for 3 h and the ITO thin film; a clear layer was present at the interface between the two materials. The interfacial layer thickness was 55 Å; this layer was thought to be NiO_2, Ni_2O_3, and Ni_3O_4, because during spraying, Ni and O can be easily combined with In or Sn in ITO. This result can be attributed to the relatively high bond energy of Ni–O (1029 kJ/mol) compared with those of Sn–O (531.8 kJ/mol) and In–O (320 kJ/mol) [47–49]. From the XPS result, it was hypothesized that the interfacial layer between the NiO and ITO was Ni_2O_3. Figure 12b shows the selected-area electron diffraction (SAED) pattern of the boundary between the L2NiO and ITO thin films. The corresponding SAED pattern exhibited (200) NiO, (211) ITO, and (002) Ni_2O_3 diffraction rings, confirming that the L2NiO (including NiO and Ni_2O_3) and ITO thin films are polycrystalline; this result also confirmed that a thin Ni_2O_3 interfacial layer was present between the L2NiO and ITO thin films. The thin Ni_2O_3 interfacial oxide layer rendered a high leakage current and low ideality factor for the L2NiO/ITO transparent heterojunction diode.

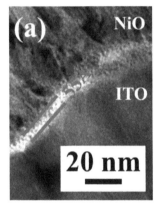 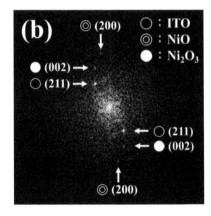

Figure 12. (a) TEM image and (b) SAED pattern of the L2NiO/ITO transparent heterojunction structure.

4. Conclusions

In this study, a modified spray method was used for the deposition of high-quality 2 % Li-doped NiO (L2NiO) thin films. The L2NiO thin films exhibited a cubic (NaCl-type) structure, and the lattice constant of the L2NiO thin films slightly decreased from 0.4178 Å to 0.4169 Å with the increase in annealing temperatures and times. As the smaller radius of Li^+ (0.6 Å) was substituted by the larger Ni^{2+} (0.69 Å), the number of substituted Li^+ increased, leading to a decrease in the lattice constant of the L2NiO thin films. According to the Burstein–Moss shift theory, the optical energy band gap (E_g) of the L2NiO thin films increased from 2.89 eV to 3.21 eV with the increase in the annealing temperatures and times because of the increase in the carrier concentration. In Hall measurements, the carrier concentration and mobility of the L2NiO thin films increased, leading to a decrease in the resistivity from 4.73 $\Omega \cdot cm$ to 1.08 $\Omega \cdot cm$ with the increase in the annealing temperatures and times. The optimum FOM (5.3×10^{-6} Ω^{-1}) was obtained for the L2NiO thin films annealed at 600 °C for 3 h, for which the resistivity and average transmittance were 1.08 $\Omega \cdot cm$ and 87.9%, respectively. Finally, the transparent heterojunction diode comprising a p-type L2NiO thin film and an n-type ITO thin film was successfully fabricated. Its properties included (1) a turn-on voltage of 1.04 V, (2) a leakage current of 1.09×10^{-4} A/cm^2 (at 1.1 V), (3) a rectification ratio of 17.3, and (4) an ideality factor of 0.46. The high leakage current resulted from the Ni_2O_3 thin layer between the heterojunction interfaces of the different crystal-type structure with the L2NiO and ITO thin films. Therefore, the L2NiO film was

Nanomaterials **2020**, *10*, 636

shown to possess satisfactory properties for applications including transparent diode, electrochromic display, and solar cell devices.

Supplementary Materials: The following are available online at http://www.mdpi.com/2079-4991/10/4/636/s1, Figure S1: Cross-section SEM images of the L2NiO thin films as a function of annealing temperatures and times: (a) 400 °C for 1 h, (b) 400 °C for 3 h, (c) 500 °C for 3 h, and (d) 600 °C for 3 h. Figure S2: (a) Cross-section SEM image and (b) optical transmittance spectra of the ITO thin film.

Author Contributions: C.-C.D. participated in I-V measurement and design the transparent heterojunction diode. C.-Y.H. participated in XRD analysis and Rietveld refinement of the XRD data. C.-F.Y. and C.-C.W. participated in the design experimental of Li doped NiO films. C.-C.W. participated in the Hall, SEM, and XPS analysis of Li doped NiO films and fabrication of Li doped NiO films. All authors have read and agreed to the published version of the manuscript.

Funding: This research received no external funding.

Acknowledgments: The authors acknowledge the financial support of the Ministry of Science and Technology (MOST 108-2221-E-143-001, MOST 108-2622-E-143-001-CC3). The authors gratefully acknowledge the use of high-resolution scanning electron microscope and multipurpose X-Ray thin-film micro area diffractometer equipment belonging to the Instrument Center of National Cheng Kung University.

Conflicts of Interest: The authors declare that they have no conflict of interests.

References

1. Wang, K.C.; Shen, P.S.; Li, M.H.; Chen, S.; Lin, M.W.; Chen, P.; Guo, T.F. Low-temperature sputtered nickel oxide compact thin film as effective electron blocking layer for mesoscopic NiO/CH$_3$NH$_3$PbI$_3$ perovskite heterojunction solar cells. *ACS Appl. Mater. Interfaces* **2014**, *6*, 11851–11858.

2. Zeng, H.; Xu, X.; Bando, Y. Template deformation-tailored ZnO nanorod/nanowire arrays: Full growth control and optimization of field-emission. *Adv. Functional Mater.* **2009**, *19*, 3165–3172. [CrossRef]

3. Song, J.; He, Y.; Chen, J. Bicolor light-emitting diode based on zinc oxide nanorod arrays and poly(2-methoxy,5-octoxy)-1,4-phenylenevinylene. *J. Electron. Mater.* **2012**, *41*, 431–436. [CrossRef]

4. Dixon, S.C.; Scanlon, D.O.; Carmalt, C.J.; Parkin, I.P. n-Type doped transparent conducting binary oxides: An overview. *J. Mater. Chem. C* **2016**, *4*, 6946–6961. [CrossRef]

5. Sun, H.; Chen, S.C.; Peng, W.C.; Wen, C.K.; Wang, X.; Chuang, T.H. The Influence of Oxygen Flow Ratio on the Optoelectronic Properties of p-Type Ni$_{1-x}$O Films Deposited by Ion Beam Assisted Sputter. *Coatings* **2018**, *8*, 168. [CrossRef]

6. Kitao, M.; Izawa, K.; Urabe, K.; Komatsu, T.; Kuwano, S.; Yamada, S. Preparation and electrochromic properties of rf-sputtered NiOx films prepared in Ar/O$_2$/H$_2$ atmosphere. *Jpn. J. Appl. Phys.* **1994**, *33*, 6656–6662. [CrossRef]

7. Kumagai, H.; Matsumoto, M.; Toyoda, K.; Obara, M. Preparation and characteristics of nickel oxide thin film by controlled growth with sequential surface chemical reactions. *J. Mater. Sci. Lett.* **1996**, *15*, 1081–1083. [CrossRef]

8. Zhang, J.Y.; Li, W.W.; Hoye, R.L.Z.; MacManus-Driscoll, J.L.; Budde, M.; Bierwagen, O.; Wang, L.; Du, Y.; Wahila, M.J.; Piper, L.F.J.; et al. Electronic and transport properties of Li-doped NiO epitaxial thin films. *J. Mater. Chem. C* **2018**, *6*, 2275–2882. [CrossRef]

9. Tiwari, S.D.; Rajeev, K.P. Magnetic properties of NiO nanoparticles. *Thin Solid Film.* **2006**, *505*, 113–117. [CrossRef]

10. Tadic, M.; Nikolic, D.; Panjan, M.; Blake, G.R. Magnetic properties of NiO (nickel oxide) nanoparticles: Blocking temperature and Neel temperature. *J. Alloys Compd.* **2015**, *647*, 1061–1068. [CrossRef]

11. Cristina, G.; Francis, L.D.; Maurizio, M.M.; Stefano, C.; Francesco, M.M.; Alfonso, P.; Claudia, I.; Claudio, S. Magneto-Plasmonic Colloidal Nanoparticles Obtained by Laser Ablation of Nickel and Silver Targets in Wate. *J. Phys. Chem. C* **2017**, *121*, 3597–3606.

12. Arif, M.; Sanger, A.; Shkir, M.; Singh, A.; Katiyar, R.S. Influence of interparticle interaction on the structural, optical and magnetic properties of NiO nanoparticles. *Phys. B* **2019**, *552*, 88–95. [CrossRef]

13. Adler, D.; Feinleib, J. Electrical and optical properties of narrow-band materials. *Phys. Rev.* **1970**, *B2*, 3112. [CrossRef]

14. Ai, L.; Fang, G.; Yuan, L.; Liu, N.; Wang, M.; Li, C.; Zhang, Q.; Li, J.; Zhao, X. Influence of substrate temperature on electrical and optical properties of p-type semitransparent conductive nickel oxide thin films deposited by radio frequency sputtering. *Appl. Surf. Sci.* **2008**, *254*, 2401–2405. [CrossRef]
15. Slack, G.A. Crystallography and domain walls in antiferromagnetic NiO crystals. *J. Appl. Phys.* **1960**, *31*, 1571–1582. [CrossRef]
16. Li, G.; Jiang, Y.; Deng, S.; Tam, A.; Xu, P.; Wong, M.; Kwo, H.S. Overcoming the limitations of sputtered nickel oxide for high-efficiency and large-area perovskite solar cells. *Adv. Sci.* **2017**, *4*, 1700463. [CrossRef]
17. Jang, W.L.; Lu, Y.M.; Hwang, W.S.; Chen, W.C. Electrical properties of Li-doped NiO films. *J. Eur. Ceram. Soc.* **2010**, *30*, 503–508. [CrossRef]
18. Jiang, D.Y.; Qin, J.M.; Wang, X.; Gao, S.; Liang, Q.C.; Zhao, J.X. Optical properties of NiO thin films fabricated by electron beam evaporation. *Vacuum* **2012**, *86*, 1083–1086. [CrossRef]
19. Dalavi, D.S.; Devan, R.S.; Patil, R.S.; Ma, Y.R.; Pati, P.S. Electrochromic performance of sol–gel deposited NiO thin film. *Mater. Lett.* **2013**, *90*, 60–63. [CrossRef]
20. Verma, V.; Katiyar, M. Effect of the deposition parameters on the structural and magnetic properties of pulsed laser ablated NiO thin films. *Thin Solid Film.* **2013**, *527*, 369–376. [CrossRef]
21. Reguig, B.A.; Khelil, A.; Cattin, L.; Morsli, M.; Bernède, J.C. Properties of NiO thin films deposited by intermittent spray pyrolysis process. *Appl. Surf. Sci.* **2007**, *253*, 4330–4334. [CrossRef]
22. Awais, M.; Dowling, D.D.; Rahman, M.; Vos, J.G.; Decker, F.; Dini, D. Spray-deposited NiOx films on ITO substrates as photoactive electrodes for p-type dye-sensitized solar cells. *J. Appl. Electrochem.* **2013**, *43*, 191–197. [CrossRef]
23. Sato, H.; Minami, T.; Takata, S.; Yamada, T. Transparent conducting p-type NiO thin films prepared by magnetron sputtering. *Thin Solid Film.* **1993**, *236*, 27–31. [CrossRef]
24. Cattin, L.; Reguig, B.A.; Khelil, A.; Morsli, M.; Benchouk, K.; Bernède, J.C. Properties of NiO thin films deposited by chemical spray pyrolysis using different precursor solutions. *Appl. Surf. Sci.* **2008**, *254*, 5814–5821. [CrossRef]
25. Antolini, E. Sintering of $Li_xNi_{1-x}O$ solid solutions at 1200 °C. *J. Mater. Sci.* **1992**, *27*, 3335.
26. Joseph, D.P.; Saravanan, M.; Muthuraaman, B.; Renugambal, P.; Sambasivam, S.; Raja, S.P.; Maruthamuthu, P.; Venkateswaran, C. Spray deposition and characterization of nanostructured Li doped NiO thin films for application in dye-sensitized solar cells. *Nanotechnology* **2008**, *19*, 485707–485717. [CrossRef]
27. Chen, X.; Zhao, L.; Niu, Q. Electrical and optical properties of p-type Li, Cu-codoped NiO thin films. *J. Electron. Mater.* **2012**, *41*, 3382–3386. [CrossRef]
28. Hasan, A.J.; Mohammad-Mehdi, B.M.; Mehrdad, S.S. Nickel–lithium oxide alloy transparent conducting films deposited by spray pyrolysis technique. *J. Alloys Comp.* **2011**, *509*, 2770–2773.
29. Desai, J.D. Nickel oxide thin films by spray pyrolysis. *J. Mater. Sci. Mater. Electron.* **2016**. [CrossRef]
30. Jarzebski, Z.M. *Oxide Semiconductors*; Pergamon Press: Oxford, UK, 1973; Volume 4, pp. 184–186.
31. Smith, N. The Structure of Thin Films of Metallic Oxides and Hydrates. *J. Am. Chem. Soc.* **1936**, *58*, 173–179. [CrossRef]
32. Lu, Y.M.; Hwang, W.S.; Yang, J.S. Effect of substrate temperature on the resistivity of non-stoichiometric sputtered NiOx films. *Surf. Coat. Technol.* **2002**, *155*, 231–235. [CrossRef]
33. Burstein, E. Anomalous optical absorption limit in InSb. *Phys. Rev.* **1954**, *93*, 632–633. [CrossRef]
34. Hamberg, I.; Granqvist, C.G.; Berggren, K.F.; Sernelius, B.E.; Engstrom, L. Band-gap widening in heavily Sn-doped In_2O_3. *Phys. Rev. B* **1984**, *30*, 3240–3249. [CrossRef]
35. Chen, X.; Guan, W.; Fang, G.; Zhao, X.Z. Influence of Substrate Temperature and Post-Treatment on the Properties of ZnO:Al Thin Films Prepared by Plused Laser Deposition. *Appl. Surf. Sci.* **2005**, *252*, 1561–1567. [CrossRef]
36. Arunodaya, J.; Sahoo, T. Effect of Li doping on conductivity and band gap of nickel oxide thin film deposited by spin coating technique. *Mater. Res. Express* **2020**, *7*, 016405.
37. Patil, P.S.; Kadam, L.D. Preparation and characterization of spray pyrolyzed nickel oxide (NiO) thin films. *App. Surf. Sci.* **2002**, *199*, 211–221. [CrossRef]
38. Yu, G.H.; Zhu, F.W.; Chai, C.L. X-ray photoelectron spectroscopy study of magnetic films. *Appl. Phys. A Mater. Sci. Process.* **2003**, *76*, 45–47. [CrossRef]
39. Oswald, S.; Bruckner, W. XPS Depth Profile Analysis of non- Stoichiometric NiO Films. *Surf. Interface Anal.* **2004**, *36*, 17–22. [CrossRef]

40. Nandy, S.; Saha, B.; Mitra, M.K.; Chattopadhyay, K.K. Effect of oxygen partial pressure on the electrical and optical properties of highly (200) oriented p-type $Ni_{1-x}O$ films by DC sputtering. *J. Mater. Sci.* **2007**, *42*, 5766–5772. [CrossRef]

41. Chang, H.L.; Lu, T.C.; Kuo, H.C.; Wang, S.C. Effect of oxygen on characteristics of nickel oxide/indium tin oxide heterojunction diodes. *J. Appl. Phys.* **2006**, *100*, 124503. [CrossRef]

42. Haacke, G. New figure of merit for transparent conductors. *J. Appl. Phys.* **1976**, *47*, 4086–4089. [CrossRef]

43. Huang, C.C.; Wang, F.H.; Wu, C.C.; Huang, H.H.; Yang, C.F. Developing high-transmittance heterojunction diodes based on NiO/TZO bilayer thin films. *Nanoscale Res. Lett.* **2013**, *8*, 1–8. [CrossRef] [PubMed]

44. Prabhua, R.R.; Sarithaa, A.C.; Shijeesha, M.R.; Jayaraj, M.K. Fabrication of p-CuO/n-ZnO heterojunction diode via sol-gel spin coating technique. *Mater. Sci. Eng. B* **2017**, *220*, 82–90. [CrossRef]

45. Ohta, H.; Hirano, M. Fabrication and photoresponse of a pn-heterojunction diode composed of transparent oxide semiconductors, p-NiO and n-ZnO. *Appl. Phys. Lett.* **2003**, *83*, 1029–1031. [CrossRef]

46. Ajimsha, R.S.; Vanaja, K.A.; Jayaraj, M.K.; Misra, P.; Dixit, V.K.; Kukreja, L.M. Transparentp-$AgCoO_2$/n-ZnO diode heterojunction fabricated by pulsed laser deposition. *Thin Solid Film.* **2007**, *515*, 7352–7356. [CrossRef]

47. Mahmoudian, M.R.; Basirun, W.J.; Alias, Y. Synthesis of polypyrrole/Ni-doped TiO_2 nanocomposites (NCs) as a protective pigment in organic coating. *Org. Coat.* **2011**, *71*, 56–64. [CrossRef]

48. Guo, X.; Zhan, Q.; Jin, G.; Li, G.; Zhan, Z. Hot-wire semiconductor metal oxide gas sensor based on F-doped SnO_2. *J. Mater. Sci. Mater. Electron.* **2015**, *26*, 860–866. [CrossRef]

49. Liu, H.F.; Chi, D.Z.; Liu, W. Layer-by-layer oxidation of InN(0001) thin films into body-center cubic In_2O_3(111) by cycle rapid thermal annealing. *CrystEngComm* **2012**, *14*, 7140–7144. [CrossRef]

nanomaterials

MDPI

Article

Electrochemical Deposition of Silicon-Carbon Films: A Study on the Nucleation and Growth Mechanism

Nina K. Plugotarenko [1], Tatiana N. Myasoedova [1,*], Mikhail N. Grigoryev [2] and Tatiana S. Mikhailova [1]

[1] Institute of Nanotechnologies, Electronics and Equipment Engineering, Southern Federal University, Chekhov str. 2, 347928 Taganrog, Russia; plugotarenkonk@sfedu.ru (N.K.P.); xelga.maks@yandex.ru (T.S.M.)
[2] Joint Stock Company, Taganrog Scientific-Research Institute of Communication, 347913 Taganrog, Russia; gregoryevmikhail@mail.ru
* Correspondence: tnmyasoedova@sfedu.ru; Tel.: +7-918-523-3488

Received: 13 November 2019; Accepted: 6 December 2019; Published: 10 December 2019

Abstract: Silicon-carbon films have been deposited on silicon and Al_2O_3/Cr-Cu substrates, making use of the electrolysis of methanol/dimethylformamide-hexamethyldisilazane (HMDS) solutions. The electrodeposited films were characterized by Raman spectroscopy and scanning electron microscopy, respectively. Moreover, the nucleation and growth mechanism of the films were studied from the experimental current transients.

Keywords: nucleation; growth; electrochemical deposition; silicon-carbon films

1. Introduction

The diamond-like carbon (DLC) films are extremely alluring for their high mechanical hardness, high electric resistivity, biocompatibility, chemical inertness, low coefficient of friction, and optical transparency in the infrared range [1–3]. The issue of stress and poor adhesion to the substrate in DLC films is a persistent problem that could be solved by incorporation of other elements (W, Ti, Al, Si, etc.) [4–6]. Therefore, the incorporation of silicon is rather promising in order to obtain amorphous silicon-carbon films.

Silicon-carbon films are very promising materials for microelectronic devices operating in aggressive environments [7]. These films are used for gas sensors, ultracapacitors, field emission devices, and other applications in aggressive environments. There are many techniques for producing these films, such as magnetron sputtering [8], ion sputtering, chemical vapor deposition, pulsed laser deposition, electrochemical deposition from molten salt, and the sol-gel method [9–11]. However, the applications of these techniques have been limited, owing to the sophisticated equipment and precise experimental conditions, including high vacuum and high temperature. It was experimentally shown that most materials that can be deposited from the vapor phase can also be deposited in a liquid phase using electrochemical techniques and inversely [10]. The application of the liquid deposition techniques is a good prospect due to such advantages as low consumption of energy, low deposition temperature, availability for large area deposition on complicated surfaces, and the simplicity of the setup. There are some reports that have demonstrated the possibility of the electrochemical deposition of DLC films from the organic liquids such as methanol [12], acetonitrile [13], dimethylsulfoxide [14], and lithium acetylide in dimethylsulfoxide [15], in ambient conditions. However, earlier, we reported the electrochemical deposition of silicon-carbon films from methanol/ethanol and hexamethyldisilazane (HMDS) solution [16,17]. However, in the development of the synthesis of a new material, the deposition kinetics is one of the first components to be studied in detail to ensure reproducibility. Currently, there is no information about the deposition mechanisms of silicon-carbon films from organic liquids onto different substrates.

Electrochemical methods allow setting and controlling the overpotential, control charge, current, the volume of the deposited solution, and a number of nuclei comparatively easily in the system, so they are suitable for the study of the nucleation and growth of a new phase. The analysis of potentiostatic current transients allows getting more information on the mechanism and kinetics of the electrodeposition [18].

The aim of the present study is to investigate the mechanisms of the nucleation and growth of silicon-carbon films onto silicon and Al_2O_3/Cr-Cu substrates through experimental potentiostatic current transients. The surface morphology, as well as structural and phase composition of the films were determined from scanning electron microscopy and Raman spectra investigations, respectively.

2. Materials and Methods

2.1. Synthesis of Silicon-Carbon Films

In this communication, the silicon-carbon films were deposited on silicon (100) (the resistivity was 4.5 Om·cm) and Al_2O_3 substrates with a size of 12×17 mm^2. In the first step, the silicon substrate was dipped in the HF solution ($\approx 15\%$) for a few minutes, and the conducting layer (Cr-Cu) was sputtered on the surface of the Al_2O_3 substrate by the magnetron technique. The substrate was mounted on the negative electrode, and graphite was mounted on the positive electrode. The distance between the substrate and the positive electrode was set to 10 mm. The deposition was done from two types of solution: (1) a methanol and HMDS solution; (2) a dimethylformamide (DMF) and HMDS solution. HMDS was dissolved in analytically pure methanol/DMF, with the volume ratio of HMDS to methanol (DMF) of 1:9. The films were deposited for 30 min. The applied potential was 180 and 500 V, for methanol-HMDS and DMF-HMDS solutions, respectively.

A schematic diagram of the experimental setup is shown in Figure 1:

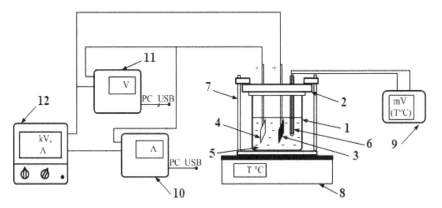

Figure 1. Schematic structure of electrolytic deposition system (1, glass cell; 2, dielectric cover; 3, graphite anode; 4, cathode substrate; 5, solution; 6, thermocouple; 7, clamps; 8, thermal table; 9, voltmeter of the thermocouple; 10, ammeter; 11, high-voltage voltmeter; 12, power supply).

2.2. Characterization

The film morphologies were investigated using scanning electron microscopy (SEM; SEM Zeiss Merlin compact VP-60-13, Stavropol, Russia). Raman spectra were recorded at ambient temperature using a Raman Microscope, Renishaw plc (Stavropol, Russia, resolution 2 cm^{-1}, 514 nm laser).

3. Results and Discussion

3.1. Characterization

During the deposition for a composite film from the DMF-HMDS solution, we found that the current density increased from 35 mA/cm² to 54–57 mA/cm² with deposition time. In the case of the methanol-HMDS solution, the current density decreased slightly from 50 mA/cm² to 44 mA/cm² and increased from 50 mA/cm² to 55 mA/cm² during the film deposition onto the silicon and Al₂O₃ substrate, respectively (Figure 2).

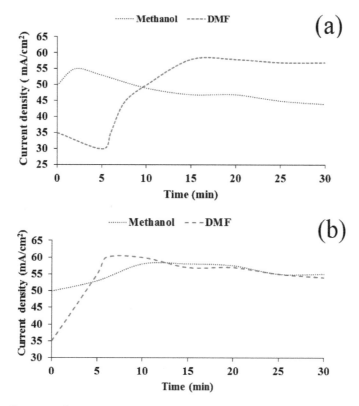

Figure 2. Experimental potentiostatic current transients for the deposition of silicon-carbon films on silicon (**a**) and Al₂O₃ (**b**) substrates.

The surface morphology of the films changes under varying technological conditions. The production of silicon-carbon materials is associated with thermodynamically nonequilibrium processes, which cause the formation of inhomogeneities as the films grow due to the self-organization of the structure. Figures 3 and 4 shows the SEM micrographs of the deposited films. From the figures, it can be seen that films deposited from the methanol-HMDS solution and DMF-HMDS solution on the silicon substrate are composed of compact grains. The average grain size was about 90, 60, and 170 nm for the films, deposited from the methanol-HMDS on the silicon substrate, from the DMF-HMDS solution on the silicon substrate, and from the methanol-HMDS on the Al₂O₃ substrate, respectively. The silicon-carbon films deposited from the DMF-HMDS solution on the Al₂O₃ substrate characterized by a powdery structure without large grains. Therefore, the histograms of the grain size distributions were built (Figure 5).

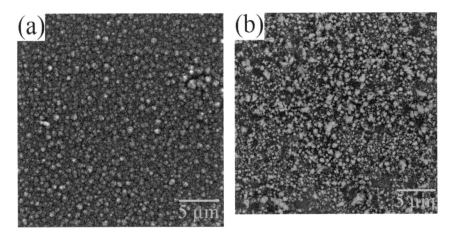

Figure 3. SEM micrographs of the silicon-carbon films deposited onto the silicon substrate from the methanol-hexamethyldisilazane (HMDS) (**a**) and DMF-HMDS (**b**) solutions.

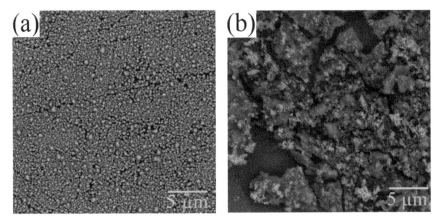

Figure 4. SEM micrographs of the silicon-carbon films deposited on the Al_2O_3 substrate from the methanol-HMDS (**a**) and DMF-HMDS (**b**) solutions.

The scatter of grain size values for the films on silicon substrates lied in the range from 20 nm to 200 nm. Grains with sizes of 50 and 80 nm predominated for the films deposited from the methanol-HMDS and DMF-HMDS solutions, respectively. For the films deposited onto the Al_2O_3 substrate, the histogram of the grain size values distribution was characterized by the absence of pronounced maxima. It was evident that the range of grain sizes for the films deposited from the methanol-HMDS solution was much narrower than for those deposited from the DMF-HMDS solution and was in the range of 60–150 nm.

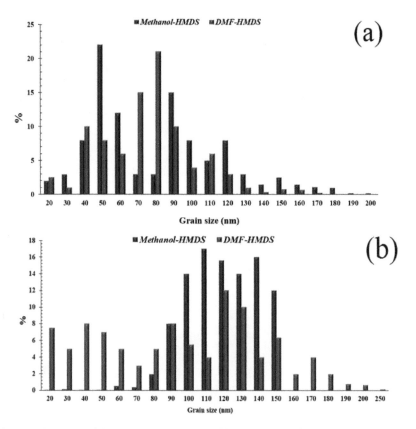

Figure 5. Histograms of the grain size distributions of the silicon-carbon films deposited on silicon (**a**) and Al$_2$O$_3$ (**b**) substrates.

The Raman spectra of the films with the deconvolution of the D and G peaks, deposited on silicon and Al$_2$O$_3$ substrates, are shown in Figure 6a,b, respectively.

The silicon-carbon films deposited from the methanol-HMDS solution investigated in this work were complex heterogeneous objects (Figure 6a). The Raman spectra contained the lines in the range that was characteristic of the SiC polytypes. The samples were characterized by the presence of the hexagonal 6H SiC polytype with the impurities of the rhombohedral 15R SiC phase. Furthermore, the bands attributed to the Si–C bond and nanocrystalline diamond (ND) were observed. The spectrum of the silicon-carbon film deposited on the Al$_2$O$_3$ substrate shifted to a lower wavenumber. The deconvolution of the Raman spectra allowed us to find out "hidden" peaks. Deconvolution was carried out on a minimum number of Gauss peak components for which their resulting curve described the experimental curve with confidence >0.99%. Therefore, in the resulting Gauss deconvolution, three peaks were observed at 1361, 1524, and 1627 cm^{-1}. The peaks centered at 1361 and 1524 cm^{-1} corresponded to the conventional D and G bands. The broadening in the G band at the higher wavenumber side was due to the presence of the D′ band at 1627 cm^{-1}. The appearance of the D′ peak proved that silicon-carbon films were highly defective structures [19]. The relative intensity ratio of the D peak to G peak (I_D/I_G) of the silicon-carbon films deposited from the methanol-HMDS solution was 1.05 for the films on both types of substrates.

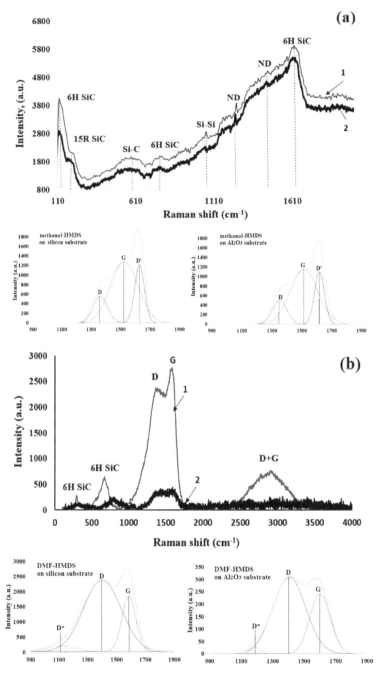

Figure 6. Raman spectra with the deconvolution of the D and G peaks (under the Raman spectra) of the silicon-carbon films deposited onto the silicon (1) and Al_2O_3 (2) substrates from the methanol-HMDS (**a**) and DMF-HMDS (**b**) solutions (D* and D′ peaks characterize disorder carbon).

Raman spectra of silicon-carbon films deposited from the DMF-HMDS solution could be characterized by the presence of the D peak and the G peak (Figure 6b). The spectrum of the silicon-carbon film on the silicon substrate was also characterized by the D + G scattering peak.

In the spectrum of the silicon-carbon film deposited on the silicon substrate, the position of the D and G peaks was 1386 and 1587 cm^{-1}, respectively (Figure 6a), while the position of the D and G peaks was 1438 and 1597 cm^{-1}, respectively, in the spectrum of silicon-carbon film, deposited on the Al$_2$O$_3$ substrate [20]. Furthermore, the G peak of the silicon-carbon film deposited on the silicon substrate shifted to a lower wavenumber, and the full width at half maximum of the G peak was also larger than that of the silicon-carbon film, deposited on the Al$_2$O$_3$ substrate. The high intensity of the D peak confirmed the existence of unsaturated hydrocarbons on the surface of SiC nanoparticles [21]. The bands attributed to the hexagonal 6H SiC polytype were observed.

The deconvolution of the D and G bands of the films deposited from the DMF-HMDS solution was also carried out as shown in Figure 6b. The D*, D, and G peaks were found. It should be noted that the D* peak has been found in disordered carbons. Some reports have attributed the D* peak to the sp^3 rich phase of disordered amorphous carbons [22]. The D and G peaks were centered at 1405 (1400) cm^{-1} and 1600 (1584) cm^{-1}.

Furthermore, it was seen that the relative intensity ratio of the D peak to G peak (I_D/I_G) of the silicon-carbon film deposited from the DMF-HMDS solution was higher than for the films deposited from the methanol-HMDS solution and reached ~1.29. The smaller ratio corresponded to smaller free carbon clusters [23].

3.2. Mechanism Study

The structure and morphology of silicon-carbon films depends on the nucleation and growth mechanism.

Potentiostatic transient measurement is an important method for studying the initial kinetics of electrocrystallization reactions [24–26].

The existing models of electrochemical deposition were based on two main ideal mechanisms for new phase nucleation on the electrode surface: instantaneous nucleation and progressive nucleation. In the case of instantaneous nucleation, all active centers are filled almost simultaneously, and further, slow growth of nuclei occurs due to the introduction of new atoms. In the presence of inhomogeneities on the surface of the substrate, germ growth first occurs at the most active centers, so with progressive nucleation, the nuclei simultaneously emerge and continue to grow. It is assumed that there is a constant supersaturation of the precursor concentration under potentiostatic conditions. Besides, both kinetic controlled and diffusion controlled growth mechanisms of a new phase on the surface are possible.

The model of 3D multiple nucleations with kinetic controlled growth was described by Isaev [18]. Instantaneous nucleation is described by:

$$\frac{j}{j_{max}} = 2.34 \frac{t}{t_{max}} \omega \left(1.50 \frac{t}{t_{max}} \right) \tag{1}$$

where j is the current density, t is time, and t_{max} is the time at the maximum current.

Progressive nucleation can be expressed as:

$$\frac{j}{j_{max}} = 2.25 \omega_2 \left(1.34 \frac{t}{t_{max}} \right) \tag{2}$$

where $\omega(x) = \exp(-x^2) \int_0^x \exp(\xi^2) d\xi$ is Dawson's integral:

$$\omega_2(y) = \exp(-y^3) \int_0^y \left(y^2 - \xi^2 \right) \exp\left(3y\xi^2 - 2\xi^3 \right) d\xi \tag{3}$$

The model of controlled nucleation was offered by Scharifker and Hills [27]. They considered the 3D nucleation model given that over time, the diffusion zones of individual nuclei overlap, which leads to a slowdown in germ growth. Instantaneous nucleation and growth are described by:

$$\left(\frac{j}{j_{max}}\right)^2 = \frac{1.9542}{(t/t_{max})}\{1 - \exp[-1.2564(t/t_{max})]\}^2 \tag{4}$$

Progressive nucleation can be expressed as:

$$\left(\frac{j}{j_{max}}\right)^2 = \frac{1.2254}{(t/t_{max})}\left\{1 - \exp\left[-2.3367(t/t_{max})^2\right]\right\}^2 \tag{5}$$

The experimental current–time transients shown in Figure 2 were analyzed using these expressions and experimentally obtained values for j_{max} and t_{max}. First, the dependences of $\ln\left(1 - \frac{j\sqrt{t}}{(j\sqrt{t})_{max}}\right)$ from t and t^2 were built in order to determine instantaneous or progressive nucleation.

Figure 7 shows graphs of electrodeposition transients characteristic of instantaneous nucleation.

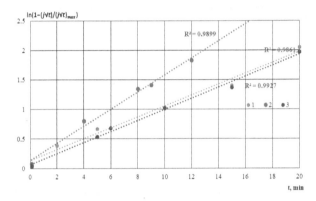

Figure 7. Semilogarithmic dependencies calculated from the current transients for the film deposition solution on the silicon substrate from the methanol-HMDS solution (1); on the Al$_2$O$_3$ substrate from the methanol-HMDS solution (2) and DMF-HMDS solution (3).

Progressive nucleation is described by Figure 8.

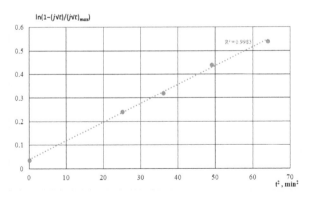

Figure 8. Semilogarithmic dependencies calculated from the current transients for the film deposition from the DMF-HMDS solution onto the silicon substrate.

As shown in the figures, for all straight lines, a high approximation confidence value was set. The comparison was based on the standard error value. Deviations from linearity were caused by concurrent processes in the solution and on the substrate: molecules' dissociation, heating of the solution, and the formation of silicon-carbon and carbon bonds, characterizing the different growth rates.

In Figure 9, the model and experimental dependencies are presented. The analysis of the semilogarithmic and (j/j_m) vs. (t/t_m) dependencies showed that the mechanism of nucleation and growth of silicon-carbon films from the methanol-HMDS and DMF-HMDS solutions on the Al_2O_3 substrate was well described by Equation 1 for instantaneous nucleation (Figure 9b,d). The experimental current transients represented in the coordinates (j/j_m) vs. (t/t_m) for the deposition of silicon-carbon film from the DMF-HMDS solution on the silicon substrate demonstrated the characteristic features of the diffusion controlled growth model (Figure 9c).

The deposition of the silicon-carbon films from the methanol-HMDS solution onto the silicon substrate was characterized by the instantaneous nucleation with kinetically controlled growth (Figure 9a), while the model for instantaneous nucleation with diffusion controlled growth fit the growth mechanisms of the new phase from the methanol-HMDS and DMF-HMDS solutions on the Al_2O_3 substrate.

All the experimental dependences of (j/j_m) vs. (t/t_m) except the deposition from DMF-HMDS solution on the silicon substrate demonstrated the higher current density compared to the model for first two minutes due to the dissociation of molecules in precursors.

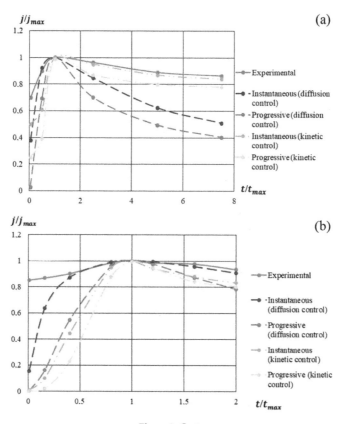

Figure 9. *Cont.*

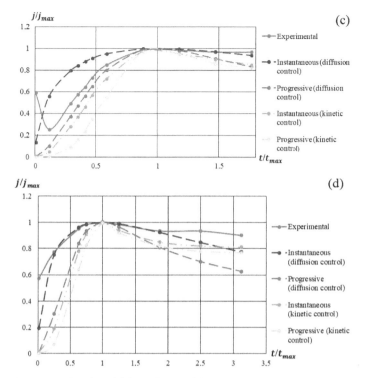

Figure 9. Experimental and model dependences of the current density on the deposition time of silicon-carbon films from: (**a**) methanol-HMDS solution on a silicon substrate; (**b**) methanol-HMDS solution on a Al_2O_3 substrate; (**c**) DMF-HMDS solution on a silicon substrate; (**d**) DMF-HMDS solution on a Al_2O_3 substrate.

4. Conclusions

The silicon-carbon films were successfully deposited on silicon and Al_2O_3/Cu-Cr substrates from organic solutions. The films deposited from the methanol-HMDS solution were mostly characterized by the presence of the hexagonal 6H SiC polytype with the impurities of the rhombohedral 15R SiC phase. Raman spectra of silicon-carbon films, deposited from the DMF-HMDS, solution can be characterized by the presence of the D peak, G peak, and D + G scattering peaks of carbon and 6H SiC polytype peaks. It was shown that the nucleation and growth mechanisms depend on the nature of the solution and substrate.

Author Contributions: Conceptualization, N.K.P. and T.N.M.; data curation, N.K.P. and T.N.M.; formal analysis, N.K.P. and T.N.M.; funding acquisition, T.N.M. and N.K.P.; investigation, M.N.G. and T.S.M.; methodology, N.K.P., T.N.M., and M.N.G.; project administration, T.N.M. and N.K.P.; resources, T.N.M., M.N.G. and N.K.P.; software, N.K.P.; supervision, T.N.M.; validation, T.N.M. and N.K.P.; visualization, T.N.M., M.N.G., N.K.P., and T.S.M.; writing, original draft preparation, N.K.P. and T.N.M.; writing, article and editing, T.N.M. and T.S.M.

Funding: This work was financially supported by the Ministry of Education of Russia, under Contract No. 14.575.21.0126 (the unique identifier for the contract is RFMEFI57517X0126).

Acknowledgments: The authors acknowledge "The Center for the Collective Use of Scientific Equipment" of the North-Caucasus Federal University (Stavropol, Russia) for Raman spectroscopy and SEM investigations.

Conflicts of Interest: The authors declare no conflict of interest. The funders had no role in the design of the study in the collection, analyses, or interpretation of data; in the writing of the manuscript; nor in the decision to publish the results.

References

1. Basa, D.K.; Ambrosone, G.; Coscia, U.; Setaro, A. Crystallization of hydrogenated amorphous silicon-carbon films with laser and thermal annealing. *Appl. Surf. Sci.* **2009**, *255*, 5528–5531. [CrossRef]
2. Wang, H.; Shen, M.R.; Ning, Z.Y.; Ye, C.; Cao, C.B.; Dang, H.Y.; Zhu, H.S. Deposition of diamond-like carbon films by electrolysis of methanol solution. *Appl. Phys. Lett.* **1996**, *69*, 1074–1076. [CrossRef]
3. Zhang, Q.; Wang, Y.; Wang, W.; Mitsuzak, N.; Chen, Z. Low voltage and ambient temperature electrodeposition of uniform carbon films. *Electrochem. Commun.* **2016**, *63*, 22–25. [CrossRef]
4. Venkatraman, C.; Brodbeck, C.; Lei, R. Tribological properties of diamond-like nanocomposite coatings at high temperatures. *Surf. Coat. Technol.* **1999**, *115*, 215–221. [CrossRef]
5. Ji, Y.; Ma, M.; Ji, X.; Xiong, X.; Sun, X. Nickel-carbonate nanowire array: An efficient and durable electrocatalyst for water oxidation under nearly neutral conditions. *Front. Chem. Sci. Eng.* **2018**, *12*, 462–467. [CrossRef]
6. Ding, X.Z. Structural and mechanical properties of Ti-containing diamond-like carbon films deposited by filtered cathodic vacuum arc. *Thin Solid Films* **2002**, *408*, 183–187. [CrossRef]
7. Silva, J.A.; Quoizola, S.; Hernandez, E.; Thomas, L.; Massines, F. Silicon carbon nitride films as passivation and antireflective coatings for silicon solar cells. *Surf. Coat. Technol.* **2014**, *242*, 157–163.
8. Hoche, H.; Pusch, C.; Riedel, R. Properties of SiCN coatings for high temperature applications—Comparison of RF-, DC- and HPPMS-sputtering. *Surf. Coat. Technol.* **2010**, *205*, S21–S27. [CrossRef]
9. Peng, Y.; Zhou, J. The influence of radiofrequency power on compositional, structural and optical properties of amorphous silicon. *Appl. Surf. Sci.* **2010**, *256*, 2189–2192. [CrossRef]
10. Ambrosone, G.; Basa, D.K.; Coscia, U. Structural and electrical properties of nanostructured silicon carbon films. *Energy Procedia* **2010**, *2*, 3–7. [CrossRef]
11. Manocha, S.; Ankur, D.; Manocha, L.M. Formation of silicon carbide whiskers from organic precursors via sol-gel method. *Eurasian Chem. Technol. J.* **2011**, *13*, 27–33. [CrossRef]
12. Yan, B.; Tay, B.K.; Chen, G.; Yang, S.R. Synthesis of silicon carbide nitride nanocomposite films by a simple electrochemical method. *Electrochem. Commun.* **2006**, *8*, 737–740. [CrossRef]
13. Guo, D.; Cai, K.; Li, L.T.; Zhu, H.S. Preparation of hydrogenated diamond-like carbon films on conductive glass from an organic liquid using pulsed power. *Chem. Phys. Lett.* **2000**, *325*, 499–502. [CrossRef]
14. Jiang, H.Q.; Huang, L.N.; Zhang, Z.J.; Xu, T.; Liu, W.M. Deposition of nanostructured diamond-like carbon films on al substrate by facile electrochemical route. *Chem. Lett.* **2004**, *33*, 378–379. [CrossRef]
15. Kulak, A.I.; Kokorin, A.I.; Meissner, D.; Ralchenko, V.G.; Vlasov, I.I.; Kondratyuk, A.V.; Kulak, T.I. Electrodeposition of nanostructured diamond-like films by oxidation of lithium acetylide. *Electrochem. Commun.* **2003**, *5*, 301–305. [CrossRef]
16. Myasoedova, T.N.; Grigoryev, M.N.; Plugotarenko, N.K.; Mikhailova, T.S. Fabrication of gas-sensor chips based on silicon–carbon films obtained by electrochemical deposition. *Chemosensors* **2019**, *7*, 52. [CrossRef]
17. Grigoryev, M.N.; Myasoedova, T.N.; Mikhailova, T.S. The electrochemical deposition of silicon-carbon thin films from organic solution. *J. Phys. Conf. Ser.* **2018**, *1124*, 081043. [CrossRef]
18. Isaev, V.A.; Grishenkova, O.V.; Zaykov, Y.P. On the theory of 3D multiple nucleation with kinetic controlled growth. *J. Electroanal. Chem.* **2018**, *818*, 265–269. [CrossRef]
19. Kaniyoor, A.; Ramaprabhu, S. A raman spectroscopic investigation of graphite oxide derived graphene. *Aip Adv.* **2012**, *2*, 032183. [CrossRef]
20. Ferrari, A.C.; Robertson, J. Raman spectroscopy of amorphous, nanostructured, diamond-like carbon, and nanodiamond. *Philos. Trans. R. Soc. Lond. A* **2004**, *362*, 2477–2512. [CrossRef]
21. Mehr, M.; Moore, D.T.; Esquivel-Elizondo, J.R.; Nino, J.C. Mechanical and thermal properties of low temperature sintered silicon carbide using a preceramic polymer as binder. *J. Mater. Sci.* **2015**, *50*, 7000–7009. [CrossRef]
22. Schwan, J.; Ulrich, S.; Batori, V.; Ehrhardt, H. Raman spectroscopy on amorphous carbon films. *J. Appl. Phys.* **1996**, *80*, 440. [CrossRef]
23. Iijima, M.; Kamiya, H. Surface modification of silicon carbide nanoparticles by azo radical initiators. *J. Phys. Chem. C* **2008**, *112*, 11786–11790. [CrossRef]
24. Bijani, S.; Schrebler, R.E.; Dalchiele, A.; Gab, M.; Martínez, L.; Ramos-Barrado, J.R. Study of the nucleation and growth mechanisms in the electrodeposition of micro- and nanostructured Cu_2O thin films. *J. Phys. Chem. C* **2011**, *115*, 21373–21382. [CrossRef]

25. Greef, R.; Peat, R.; Peter, L.M.; Pletcher, D. *Instrumental Methods in Electrochemistry*; Robinson, J., Ed.; Ellis Horwood: Chichester, UK, 1985; p. 283.

26. Quayum, M.E.; Ye, S.; Uosaki, K. Mechanism for nucleation and growth of electrochemical palladium deposition on an Au (111) electrode. *J. Electroanal. Chem.* **2002**, *520*, 126–132. [CrossRef]

27. Scharifker, B.R.; Hills, G.J. Theoretical and experimental studies of multiple nucleation. *Electrochim. Acta* **1983**, *28*, 879–889. [CrossRef]

 *nanomaterials*

MDPI

Article

Influence of Sintering Strategy on the Characteristics of Sol-Gel Ba$_{1-x}$Ce$_x$Ti$_{1-x/4}$O$_3$ Ceramics

Cătălina A. Stanciu [1,2], Ioana Pintilie [3], Adrian Surdu [1], Roxana Truşcă [1], Bogdan S. Vasile [1], Mihai Eftimie [1] and Adelina C. Ianculescu [1,*]

[1] Department of Science and Enginnering of Oxide Materials and Nanomaterials, University POLITEHNICA of Bucharest, Bucharest 060042, Romania; catalina.a.stanciu@gmail.com (C.A.S.); adrian.surdu@upb.ro (A.S.); truscaroxana@yahoo.com (R.T.); bogdan.vasile@upb.ro (B.S.V.); mihai.eftimie@yahoo.com (M.E.)
[2] National Institute for Lasers, Plasma and Radiation Physics, P.O. Box MG54, Bucharest-Magurele 077125, Romania
[3] National Institute of Materials Physics, P.O. Box MG-7, Bucharest-Magurele 077125, Romania; ioana@infim.ro
* Correspondence: a_ianculescu@yahoo.com; Tel.: +40-21-4023960 or +40-721487396; Fax: +40-21-3181010

Received: 22 October 2019; Accepted: 18 November 2019; Published: 23 November 2019

Abstract: Single-phase Ce^{3+}-doped BaTiO$_3$ powders described by the nominal formula Ba$_{1-x}$Ce$_x$Ti$_{1-x/4}$O$_3$ with $x = 0.005$ and 0.05 were synthesized by the acetate variant of the sol-gel method. The structural parameters, particle size, and morphology are strongly dependent on the Ce^{3+} content. From these powders, dense ceramics were prepared by conventional sintering at 1300 °C for 2 h, as well as by spark plasma sintering at 1050 °C for 2 min. For the conventionally sintered ceramics, the XRD data and the dielectric and hysteresis measurements reveal that at room temperature, the specimen with low cerium content ($x = 0.005$) was in the ferroelectric state, while the samples with significantly higher Ce^{3+} concentration ($x = 0.05$) were found to be in the proximity of the ferroelectric–paraelectric phase transition. The sample with low solute content after spark plasma sintering exhibited insulating behavior, with significantly higher values of relative permittivity and dielectric losses over the entire investigated temperature range relative to the conventionally sintered sample of similar composition. The spark-plasma-sintered Ce-BaTiO$_3$ specimen with high solute content ($x = 0.05$) showed a fine-grained microstructure and an almost temperature-independent colossal dielectric constant which originated from very high interfacial polarization.

Keywords: sol-gel; Ce^{3+}-doped BaTiO$_3$; dielectric permittivity; ferroelectric–paraelectric phase transition; spark plasma sintering

1. Introduction

Due to their environmentally friendly character, along with multiple unique and useful properties such as high dielectric permittivity, piezoelectric, pyroelectric and ferroelectric behavior, and PTCR (positive temperature coefficient of resistivity) effect and high endurance under DC field stress, BaTiO$_3$-based ceramics are widely used in microelectronics applications as multilayer ceramic capacitors (MLCC), ferroelectric memory (FeRAM), energy harvester actuators, transducers, IR sensors, thermistors, controllers, and phase shifters [1–5].

Taking into account the flexibility of the perovskite lattice, a proper approach to tailoring the functional behavior of BaTiO$_3$-based ceramics consists of adopting an adequate compositional design, which involves doping strategies or the formation of solid solutions.

It is well known that A-site donor doping in BaTiO$_3$ involving the incorporation of larger rare-earth ions such as La^{3+} and Nd^{3+} [6–12] in the perovskite lattice is more effective in shifting the Curie temperature toward lower values than the effect induced by homovalent species such as Sr^{2+} and Ca^{2+} [13–17]. The content of donor dopant also exhibits a complicated influence on the electrical properties. Thus, a higher content of donor solutes, exceeding a critical composition of ~0.5 atom %, but found within their solubility range, determines a typical insulating behavior determined by a compensating mechanism via cation vacancies, in order to maintain electroneutrality [18]. For concentrations below the already mentioned critical value, such donor-doped $BaTiO_3$ systems behave at room temperature like semiconductors, with a well-marked PTCR effect just above the Curie temperature due to the compensation of the donor dopant's positive "extra-charge" inside the ceramic grains via electrons [19]. However, a previous study revealed that even in the case of lightly La-doped $BaTiO_3$ ceramics, changes in the defect chemistry by modifying the preparation and sintering strategy can induce a typical insulating, nonlinear behavior at room temperature [9].

Because of its variable oxidation state, cerium can be incorporated as Ce^{3+} on Ba sites, acting as a donor dopant [20–22]; as Ce^{4+} on Ti^{4+}, forming homovalent solid solutions [23–26]; or simultaneously as Ce^{3+} on *A* sites and Ce^{4+} on *B* sites of the ABO_3 perovskite lattice [27,28]. The valence of cerium and, consequently, its site occupancy and solubility in the host $BaTiO_3$ lattice are strongly dependent on the starting Ba/Ti ratio and oxygen partial pressure [28].

According to Equation (1), "built-in" titanium vacancies as compensating defects for the effective positive extra-charge of the donor dopant have to be stipulated in the starting formula, and sintering in air at temperatures below 1350 °C followed by slow cooling has to be carried out in order to obtain single-phase, insulating ceramics [6–8,29,30].

$$4CeO_2 + 3TiO_2 \rightarrow 4Ce_{Ba}^{\bullet} + 3Ti_{Ti}^{X} + V_{Ti}^{''''} + 12O_O^{X} + O_2(g) \tag{1}$$

In the present work, we considered two compositions, corresponding to the nominal formula $Ba_{1-x}Ce_xTi_{1-x/4}O_3$, with $x = 0.005$ and 0.05, respectively. As already mentioned, a Ce^{3+} doping level of 0.5 atom % ($x = 0.005$) is considered the "critical" threshold for which a semiconductor–insulator transition accompanied by the so-called "grain growth anomaly" takes place, while a one-order-of-magnitude-higher solute concentration of 5 atom % ($x = 0.05$) represents the "morphotropic" concentration reported in the literature for shifting the Curie temperature in the proximity of room temperature [20–22].

Various techniques such as the classical solid-state reaction method [20,31,32], the sol-gel route [33], and the Pechini procedure [21,22,34,35] have been reported for the preparation of Ce^{3+}-doped $BaTiO_3$ ceramics. The "acetate" variant of the sol-gel route is known as an excellent preparation technique for controlling the stoichiometry and microstructure homogeneity of doped $BaTiO_3$-based products [33,36–38], so we considered it as adequate for our study. The sol-gel powders were labelled after their composition as BCT-005 and BCT-05, respectively.

In the last several years, spark plasma sintering has quite often been used as an alternative method to the classical sintering procedure in order to obtain highly densified, nanocrystalline ceramics from powders prepared by various wet chemical methods [39–42]. $BaTiO_3$ ceramic bodies with minimum grain growth and therefore exhibiting grain size (GS) down to 30 nm and relative density of 95–99% have been produced by this procedure [43,44]. Unlike pure $BaTiO_3$, spark-plasma-sintered ceramics derived from substituted barium titanate solid solutions are much less studied from the point of view of their microstructure and electrical characteristics [45–48]. Regarding the Ce^{3+}-doped $BaTiO_3$ system, only results describing the characteristics of the ceramics consolidated by conventional sintering are reported in the literature. No data referring to the microstructure and electrical behavior of cerium-doped barium titanate ceramics obtained by field-assisted sintering techniques were found. Therefore, in order to obtain fine-grained ceramics and to investigate the influence of grain size on the electrical behavior, the sol-gel Ce^{3+}-doped $BaTiO_3$ powders were also consolidated by spark plasma sintering. The structural parameters, microstructure, and functional properties of the resulting ceramics after conventional sintering and spark plasma sintering are comparatively discussed. The ceramic specimens were differentiated after the sintering procedure by adding to the powder labels the indicative "CS" for conventional sintering and "SPS" for spark plasma sintering.

2. Materials and Methods

2.1. Sample Preparation

Barium acetate (Ba(CH$_3$COO)$_2$, 99%, Sigma Aldrich, Saint Louis, MO, USA), titanium (IV) isopropoxide 97% solution in 2-propanol (Ti[OCH(CH$_3$)$_2$]$_4$, Sigma Aldrich, Saint Louis, MO, USA), and cerium acetate (Ce(CH$_3$CO$_2$)$_3$, 99.9%, Sigma Aldrich, Saint Louis, MO, USA) were used as starting reagents. Two different solutions were prepared by dissolving appropriate amounts of barium acetate in acetic acid and cerium acetate in acetic acid, at 70 °C, under continuous stirring. As stabilizers for the sol, 2-methoxyethanol and acetylacetone in a 2:1 volume ratio were used. Another solution was formed by mixing titanium isopropoxide in 2-propanol. The barium acetate solution was added to the titanium isopropoxide solution under continuous stirring. Then, the cerium acetate solution was added to the barium and titanium mixture solution. In a subsequent step, acetylacetone (CH$_3$COCH$_2$COCH$_3$, 97%, Aldrich, Saint Louis, MO, USA) was added to the as-obtained solution. The resulting clear, yellowish sols were then heated on a hotplate under magnetic stirring at 80 °C for 2 h. During this process, the solutions became increasingly viscous, until yellow-brown gels were obtained. These gels were thermally treated at 100 °C for 12 h, resulting in brown-orange glassy resins which were lightly ground in a mortar. The amorphous precursor powders were then thermally treated in static air at 900 °C for 2 h in a muffle furnace, using a heating rate of 5 °C min^{-1}. An intermediate plateau of 2 h at a lower temperature of 450 °C was also carried out in order to ensure the complete removal of organic material.

The as-synthesized oxide powders with compositions described by the nominal formula Ba$_{1-x}$Ce$_x$Ti$_{1-x/4}$O$_3$ ($x = 0.005$ and 0.05) were milled and shaped by uniaxial die pressing at 174 MPa into pellets with a diameter of ~13 mm and a thickness of ~1.2–1.6 mm, using a small amount of organic binder (5% PVA aqueous solution). The green bodies were sintered in a muffle furnace in static air at 1300 °C for a 4 h plateau, using a heating rate of 5 °C·min^{-1}, and then they were slowly cooled (at the normal cooling rate of the furnace) to room temperature in order to obtain dense Ce^{3+}-doped BaTiO$_3$ ceramic samples. After sintering, the color of the ceramic samples varied from light orange to reddish with varying Ce^{3+} content from 0.5 atom % to 5 atom %.

Amounts of sol-gel powders with the abovementioned compositions were also used to prepare dense ceramics by spark plasma sintering (SPS). The powders were poured into a graphite die and then sintered under vacuum to create dense ceramic pellets of 10 mm diameter and ~1 mm thickness, using commercial SPS equipment (FCT Systeme GmbH, Rauenstein, Germany, Spark Plasma Sintering Furnace type HP D 1.25). The temperature was raised at a fixed heating rate of 100 °C·min^{-1} under a constant applied pressure of 50 MPa and then held at a constant temperature of 1050 °C for 2 min. Rapid heating was provided by a pulsed DC current. After polishing, all the ceramic samples were annealed in air at 1000 °C for 16 h, with a heating rate of 10 °C min^{-1}. The aims of this post-sintering thermal treatment were (*i*) to reduce the concentration of oxygen vacancies originating from the reducing conditions of the SPS process and to ensure the re-oxidation of the Ti^{3+} species to Ti^{4+}; (*ii*) to remove possible surface carbon contamination; and (*iii*) to relieve the residual stresses arising either from the SPS process or from polishing.

2.2. Sample Characterization

To monitor the decomposition process of the precursors, thermal analysis investigations were performed in a static air atmosphere up to 1200 °C, with a heating rate of 10 °C min^{-1}, by using a NETZSCH STA 409 PC LUXX (Selb, Germany) thermal analyzer.

The phase purity and crystal structure of the Ce^{3+}-doped BaTiO$_3$ powders and related ceramics were determined by X-ray diffraction (XRD) investigations, performed at room temperature (23 °C) by means of a SHIMADZU XRD 6000 diffractometer (Kyoto, Japan), using Ni-filtered Cu Kα radiation ($\lambda = 1.5406$ Å). The measurements were performed in θ–2θ mode with a scan step increment of 0.02° and a counting time of 1 s/step in the 2θ range of (20°–80°). Phase identification was performed using HighScore Plus 3.0e software (PANalytical, Almelo, The Netherlands), PANalytical, Almelo,

The Netherlands connected to the ICDD PDF-4+ 2017 database. To estimate the structural characteristics, the same step increment but with a counting time of 10 s/step in the same 2θ range was used. Lattice parameters were refined by the Rietveld method. After removing the instrumental contribution, the full width at half-maximum (FWHM) of the diffraction peaks can be interpreted in terms of the crystallite size and lattice strain. A pseudo-Voigt function was used to refine the shapes of the Ce^{3+}-doped $BaTiO_3$ peaks.

For a high-accuracy estimation of the morphology and crystallinity degree of the constitutive Ce^{3+}-doped $BaTiO_3$ particles, transmission electron microscopy (TEM/HRTEM) and selected-area electron diffraction (SAED) investigations were performed. The bright-field and high-resolution images were collected by using a TecnaiTM G^2 F30 S-TWIN transmission electron microscope (FEI Co., Hillsboro, OR, USA). For these purposes, small amounts of powdered samples were suspended in ethanol by 15 min ultrasonication. A drop of suspension was put onto a 400 mesh, holey carbon-coated film Cu grid and dried. The average particle size for each Ce^{3+}-doped $BaTiO_3$ powder was calculated using Origin Pro 9.0 software (OriginLab, Northampton, MA, USA) by taking into account size measurements on ~60 particles (from images of appropriate magnifications obtained from various microscopic fields) performed using the microscope software Digital Micrograph 1.8.0 (Gatan, Sarasota, FL, USA).

A high-resolution FEI QUANTA INSPECT F microscope with field emission gun (FEI Co., Hillsboro, OR, USA), coupled with energy-dispersive X-ray spectroscopy (EDX), was used to analyze the microstructure and the elemental composition of the ceramics obtained by both conventional sintering and spark plasma sintering. The grain size of the ceramics was determined as the mean intercept length by taking into account measurements on ~70–80 grains. The relative density values of the sintered ceramic pellets were roughly determined as the ratio between the apparent density measured by Archimedes' principle and their crystallographic (theoretical) density calculated from the diffraction data.

The dielectric behavior was studied by performing electrical measurements of capacitance and dielectric losses, with the aid of an HP 4194A impedance/gain analyzer, on samples having a parallel-plate capacitor configuration (by applying Ag–Ag electrodes on both polished surfaces of the sintered ceramic disks of about 1 mm thickness). The measurements were performed in a Janis cryostat, in a temperature range of −250 to +200 °C, and at different fixed frequencies of the small-amplitude AC signal, in the range of 10^3–10^5 Hz. All these measurements were performed in vacuum, during heating up, with a heating rate of $\beta = 0.01$ K/s (0.6 K/minute). It is thought that these measurement conditions ensure a good thermal equilibrium for the sample during the capacitance measurements.

Hysteresis measurements, revealing the total (dynamic) polarization versus the electric field, were performed at ambient temperature (22 °C) using a "Premier II" ferritester together with the "Vision" software (Radiant Technologies, Inc., Albuquerque, NM, USA, version 4.2.0) package provided by Radiant Technologies Inc (Radiant Technologies, Inc., Albuquerque, NM, USA). Measurements were performed using a 1 s pulse (1 Hz measurement frequency).

3. Results and Discussion

3.1. Thermal Behaviour of Precursors

Thermal analysis methods allowed us to investigate the transformation of the amorphous precursors into oxide powders during heating. Monitoring the thermal behavior of the gel precursor with the highest cerium content showed that the decomposition process is complex, involving three main decomposition stages showed on the derivative thermogravimetry (DTG) curve in the temperature ranges 100–250 °C, 250–540 °C, and 540–785 °C, respectively. Each of these main stages consists of several decomposition steps indicated by corresponding thermal effects recorded on the differential thermal analysis (DTA) curve and accompanied by mass losses observed on the thermogravimetric (TG) curve, as shown for the gel precursor with the highest cerium content in Figure 1.

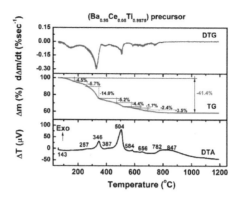

Figure 1. Thermal analysis curves of the precursor powder with the highest cerium content (x = 0.05).

In the low temperature range, the DTA curve exhibited a weak endothermic effect with a maximum centered around ~143 °C and accompanied by a slight mass loss of 4.5% on the TG curve, which was attributed to the release of adsorbed water and to the evaporation of the solvents. At temperatures ranging between 250 °C and 540 °C, combustion of the organic groups took place in four different steps as indicated by the exothermic effects whose maxima are located at 257, 346, 387, and 504 °C, respectively. For these processes, a cumulative mass loss of 24.9% was estimated from the thermogravimetric measurements. Further, a succession of weak, exothermic effects was recorded in the temperature range of 540–785 °C, with the maxima centered at 584 °C, 656 °C, and 782 °C on the DTA curve. These effects might appear to result from two opposite processes, i.e., endothermic decomposition of some intermediate carbonated phases such as $BaCO_3$ and $Ba_2Ti_2O_5 \cdot CO_3$ due to combustion, accompanied by a mass loss of 12%, occurring simultaneously with the exothermic formation of the perovskite skeleton, which seems to prevail. The last decomposition step (~782 °C) is clearly overlapped with the formation of the perovskite Ce-BaTiO$_3$ phase by a complex process involving crystallization and/or solid state reaction between small amounts of barium- and titanium-rich secondary phases (Ba_2TiO_4 and $BaTi_2O_5$), which usually originate from some residual chemical heterogeneities in the precursor gels. This overlapping results in a broad, flattened, high-temperature feature on the DTA curve. The formation/crystallization of the Ce-BaTiO$_3$ solid solution was completed at ~950 °C.

3.2. Ce^{3+}-Doped BaTiO$_3$ Powders

3.2.1. Phase Composition and Crystalline Structure

The X-ray diffraction patterns recorded at room temperature for the oxide powders obtained after annealing at 900 °C for 2 h show the presence of well-crystallized Ce-BaTiO$_3$ perovskite as a unique phase, irrespective of cerium content (Figure 2a).

The analysis of the positions and profiles of the main diffraction peaks revealed the presence of two features: (*i*) a shift in the main reflections toward higher diffraction angles with increasing Ce^{3+} content (Figure 2a), induced by the difference in the ionic radius values of the *A*-site substituting and substituted species ($r(Ce^{3+})$ = 1.34 Å compared to $r(Ba^{2+})$ = 1.61 Å [49]), proving the incorporation of the solute on the barium site; and (*ii*) the absence of splitting of the (200) reflection in the XRD patterns of the powders under investigation, which seems to indicate a cubic structure (Figure 2b).

However, Rietveld refinement revealed that only the heavily doped powder (x = 0.05) exhibited cubic Pm3m symmetry of the unit cell (ICDD card no. 01-083-3859). For the powder with lower Ce^{3+} addition (x = 0.005), the best fit for the corresponding XRD pattern (the lowest values of the Rietveld parameters R_p, R_{wp}, and χ^2) was obtained by using the P4mm space group, specific to the tetragonal structure (ICDD card no. 01-081-8524). However, the low *c/a* ratio suggests that the size of the particles is small enough to induce internal strains which seem to affect tetragonality in the $Ba_{0.995}Ce_{0.005}Ti_{0.99875}O_3$

powder. The stabilization of the cubic structure in the heavily doped $Ba_{0.95}Ce_{0.05}Ti_{0.9875}O_3$ powder is determined by several factors: (*i*) a significant increase in the concentration of smaller Ce^{3+} ions on barium sites; (*ii*) a consequent increase in the concentration of titanium vacancies [35]; and (*iii*) an increase of the internal strains simultaneous to the decrease in crystallite size induced by the greater addition of donor dopant.

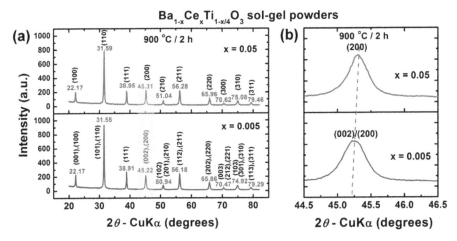

Figure 2. (**a**) XRD patterns recorded at room temperature for the 0.5% and 5% Ce^{3+}-doped $BaTiO_3$ powders thermally treated at 900 °C /2 h, and (**b**) details (light blue rectangle of Figure 2a) of the region corresponding to diffraction angles $2\theta = 44.5°–46.5°$.

The obvious decrease in the lattice parameters and the contraction of the unit cell volume are also related to the increasing cerium content incorporated on barium sites, taking into account the lower ionic radius of Ce^{3+} with respect to Ba^{2+}, as mentioned above. The structural parameters of the perovskite phases are presented in Table 1.

Table 1. Phase composition and structural parameters obtained by Rietveld refinement and average particle size estimated from TEM investigations for the sol-gel Ce^{3+}-doped $BaTiO_3$ powders annealed at 900 °C/2 h.

Formula		$Ba_{0.995}Ce_{0.005}Ti_{0.99875}O_3$	$Ba_{0.95}Ce_{0.05}Ti_{0.9875}O_3$
Sample Symbol		BCT-005	BCT-05
Phase composition		BCTss-100%	BCTss-100%
BCTss structure		Tetragonal, P4mm	Cubic, Pm3m
Unit cell parameters	a (Å)	4.003105 ± 0.000113	4.001052 ± 0.001342
	b (Å)	4.003105 ± 0.000113	4.001052 ± 0.001342
	c (Å)	4.016862 ± 0.000156	4.001052 ± 0.001342
Tetragonality, c/a		1.0034	1.0000
Unit cell volume, V (Å^3)		64.36961	64.05051
Theoretical density, ρ_t (g/cm^3)		6.014	6.032
R profile, R_p		5.04481	4.89943
Weighted R profile, R_{wp}		6.56691	6.38165
Goodness of fit, χ^2		0.01703	0.01539
Crystallite size, $<D>$ (nm)		33.66 ± 10.03	25.98 ± 5.35
Internal strains, $<S>$ (%)		0.27 ± 0.05	0.35 ± 0.07
Particle size, $<d_{TEM}>$ (nm)		109.25 ± 30.68	60.93 ± 13.71

BCTss, Ce^{3+}-$BaTiO_3$ solid solution: Tetragonal, P4mm (ICDD card no. 01-081-8524); Cubic, Pm3m (ICDD card no. 01-083-3859).

3.2.2. Morphology

TEM/HRTEM analyses coupled with SAED investigations revealed that the Ce^{3+}-doped $BaTiO_3$ powders consist of well-crystallized, polyhedral particles with well-defined edges and rounded corners (Figures 3a–d and 4a–c). A high agglomeration tendency was observed for the particles of the sample BCT-05, while partially sintered blocks were noticed in the case of the powder BCT-005. Average particle sizes of 109.25 nm and 60.93 nm were estimated for the powders BCT-005 and BCT-05, respectively (Figures 3b and 4b, Table 1). In the analysis of the influence of Ce^{3+} content on the powder morphology, decreases in both the average particle size and the aggregation tendency when the dopant content increased from 0.5 atom % to 5 atom % were indicated by the TEM images of Figures 3a and 4a. When comparing the values of the average crystallite size calculated from the diffraction data with the values corresponding to the average particle size estimated from TEM analysis, one can assume that in both cases most of the particles are not single crystals. Thus, the grains inside the aggregates observed in the powdered sample BCT-005 seem to consist of ~2–5 crystallites (Figure 3b). A more complicated particle size distribution was determined for the powdered sample BCT-05, in which a major proportion of the polycrystalline particles (consisting of ~2–4 crystallites) coexists with a smaller fraction of single-crystal particles (Figure 4b).

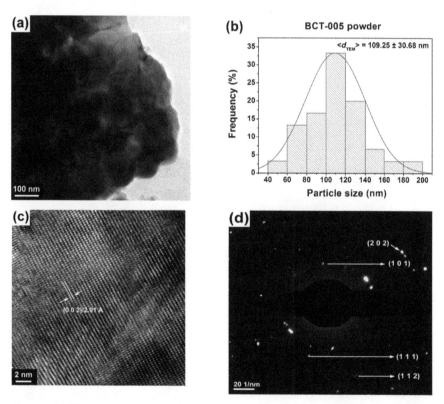

Figure 3. (a) TEM image, (b) histogram indicating the particle size distribution, (c) HRTEM image, and (d) selected-area electron diffraction (SAED) pattern of the sol-gel $Ba_{0.995}Ce_{0.005}Ti_{0.99875}O_3$ (BCT-005) powder.

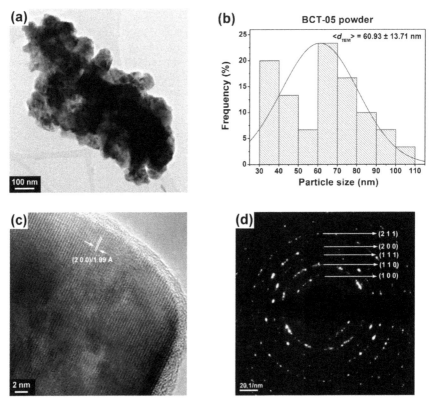

Figure 4. (**a**) TEM image, (**b**) histogram indicating the particle size distribution, (**c**) HRTEM image, and (**d**) SAED pattern of the sol-gel $Ba_{0.95}Ce_{0.05}Ti_{0.9875}O_3$ (BCT-05) powder.

The HRTEM images in Figures 3c and 4c clearly show long-range highly ordered fringes spaced at 2.01 and 1.99 Å, corresponding to the (002) and (200) crystalline planes of the tetragonal and cubic structures of the powdered samples BCT-005 and BCT-05, respectively. The high crystallinity degree of the randomly oriented particles/grains was also indicated by the bright spots, forming well-defined diffraction rings assigned to several crystalline planes of the perovskite phase in the SAED patterns in Figures 3d and 4d. The results of the SAED investigations are in good agreement with the XRD data, indicating the tetragonal distortion of the specimen BCT-005 and cubic symmetry of the unit cell for the powder BCT-05.

3.3. Ce^{3+}-Doped $BaTiO_3$ Ceramics

3.3.1. Phase Composition and Crystalline Structure

The XRD patterns of the ceramics derived from sol-gel powders and conventionally sintered at 1300 °C/4 h show, irrespective of the cerium content, the obtaining of single-phase compositions consisting of well-crystallized Ce^{3+}-$BaTiO_3$ solid solutions as identified by the main reflections of the perovskite structure (Figure 5a). These observations were also sustained by the results obtained from the Rietveld analysis, which indicated the presence of 100% Ce^{3+}-$BaTiO_3$ in both ceramics BCT-005_CS and BCT-05_CS (Table 2 and Figure 5c,d). As in the case of the starting powders, the formation of single-phase ceramics, free of any barium- or titanium-rich secondary phases, suggests the effectiveness of the "built-in" tetra-ionized titanium vacancies in compensating the supplementary positive electrical charge induced by the presence of the Ce^{3+} solute on Ba sites.

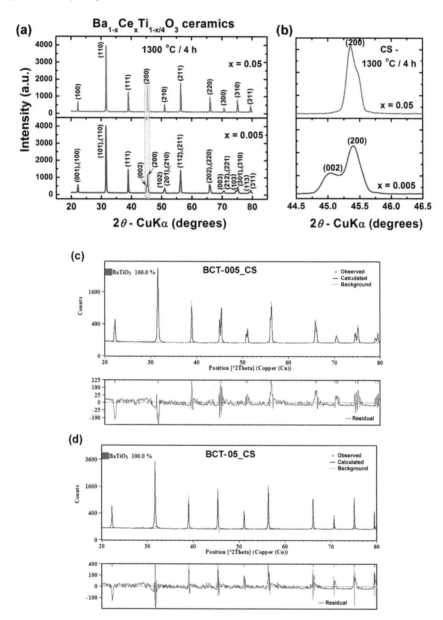

Figure 5. (a) XRD patterns recorded at room temperature for $Ba_{1-x}Ce_xTi_{1-x/4}O_3$ ceramics obtained by conventional sintering (CS) at 1300 °C/4 h; (b) details (light blue rectangle of Figure 5a) of the region corresponding to diffraction angles $2\theta = 44.5°–46.5°$; (c,d) results of Rietveld analyses of X-ray diffraction data for the conventionally sintered ceramics: (c) BCT-005_CS and (d) BCT-05_CS.

Table 2. Structural/microstructural parameters of Ce^{3+}-doped $BaTiO_3$ ceramics derived from sol-gel powders and consolidated by conventional sintering or spark plasma sintering.

Formula		$Ba_{0.995}Ce_{0.005}Ti_{0.99875}O_3$		$Ba_{0.95}Ce_{0.05}Ti_{0.9875}O_3$	
Sample Symbol		BCT-005_CS	BCT-005_SPS	BCT-05_CS	BCT-05_SPS
Sintering procedure/conditions		CS 1300 °C/4 h	SPS 1050 °C/2 min	CS 1300 °C/4 h	SPS 1050 °C/2 min
Phase composition		BCTss-100%	BCTss-100%	BCTss-100%	BCTss-98.2% BT_2-1.1% C-0.7%
BCTss structure		Tetragonal, P4mm	Tetragonal, P4mm	Cubic, Pm3m	Cubic, Pm3m
Unit cell parameters	a (Å)	3.994169 ± 0.000108	4.000073 ± 0.000273	3.999053 ± 0.000054	3.999968 ± 0.000167
	b (Å)	3.994169 ± 0.000108	4.000073 ± 0.000273	3.999053 ± 0.000054	3.999968 ± 0.000167
	c (Å)	4.024993 ± 0.000127	4.021890 ± 0.000344	3.999053 ± 0.000054	3.999968 ± 0.000167
Tetragonality, c/a		1.0077	1.0054	1.0000	1.0000
Unit cell volume, V (Å^3)		64.21228	64.35261	63.95455	63.99847
Theoretical density, $_t$ (g/cm^3)		6.027	6.014	6.041	6.037
R profile, R_p		6.48709	6.75345	7.53511	7.3148
Weighted R profile, R_{wp}		9.1197	10.6243	10.8429	10.79943
Goodness of fit, χ^2		0.47706	0.45168	0.62327	0.48938
Crystallite size, $<D>$ (nm)		48.72 ± 4.31	45.85 ± 6.80	171.43 ± 7.49	45.05 ± 6.37
Internal strains, $<S>$ (%)		0.19 ± 0.07	0.21 ± 0.12	0.18 ± 0.08	0.22 ± 0.11
Relative density, ρ_r (%)		91.7	95.8	97.1	98.6
Grain size, $<GS>$ (µm)		1.167 ± 0.183	0.279 ± 0.089	1.066 ± 0.324	0.146 ± 0.054

CS, conventional sintering; SPS, spark plasma sintering; BCTss, Ce^{3+}-$BaTiO_3$ solid solution: Tetragonal, P4mm (ICDD card no. 01-081-8524); Cubic, Pm3m (ICDD card no. 01-083-3859); BT_2, $BaTi_2O_5$: Monoclinic, C2 (ICDD card no. 04-012-4418); C, CeO_2: Cubic, Fm3m (ICDD card no. 00-067-0121).

From a structural point of view, unlike the starting oxide powder, in the case of the ceramic sample doped with 0.5 atom % Ce^{3+} (BCT-005_CS), the splitting of the (200) reflection in two adjacent (002) and (200) peaks is more significant, indicating higher tetragonality of the unit cell due to the microstructure coarsening induced by sintering (Figure 5b). For the $BaTiO_3$ specimen with a higher cerium concentration (5 atom %), even though the two (002) and (200) peaks merge into a single symmetric (200) peak, the profile of this peak remains complicated (Figure 5b). This seems to suggest the presence of a mixture of tetragonal and cubic modifications, which means that, for the BCT-05_CS specimen, ferroelectric–paraelectric phase transition might occur near room temperature.

Calculation of the structural parameters based on the XRD data sustained this assertion, showing a typical tetragonal structure for the lightly doped sample BCT005_CS, while a cubic structure was determined for the specimen BCT05_CS (Table 2).

In the case of the ceramics obtained by spark plasma sintering, the corresponding XRD patterns also revealed a clear tendency towards single-phase compositions (Figure 6a). For the sample with higher cerium content (BCT-05_SPS), a small amount of CeO_2 was identified at the detection limit, most likely because of incipient oxidation of Ce^{3+} to Ce^{4+} which takes place during the post-sintering thermal treatment. For this reason, the content of Ce^{3+} in the perovskite solid solution was slightly lower that that stipulated by the nominal formula. Consequently, a small amount of $BaTi_2O_5$ was also expelled from the perovskite lattice in order to preserve the stoichiometric ratio (Table 2 and Figure 6d). These data are in agreement with the results reported by Makovec and Kolar [50], who showed that after annealing at temperatures of 1000–1100 °C, internal partial oxidation of Ce^{3+} to Ce^{4+} accompanied by heterogeneous precipitation of small amounts of CeO_2 and polytitanate phases occurs in the perovskite matrix.

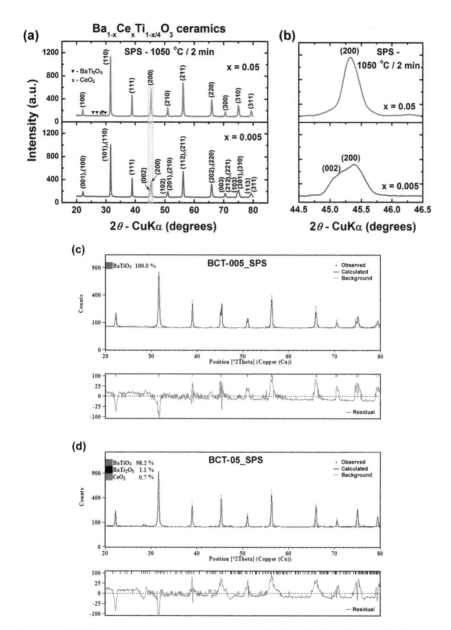

Figure 6. (a) XRD patterns recorded at room temperature for $Ba_{1-x}Ce_xTi_{1-x/4}O_3$ ceramics obtained by spark plasma sintering (SPS) at 1050 °C/2 min, followed by post-sintering re-oxidation thermal treatment at 1000 °C/16 h; (b) details (light blue rectangle of Figure 6a) of the region corresponding to diffraction angles $2\theta = 44.5$–46.5°; (c,d) results of Rietveld analyses of X-ray diffraction data for the SP-sintered ceramics: (c) BCT-005_SPS and (d) BCT-05_SPS.

For the ceramic specimen with low Ce^{3+} content (BCT-005_SPS), the oxidation process was negligible, so quantitative formation of the perovskite Ce^{3+}-$BaTiO_3$ solid solution as a unique phase was indicated by the Rietveld analysis (Table 2 and Figure 6c).

The details (light blue rectangle of Figure 6a) of the region corresponding to diffraction angles 2θ = 44.5°–46.5° revealed similar structural features to those in the case of the ceramics obtained by conventional sintering (Figure 6b). For the ceramic sample BCT-005_SPS, even if the splitting of the (200) reflection is less pronounced, the left-side asymmetry of the profile of the diffraction peak suggests a tetragonal structure but with a lower tetragonality degree than that corresponding to the conventionally sintered specimen of similar composition (BCT-005_CS).

In the case of the ceramic with high cerium content (BCT-05_SPS), the symmetric profile of the (200) reflection indicates a cubic structure (Figure 6b). These observations were sustained by the data provided by the Rietveld refinement (Table 2).

A reduction of the unit cell volume with increasing Ce^{3+} content was revealed for all the ceramic samples, no matter the sintering procedure. The slightly higher values of unit cell volume specific to the perovskite phase in the spark-plasma-sintered ceramics relative to the conventionally sintered specimens of similar composition could be explained in terms of higher inter-ionic distances induced by the presence of smaller grains, with lower crystallinity (Table 2).

3.3.2. Microstructure

The FE-SEM image of the ceramic sample with low cerium content obtained after conventional sintering at 1300 °C/4 h (BCT-005_CS) shows a fine-grained, homogeneous microstructure consisting of polyhedral, well-faceted grains (with truncated edges and corners) uniform in shape and size and with a small amount of intergranular porosity (Figure 7a). An average grain size <*GS*> of 1.167 μm (Table 2, Figure 7b), lower than the 5 μm reported in the case of sol-gel La^{3+}-doped $BaTiO_3$ ceramics of similar composition and sintered under similar conditions [10], was estimated in this case.

An increase in Ce^{3+} content only slightly affected the average grain size, but it seems to have promoted densification and induced a change in the morphology of the grains, which became rounded, without faces, edges, or corners. However, the grains exhibited well-defined boundaries and perfect triple junctions, while the intergranular pores were almost entirely missing, as the FE-SEM image in Figure 7c revealed. An average grain size of 1.066 μm was estimated for the specimen BCT-05_CS (Table 2, Figure 7d).

The ceramics obtained by spark plasma sintering exhibited denser microstructures composed of grains with significantly lower sizes but with similar morphologies relative to the samples resulting from conventional sintering (Figure 8a–c). Relative density values of 97.1% and 98.6% and average grain size values of 0.279 and 0.146 μm were determined for the ceramics denoted BCT-005_SPS and BCT-05_SPS, respectively (Table 2, Figure 8b,d). As in the case of the conventionally sintered sample, the SP-sintered specimen with low Ce^{3+} content (BCT-005_SPS) exhibited faceted polyhedral grains, tightly welded together, generating in some regions larger clusters composed of nanometric subgrains separated by dislocation networks (Figure 8a). An increase in cerium concentration not only induced a change in the grain morphology but also determined an obvious decrease in the average grain size. Thus, the pore-free specimen with high Ce^{3+} content (BCT-05_SPS) consisted of small, equiaxial grains and exhibited slightly higher densification and an almost unimodal grain size distribution (Figure 8c,d). FE-SEM investigations in back-scattered electrons (BSE) mode did not reveal the presence of secondary phases detected by XRD measurements, most likely because of their small amounts. It is worth mentioning that spark plasma sintering performed at 1050 °C for 2 min followed by re-oxidation thermal treatment at 1000 °C for 16 h induced an increase in the average grain size by 2.4–2.6 times compared to the average particle size of the starting sol-gel Ce^{3+}-doped $BaTiO_3$ powders. On the other hand, regarding the conventional sintering, a higher grain growth rate was estimated for the sample with higher cerium content (BCT-05_CS) than for BCT-005_CS, taking into account the values of the average particle sizes of the related powders.

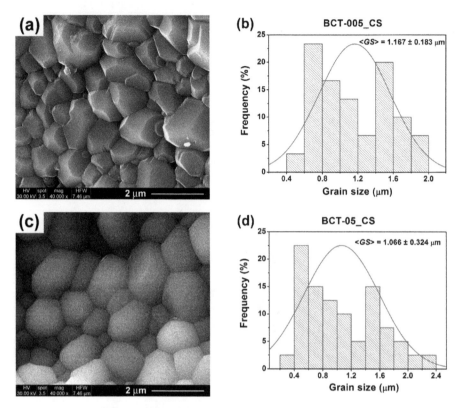

Figure 7. (**a,c**) FE-SEM images and (**b,d**) histograms indicating the grain size distribution for the conventionally sintered ceramics: (**a,b**) $Ba_{0.995}Ce_{0.005}Ti_{0.99875}O_3$ (BCT-005_CS) and (**c,d**) $Ba_{0.95}Ce_{0.05}Ti_{0.9875}O_3$ (BCT-05_CS).

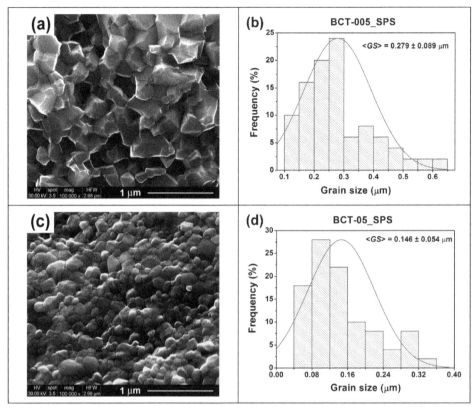

Figure 8. (**a,c**) FE-SEM images and (**b,d**) histograms indicating the grain size distribution for the spark-plasma-sintered ceramics: (**a,b**) $Ba_{0.995}Ce_{0.005}Ti_{0.99875}O_3$ (BCT-005_SPS) and (**c,d**) $Ba_{0.95}Ce_{0.05}Ti_{0.9875}O_3$ (BCT-05_SPS).

3.3.3. Dielectric and Ferroelectric Properties

Figure 9a–d presents the temperature dependence of the dielectric properties recorded for five frequencies, in the range of 1 kHz–500 kHz, for the ceramics obtained by conventional sintering. Both samples showed a well-defined, sharp, and frequency-independent maximum of permittivity assigned to the ferroelectric–paraelectric phase transition at the Curie temperature (T_C) (Figure 9a,c). A small frequency dispersion was observed only in the ferroelectric phase, mainly for the specimen with low cerium content (BCT-005_CS), whereas in the paraelectric state (above T_C), the permittivity was almost frequency-independent for both samples, irrespective of the amount of Ce^{3+} solubilized in the perovskite lattice.

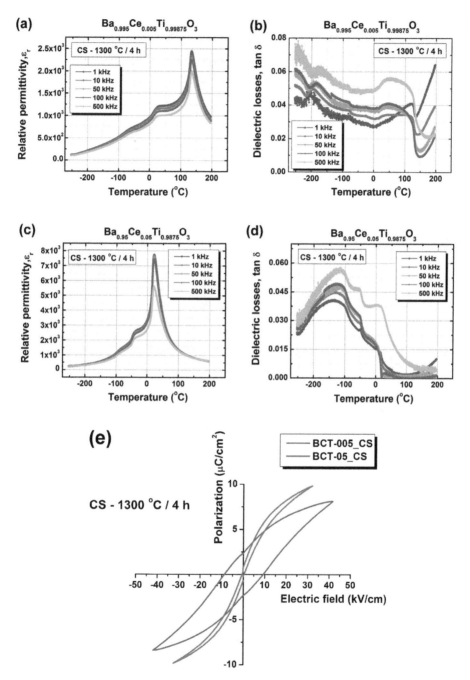

Figure 9. (**a,c**) Relative permittivity and (**b,d**) dielectric losses vs. temperature for the conventionally sintered ceramics described by the nominal formula $Ba_{1-x}Ce_xTi_{1-x/4}O_3$: (**a,b**) BCT-005_CS; (**c,d**) BCT-05_CS. (**e**) *P–E* dependence for the Ce^{3+}-doped $BaTiO_3$ ceramics obtained by conventional sintering.

The samples showed good dielectric properties, with high values of relative permittivity and dielectric losses below 0.07 in the temperature range of −200 to +200 °C. The BaTiO$_3$ sample doped with 5 atom % Ce^{3+} (BCT-05_CS) showed a drop of the loss tangent to below 0.01 at temperatures above the Curie temperature (Figure 9d). This insulating behavior is consistent with full electrical compensation by cation vacancies, homogeneously distributed inside the perovskite grains. These results are in agreement with the data reported earlier for ceramics with similar composition derived from powders synthesized using the modified Pechini method [35]. On the contrary, the BaTiO$_3$ ceramic sample doped with a significantly lower Ce^{3+} content (0.5 atom %) exhibited an increase in the tangent loss, particularly at higher temperatures (above T_C) and at lower frequencies (Figure 9b). Higher values of dielectric losses in the low frequency range are most likely associated with thermally activated space charge effects (Maxwell–Wagner phenomena), which commonly occur in lightly donor-doped BaTiO$_3$ due to the grain boundary contribution or to the ionization of oxygen vacancies in the sintered ceramics [9,10,51].

For the lightly doped BCT-005_CS sample, the maximum permittivity at 1 kHz reached a value of 2448 at a Curie temperature of 133 °C, close to that specific to the undoped BaTiO$_3$ ceramic, indicating that at room temperature this specimen was found in its ferroelectric state (Figure 9a and Table 3). The maximum permittivity was lower than that reported by Hwang and Han for their Ba$_{0.9995}$Ce$_{0.005}$TiO$_3$ ceramic sample prepared by the Pechini method but sintered in air for 5 h at a higher temperature of 1380 °C [21]. The other two strongly flattened permittivity maxima recorded at $T_1 = 25$ °C and $T_2 = -69$ °C correspond to the tetragonal–orthorhombic (T–O) and orthorhombic–rhombohedral (O–R) phase transitions (Figure 9a). It must be noted that even though the best fit obtained by Rietveld analysis of the diffraction data indicated a tetragonal structure, the results of the dielectric measurements show that, actually, at room temperature, a mixture of tetragonal and orthorhombic modifications was found in the BCT-005_CS sample. The higher values of T_1 and T_2 than those corresponding to the undoped BaTiO$_3$ single crystals (for which $T_1 = 5$ °C and $T_2 = -90$ °C) seem to be related to the small grain size of this polycrystalline ceramic sample.

Table 3. Dielectric properties at 1 kHz for the Ce^{3+}-doped BaTiO$_3$ ceramics derived from sol-gel powders and consolidated by conventional sintering or spark plasma sintering.

Formula	Ba$_{0.995}$Ce$_{0.005}$Ti$_{0.99875}$O$_3$		Ba$_{0.95}$Ce$_{0.05}$Ti$_{0.9875}$O$_3$	
Sample Symbol	BCT-005_CS	BCT-005_SPS	BCT-05_CS	BCT-05_SPS
Sintering procedure/conditions	CS 1300 °C/4 h	SPS 1050 °C/2 min	CS 1300 °C/4 h	SPS 1050 °C/2 min
ε'_{max}	2448	19782	7758	3.67 × 10^6
T_C	133	91	21	−
T_1	25	−49	−44	−
T_2	−69	−214	−93	−
ε'_{RT}	1163	18367	7695	3.35 × 10^6
tan δ_{RT}	0.0287	0.2498	0.0098	~10

The specimen with high Ce^{3+} addition (BCT-05_CS) showed a ferroelectric–paraelectric phase transition in the proximity of room temperature. As in the case of the BCT-005_CS ceramic, the sharp permittivity maximum suggests a first-order phase transition, specific to a typical ferroelectric material. A Curie temperature of 21 °C with a permittivity maximum of 7758 was determined in this case (Figure 9c and Table 3). This result is in agreement with the structural data, which showed cubic symmetry of the unit cell for the sample BCT-05_CS, proving that this ceramic was already found in the paraelectric state. The evolution of the Curie temperature with increasing donor dopant concentration shows that the Ce^{3+} solute was exclusively incorporated on Ba sites in the perovskite lattice. It has to be mentioned that for a ceramic sample with the same composition, derived from a powder prepared by the Pechini method and sintered in similar conditions, a slightly higher Curie temperature of 25 °C was reported in our previous work [35], which suggests that the wet chemical synthesis procedure of the starting oxide powder could play a certain role in stabilizing the paraelectric state at room temperature. A decreasing rate of −24.9 °C/atom % of Ce^{3+} was determined for the sol-gel ceramics described by the nominal formula $Ba_{1-x}Ce_xTi_{1-x/4}O_3$ investigated in this study. This value is higher compared to the −18 °C/atom % reported in the literature by Jing et al. for $Ba_{1-x}Ce_xTi_{1-x/4}O_3$ specimens prepared by the Pechini procedure [31].

A decreasing trend with increasing solute content, specific to *A*-site doping in $BaTiO_3$, was also recorded for the T–O and O–R phase transition temperatures, which showed values of −44 and −93 °C, respectively (Table 3).

Measurements of *P–E* dependence were performed only for the ceramics obtained by conventional sintering. The higher leakage currents in spark-plasma-sintered ceramics did not allow us to measure the polarization versus the electric field, even for the single-phase specimen BCT-005_SPS.

In the case of the conventionally sintered sample with a small cerium addition, the hysteretic *P–E* dependence at fields ranging between −40 and +40 kV/cm clearly indicated the ferroelectric state. The sample labelled BCT-005_CS exhibited a tilted hysteresis loop, characterized by coercive fields of 8.7 kV/cm, saturation polarization of 8.1 μC/cm² , and a low remnant polarization value of 2.3 μC/cm² (Figure 9e).

Regarding the specimen with high Ce^{3+} content (BCT-05_CS), the drop in coercivity and remnant polarization was consistent with the proximity of the paraelectric state at room temperature (Figure 9e). However, the "*S*" shape of the hysteresis loop seems to indicate a superparaelectric state, like that specific to the relaxor systems, rather than a typical paraelectric state, characterized by linear *P–E* dependence [52–55]. This suggests that at room temperature, a more complicated local structure, involving polar nano-regions consisting of unit cells with low tetragonality, embedded in a matrix of cubic unit cells, might be present in this sample. This kind of structure cannot be detected by XRD measurements, which involves averaging over at least 10,000 unit cells in the calculation of unit cell parameters.

The spark-plasma-sintered ceramics with the nominal formula $Ba_{0.995}Ce_{0.005}Ti_{0.99875}O_3$ exhibited much higher permittivity values over the whole investigated temperature range when compared to the conventionally sintered samples with similar composition. The temperature dependence of relative permittivity was flattened, showing at 1 kHz a broad feature centered around the temperature of 91 °C (Figure 10a and Table 3). This might correspond to a diffuse ferroelectric–paraelectric phase transition caused by the low grain size, as reported in the case of undoped $BaTiO_3$ ceramics consolidated by spark plasma sintering [56–61]. The T–O phase transition was barely detected as an asymmetric feature centered at −49 °C, while the shoulder at approximately −214 °C was assigned to the O–R phase transition. These data show a much more pronounced shift of the phase transition temperatures T_1 and T_2 towards lower temperature values compared to that specific to the Curie temperature, T_C. Very high permittivity values of 800–20,000 were recorded at 1 kHz over the entire temperature range of −250 to +200 °C, and a maximum permittivity of 19,782 was found at 91 °C. Increasing frequency induced a strong flattening of the temperature dependence of the relative permittivity, so that at 500 kHz, only the ferroelectric–paraelectric phase transition could be identified, with the maximum permittivity reaching 9915 at 107 °C (Figure 10a). Regarding the dielectric losses, higher tan δ values of 0.27–0.42 and 0.07–0.13 were recorded in the temperature range of −100 to 200 °C for lower frequencies of 1 kHz and 10 kHz, respectively (Figure 10b, Table 3). At higher frequencies (of 50–500 kHz), in the same temperature range, the dielectric tangent exhibited values below 0.07, similar to those obtained for the conventionally sintered bulk ceramics with similar Ce^{3+} content. A strong anomaly in the temperature dependence of the dielectric loss tangent, the origin of which is still not clear but seems to be related to the (O–R) phase transition, was found in the low temperature range of −250 to +200 °C. This anomaly shows a certain frequency dispersion, reflected by the decrease and concurrent shift towards slightly higher temperature values of the tangent loss maximum with increasing frequency, as in the case of the relaxor systems [52,53] (Figure 10b). The higher values of dielectric losses in the low frequency region in spark-plasma-sintered $BaTiO_3$ ceramics, relative to those in conventionally sintered samples, originated from the higher leakage currents induced by the higher density of interface states associated with the lower grain size. Besides this, one cannot exclude the presence of a very small concentration of oxygen vacancies which can persist even after prolonged re-oxidation thermal treatments; these might induce a conductive contribution via a mechanism involving small polarons hopping [62]. However, the white color in the cross section of the sample indicates that the insulating behavior prevails.

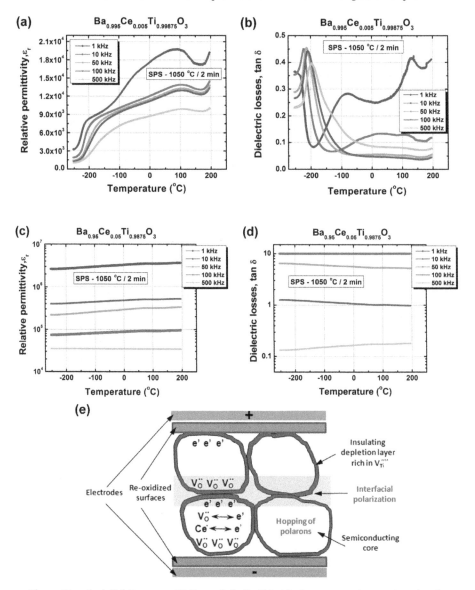

Figure 10. (**a,c**) Relative permittivity and (**b,d**) dielectric losses vs. temperature for the spark-plasma-sintered ceramics described by the nominal formula $Ba_{1-x}Ce_xTi_{1-x/4}O_3$: (**a,b**) x = 0.005 (BCT-005_SPS); (**c,d**) x = 0.05 (BCT-05_SPS). (**e**) A schematic representation of the processes generating colossal permittivity in the BTC-05_SPS ceramic sample.

The sample with high cerium content, described by the nominal formula $Ba_{0.95}Ce_{0.05}Ti_{0.9875}O_3$, showed colossal values of effective permittivity, almost temperature independent, which decreased from 3×10^6 to 2×10^4 when the frequency increased from 1 kHz to 500 kHz (Figure 10c). The same temperature invariance was recorded for dielectric losses, which decreased from $\tan \delta \gg 1$ to $\tan \delta = 0.2$ when the frequency increased from 1 kHz to 500 kHz (Figure 10d). Such large values of effective permittivity and dielectric losses can be explained in terms of Maxwell–Wagner phenomena and rather indicate a semiconducting behavior. Similar characteristics were also reported for a fine-grained $Ba_{0.95}La_{0.05}TiO_3$ ceramic sample prepared from nano-sized powders and consolidated by spark plasma sintering, where the colossal permittivity was found to arise from a conjugated effect of the re-oxidized insulating surface and high interfacial polarization inside the ceramic specimen [63,64]. This large interfacial polarization seems to be determined by a "brick-wall"-type electric microstructure due to the presence of highly semiconducting grain cores, with a high concentration of carriers (polarons) induced by both reducing conditions during the SPS process and a high concentration of donor dopant in association with the high density of very thin grain boundary depletion layers, rich in electron traps (cation vacancies) and consequently exhibiting insulating properties (Figure 10e). In a highly donor-doped, fine-grained barium titanate sample of grey color such as this, the semiconducting behavior prevails and can be explained in terms of a combination of the two alternative models proposed earlier by Takeuchi et al. to explain the presence of the conductive contribution induced by the oxygen deficiency in partially reduced undoped $BaTiO_3$ ceramics [39].

4. Conclusions

Ce^{3+}-doped $BaTiO_3$ powders described by the nominal formula $Ba_{1-x}Ce_xTi_{1-x/4}O_3$ with x = 0.005 and 0.05 were synthesized by the acetate variant of the sol-gel method. The structural parameters, particle size, and morphology are strongly dependent on the Ce^{3+} content. From these powders, ceramics were prepared by conventional sintering at 1300 °C/4 h or by spark plasma sintering at 1050 °C/2 min. For all the investigated ceramics, irrespective of their composition and sintering procedure, the solute was exclusively incorporated as Ce^{3+} on Ba^{2+} sites, acting as a donor dopant. The results of the dielectric and P–E measurements were in good agreement with the values of the structural parameters calculated based on the XRD data, revealing that at room temperature, the ceramic with low cerium content (x = 0.005) was in the ferroelectric state, while the sample with a significantly higher Ce^{3+} concentration (x = 0.05) was found to be in the proximity of the ferroelectric–paraelectric phase transition. Curie temperatures of 133 °C and 21 °C and values of maximum permittivity of 2448 and 7758 were determined for the conventionally sintered ceramics with x = 0.005 and x = 0.05, respectively. The fine-grained and dense ceramic sample with low solute content resulting from spark plasma sintering exhibited insulating behavior at 1 kHz, with significantly higher values of relative permittivity (ε_r ~20,000) and dielectric losses ($\tan \delta$ = 0.2–0.4) over the entire investigated temperature range relative to the conventionally sintered sample with similar composition. The spark-plasma-sintered Ce^{3+}-$BaTiO_3$ specimen with high cerium content (x = 0.05) showed a fine-grained microstructure and an almost temperature-independent colossal dielectric constant (ε_r ~ 3.35×10^6). However, in this case, the enormous dielectric losses rather indicated a semiconducting behavior.

To conclude, in this work we revealed that by reducing the grain size from micro- toward nano-scale, the interplay of the influences of both the composition and the microstructural features determined opposing evolution in the two BCT ceramics. Thus, a dielectric of average performance (the BCT ceramic sample with x = 0.005) acquired significantly higher permittivity, more stable with temperature, while a better dielectric (the BCT sample with x = 0.05) was converted into a typical semiconductor, due to the high concentration of polarons.

Author Contributions: Formal analysis, A.S. and M.E.; Investigation, C.A.S., I.P., R.T. and B.S.V.; Methodology, C.A.S., A.S., R.T., B.S.V. and M.E.; Project administration, A.C.I.; Supervision, A.C.I.; Writing—original draft, A.C.I., C.A.S. and I.P.; Writing—review and editing, A.C.I.

Funding: The financial support of the Romanian CNCS-UEFISCDI Project No. PN-III-P4-ID-PCE-2016-0072 is gratefully acknowledged.

Conflicts of Interest: The authors declare no conflict of interest.

References

1. Acosta, M.; Novak, N.; Rojas, V.; Patel, S.; Vaish, R.; Koruza, J.; Rossetti, G.A.; Rödel, J. BaTiO$_3$-based piezoelectrics: Fundamentals, current status, and perspectives. *Appl. Phys. Rev.* **2017**, *4*, 041305. [CrossRef]
2. Lines, M.E.; Glass, A.M. *Principles and Applications of Ferroelectrics and Related Materials*; Clarendon, Oxford University Press: Oxford, UK, 1977.
3. Haertling, G.H. Ferroelectric Ceramics: History and Technology. *J. Am. Ceram. Soc.* **1999**, *82*, 797–818. [CrossRef]
4. Rae, A.; Chu, M.; Ganine, V. Dielectric Ceramic Materials. In *Ceram Trans Barium Titanate: Past, Present and Future*; Nair, K.M., Bhalla, A.S., Eds.; The American Ceramic Society: Westerville, OH, USA, 2007; Volume 100, pp. 1–12.
5. Randall, C.A. Scientific and Engineering Issues of the State-of-the-Art and Future Multilayer Capacitors. *J. Ceram. Soc. Jpn.* **2001**, *109*, S2–S6. [CrossRef]
6. Morrison FDSinclair, D.C.; West, A.R. Doping mechanisms and electrical properties of La-doped BaTiO$_3$ ceramics. *Int. J. Inorg. Mater.* **2001**, *3*, 1205–1210. [CrossRef]
7. Morrison, F.D.; Sinclair, D.C.; West, A.R. Electrical and structural characteristics of lanthanum-doped barium titanate ceramics. *J. Appl. Phys.* **1999**, *86*, 6355–6366. [CrossRef]
8. Morrison, F.D.; Sinclair, D.C.; Skakle, J.M.S.; West, A.R. Novel Doping Mechanism for Very-High-Permittivity Barium Titanate Ceramics. *J. Am. Ceram. Soc.* **1998**, *81*, 1957–1960. [CrossRef]
9. Ianculescu, A.; Mocanu, Z.V.; Curecheriu, L.P.; Mitoseriu, L.; Padurariu, L.; Trusca, R. Dielectric and tunability properties of La-doped BaTiO$_3$ ceramics. *J. Alloy. Compd.* **2011**, *509*, 10040–10049. [CrossRef]
10. Ianculescu, A.C.; Vasilescu, C.A.; Crisan, M.; Raileanu, M.; Vasile, B.S.; Calugaru, M.; Crisan, D.; Dragan, N.; Curecheriu, L.; Mitoseriu, L. Formation mechanism and characteristics of lanthanum-doped BaTiO$_3$ powders and ceramics prepared by the sol-gel process. *Mater. Charact.* **2015**, *106*, 195–207. [CrossRef]
11. Murugaraj, P.; Kutty, T.R.N.; Subba, R.M. Diffuse phase transformations in neodymium-doped BaTiO$_3$ ceramics. *J. Mater. Sci.* **1986**, *21*, 3521–3527. [CrossRef]
12. Yao, Z.; Liu, H.; Liu, Y.; Wu, Z.; Shen, Z.; Liu, Y.; Cao, M. Structure and dielectric behavior of Nd-doped BaTiO$_3$ perovskites. *Mater. Chem. Phys.* **2008**, *109*, 475–481. [CrossRef]
13. Tagantsev, A.K.; Sherman, V.O.; Astafiev, K.F.; Venkatesh, J.; Setter, N. Ferroelectric Materials for Microwave Tunable Applications. *J. Electroceramics* **2003**, *11*, 5–66. [CrossRef]
14. Ianculescu, A.; Berger, D.; Viviani, M.; Ciomaga, C.; Mitoseriu, L.; Vasile, E.; Drăgan, N.; Crişan, D. Investigation of Ba$_{1-x}$Sr$_x$TiO$_3$ ceramics prepared from powders synthesized by the modified Pechini route. *J. Eur. Ceram. Soc.* **2007**, *27*, 3655–3658. [CrossRef]
15. Ianculescu, A.; Berger, D.; Mitoşeriu, L.; Curecheriu, L.P.; Drăgan, N.; Crişan, D.; Vasile, E. Properties of Ba$_{1-x}$Sr$_x$TiO$_3$ ceramics prepared by the modified Pechini method. *Ferroelectrics* **2008**, *369*, 22–34. [CrossRef]
16. Kadira, L.; Elmesbahi, A.; Sayouri, S. Dielectric study of calcium doped barium titanate Ba$_{1-x}$Ca$_x$TiO$_3$ ceramics. *Int. J. Phys. Sci.* **2016**, *11*, 71–79.
17. Zhang, L.; Thakur, O.P.; Feteira, A.; Keith, G.M.; Mould, A.G.; Sinclair, D.C.; West, A.R. Comment on the use of calcium as a dopant in X8R BaTiO$_3$-based ceramics. *Appl. Phys. Lett.* **2007**, *90*, 142914. [CrossRef]
18. Chan, N.H.; Harmer, M.P.; Smyth, D.M. Compensating defects in highly donor-doped BaTiO$_3$. *J. Am. Ceram. Soc.* **1986**, *69*, 507–510. [CrossRef]
19. Chan, N.H.; Smyth, D.M. Defect chemistry of donor-doped BaTiO$_3$. *J. Am. Ceram. Soc.* **1984**, *67*, 285–288. [CrossRef]
20. Hennings, D.F.K.; Schreinemacher, B.; Schreinemacher, H. High-permittivity dielectric ceramics with high endurance. *J. Eur. Ceram. Soc.* **1994**, *13*, 81–88. [CrossRef]
21. Hwang, J.H.; Han, Y.H. Dielectric properties of (Ba$_{1-x}$Ce$_x$)TiO$_3$. *Jpn. J. Appl. Phys.* **2000**, *39*, 2701–2704. [CrossRef]
22. Hwang, J.H.; Han, Y.H. Electrical properties of cerium-doped BaTiO$_3$. *J. Am. Ceram. Soc.* **2001**, *84*, 1750–1754. [CrossRef]

23. Guha, J.P.; Kolar, D. Subsolidus equilibria in the system BaO-CeO$_2$-TiO$_2$. *J. Am. Ceram. Soc.* **1973**, *56*, 5–6. [CrossRef]

24. Jing, Z.; Yu, Z.; Ang, C. Crystalline structure and dielectric properties of Ba(Ti$_{1-y}$Ce$_y$)O$_3$. *J. Mater. Sci.* **2003**, *38*, 1057–1061. [CrossRef]

25. Ang, C.; Jing, Z.; Yu, Z. Ferroelectric relaxor Ba(Ti,Ce)O$_3$. *J. Phy. Condens. Matter.* **2002**, *14*, 8901–8912. [CrossRef]

26. Canu, G.; Confalonieri, G.; Deluca, M.; Curecheriu, L.; Buscaglia, M.T.; Asandulesa, M.; Horchidan, N.; Dapiaggi, M.; Mitoseriu, L.; Buscaglia, V. Structure-property correlations and origin of relaxor behaviour in BaCe$_x$Ti$_{1-x}$O$_3$. *Acta Mater.* **2018**, *152*, 258–268. [CrossRef]

27. Makovec, D.; Samardžija, Z.; Kolar, D. Incorporation of cerium into the BaTiO$_3$ lattice. In *Third Euro-Ceramics, Proceedings of the 3rd European Ceramic Society Conference, Madrid, Spain, 12–17 September 1993*; Durán, P., Fernández, J.P., Eds.; Faenza Editrice Ibérica, S. L.: Faenza, Italy, 1993; Volume 1, pp. 961–966.

28. Makovec, D.; Samardžija, Z.; Kolar, D. Solid solubility of cerium in BaTiO$_3$. *J. Solid State Chem.* **1996**, *123*, 30–38. [CrossRef]

29. Jonker, G.H.; Havinga, E.E. The influence of foreign ions on the crystal lattice of BaTiO$_3$. *Mater. Res. Bull.* **1982**, *17*, 345–350. [CrossRef]

30. Lewis, C.V.; Catlow, C.R.A. *Defect studies of doped and undoped* barium titanate using *computer simulation techniques*. *J. Phys. Chem. Solids* **1986**, *47*, 89–97. [CrossRef]

31. Jing, Z.; Yu, Z.; Ang, C. Crystalline structure and dielectric behavior of (Ce,Ba)TiO$_3$ ceramics. *J. Mater. Res.* **2002**, *17*, 2787–2793. [CrossRef]

32. Yasmin, S.; Choudhury, S.; Hakim, M.A.; Bhuiyan, A.H.; Rahman, M.J. Effect of cerium doping on microstructure and dielectric properties of BaTiO$_3$ ceramics. *J. Mater. Sci. Technol.* **2011**, *27*, 759–763.

33. Cernea, M.; Monereau, O.; Llewellyn, P.; Tortet, L.; Galassi, C. Sol–gel synthesis and characterization of Ce doped-BaTiO$_3$. *J. Eur. Ceram. Soc.* **2006**, *26*, 3241–3246. [CrossRef]

34. Pechini, M.P. Method of Preparing Lead and Alkaline Earth Titanates and Niobates and Coating Method using the Same for a Capacitor. U.S. Patent No. 3,330,697, 11 July 1967.

35. Ianculescu, A.; Berger, D.C.; Vasilescu, C.A.; Olariu, M.; Vasile, B.S.; Curecheriu, L.P.; Gajovic, A.; Trusca, R. Incorporation mechanism and functional properties of Ce-doped BaTiO$_3$ ceramics derived from nanopowders prepared by the modified-Pechini method. In *Nanoscale Ferroelectrics and Multiferroics: Key Processing and Characterization Issues, and Nanoscale Effects*; John Wiley & Sons, Ltd.: Hoboken, NJ, USA, 2016; Volume 1, pp. 13–43.

36. Ianculescu, A.; Pintilie, I.; Vasilescu, C.A.; Botea, M.; Iuga, A.; Melinescu, A.; Drăgan, N.; Pintilie, L. Intrinsic pyroelectric properties of thick, coarse grained Ba$_{1-x}$Sr$_x$TiO$_3$ ceramics. *Ceram. Int.* **2016**, *42*, 10338–10348. [CrossRef]

37. Vasilescu, C.A.; Trupina, L.; Vasile, B.S.; Trusca, R.; Cernea, M.; Ianculescu, A.C. Characteristics of 5 mol% Ce^{3+}-doped barium titanate nanowires prepared by a combined route involving sol-gel chemistry and polycarbonate membrane-templated process. *J. Nanopart. Res.* **2015**, *17*, 434. [CrossRef]

38. Ianculescu, A.C.; Vasilescu, C.A.; Trupină, L.; Vasile, B.S.; Truşcă, R.; Cernea, M.; Pintilie, L.; Nicoară, A. Characteristics of Ce^{3+}-doped barium titanate nanoshell tubes prepared by template-mediated colloidal chemistry. *J. Eur. Ceram. Soc.* **2016**, *36*, 1633–1642. [CrossRef]

39. Takeuchi, T.; Suyama, Y.; Sinclair, D.C.; Kageyama, H. Spark-plasma-sintering of fine BaTiO$_3$ powder prepared by a sol-crystal method. *J. Mater. Sci.* **2001**, *36*, 2329–2334. [CrossRef]

40. Li, B.; Wang, X.; Cai, M.; Hao, L.; Li, L. Densification of uniformly small-grained BaTiO$_3$ using spark-plasma-sintering. *Mater. Chem. Phys.* **2003**, *82*, 173–180. [CrossRef]

41. Licheri, R.; Fadda, S.; Orrù, R.; Cao, G.; Buscaglia, V. Self-propagating high-temperature synthesis of barium titanate and subsequent densification by spark plasma sintering (SPS). *J. Eur. Ceram. Soc.* **2007**, *27*, 2245–2253. [CrossRef]

42. Luan, W.; Gao, L.; Kawaoka, H.; Sekino, T.; Niihara, K. Fabrication and characteristics of fine-grained BaTiO$_3$ ceramics by spark plasma sintering. *Ceram. Int.* **2004**, *30*, 405–410. [CrossRef]

43. Mitoşeriu, L.; Harnagea, C.; Nanni, P.; Testino, A.; Buscaglia, M.T.; Buscaglia, V.; Viviani, M.; Zhao, Z.; Nygren, M. Local switching properties of dense nanocrystalline BaTiO$_3$ ceramics. *Appl. Phys. Lett.* **2004**, *84*, 2418–2420. [CrossRef]

44. Buscaglia, M.T.; Buscaglia, V.; Viviani, M.; Petzelt, J.; Savinov, M.; Mitoşeriu, L.; Testino, A.; Nanni, P.; Harnagea, C.; Zhao, Z.; et al. Ferroelectric properties of dense nanocrystalline BaTiO$_3$ ceramics. *Nanotechnology* **2004**, *15*, 1113–1117. [CrossRef]

45. Hungría, T.; Algueró, M.; Hungría, A.B.; Castro, A. Dense fine-grained Ba$_{1-x}$Sr$_x$TiO$_3$ ceramic prepared by the combination of mechanosynthesized nanopowders and spark plasma sintering. *Chem. Mater.* **2005**, *17*, 6205–6212. [CrossRef]

46. Aldica, G.; Cernea, M.; Ganea, P. Dielectric Ba(Ti$_{1-x}$Sn$_x$)O$_3$ (x = 0.13) ceramics, sintered by spark plasma and conventional methods. *J. Mater. Sci.* **2010**, *45*, 2606–2610. [CrossRef]

47. Maiwa, H. Structure and properties of Ba(Zr$_{0.2}$Ti$_{0.8}$)O$_3$ ceramics prepared by spark plasma sintering. *J. Mater. Sci.* **2008**, *43*, 6385–6390. [CrossRef]

48. Maiwa, H. Dielectric and Electromechanical Properties of Ba(Zr$_x$Ti$_{1-x}$)O$_3$ (x = 0.1 and 0.2) Ceramics Prepared by Spark Plasma Sintering. *Jpn. J. Appl. Phys.* **2007**, *46*, 7013–7017. [CrossRef]

49. Shannon, R.D. Revised effective ionic radii and systematic studies of interatomic distances in halides and chalcogenides. *Acta Cryst. A* **1976**, *32*, 751–767. [CrossRef]

50. Makovec, D.; Kolar, K. Internal oxidation of Ce^{3+}–BaTiO$_3$ solid solutions. *J. Am. Ceram. Soc.* **1997**, *80*, 45–52. [CrossRef]

51. Ramoška, T.; Banys, J.; Sobiestianskas, R.; Vijatović Petrović, M.; Bobić, J.; Stojanović, B. Dielectric investigations of La-doped barium titanate. *Proc. Appl. Ceram.* **2010**, *4*, 193–198. [CrossRef]

52. Cross, L.E. Relaxor Ferroelectrics: An Overview. *Ferroelectrics* **1994**, *151*, 305–320. [CrossRef]

53. Cross, L.E. Ferroic Materials and Composites: *Past, Present, and Future*. In *Advanced Ceramics*, 3rd ed.; Elsevier Applied Science: London, UK; New York, NY, USA, 1990; pp. 71–101.

54. Samara, G.A. The relaxational properties of compositionally disordered ABO$_3$ perovskites. *J. Phys. Condens. Matter* **2003**, *15*, R367–R411. [CrossRef]

55. Ahn, C.W.; Wang, K.; Li, J.F.; Lee, J.-S.; Kim, I.W. A Brief Review on Relaxor Ferroelectrics and Selected Issues in Lead-Free Relaxors. *J. Korean Phys. Soc.* **2016**, *68*, 1481–1494. [CrossRef]

56. Li, B.; Wang, X.; Li, L.; Zhou, H.; Liu, X.; Han, X.; Zhang, Y.; Qi, X.; Deng, X. Dielectric properties of fine-grained BaTiO$_3$ prepared by spark-plasma-sintering. *Mater. Chem. Phys.* **2004**, *83*, 23–28. [CrossRef]

57. Takeuchi, T.; Capiglia, C.; Balakrishnan, N.; Takeda, Y.; Kageyama, H. Preparation of fine grained BaTiO$_3$ ceramics by spark plasma sintering. *J. Mater. Res.* **2002**, *17*, 575–581. [CrossRef]

58. Deng, X.; Wang, X.; Wen, H.; Kang, A.; Gui, Z.; Li, L. Phase transitions in nanocrystalline barium titanate ceramics prepared by spark plasma sintering. *J. Am. Ceram. Soc.* **2006**, *89*, 1059–1064. [CrossRef]

59. Zhao, Z.; Buscaglia, V.; Viviani, M.; Buscaglia, M.T.; Mitoseriu, L.; Testino, A.; Nygren, M.; Johnsson, M.; Nanni, P. Grain-size effects on the ferroelectric behavior of dense nanocrystalline BaTiO$_3$ ceramics. *Phys. Rev. B* **2004**, *70*, 024107. [CrossRef]

60. Buscaglia, M.T.; Viviani, M.; Buscaglia, V.; Mitoşeriu, L.; Testino, A.; Nanni, P.; Zhao, Z.; Nygren, M.; Harnagea, C.; Piazza, D.; et al. High dielectric constant and frozen macroscopic polarization in dense nanocrystalline BaTiO$_3$ ceramics. *Phys. Rev. B* **2006**, *73*, 064114. [CrossRef]

61. Curecheriu, L.; Balmus, S.B.; Nica, V.; Buscaglia, M.T.; Buscaglia, V.; Ianculescu, A.; Mitoseriu, L. Grain-size dependent properties of dense nanocrystalline barium titanate ceramics. *J. Am. Ceram. Soc.* **2012**, *95*, 3912–3921. [CrossRef]

62. Yang, X.; Li, D.; Ren, Z.H.; Zeng, R.G.; Gong, S.Y.; Zhou, D.; Tian, H.; Li, J.X.; Xu, G.; Shen, Z.J.; et al. Colossal dielectric performance of pure barium titanate ceramics consolidated by spark plasma sintering. *RSC Adv.* **2016**, *6*, 75422. [CrossRef]

63. Valdez-Nava, Z.; Guillemet-Fritsch, S.; Tenailleau, C.; Lebey, T.; Durand, B.; Chane-Ching, J.Y. Colossal dielectric permittivity of BaTiO$_3$-based nanocrystalline ceramics sintered by spark plasma sintering. *J. Electroceram.* **2009**, *22*, 238–244. [CrossRef]

64. Guillemet-Fritsch, S.; Valdez-Nava, Z.; Tenailleau, C.; Lebey, T.; Durand, B.; Chane-Ching, J.Y. Colossal Permittivity in Ultrafine Grain Size BaTiO$_{3-x}$ and Ba$_{0.95}$La$_{0.05}$TiO$_{3-x}$ Materials. *Adv. Mater.* **2008**, *20*, 551–555. [CrossRef]

MDPI

Article

Vertically Aligned Single-Crystalline CoFe$_2$O$_4$ Nanobrush Architectures with High Magnetization and Tailored Magnetic Anisotropy

Lisha Fan [1,2,†], Xiang Gao [1,†,‡], Thomas O. Farmer [1], Dongkyu Lee [1], Er-Jia Guo [1], Sai Mu [1], Kai Wang [1], Michael R. Fitzsimmons [1,3], Matthew F. Chisholm [1], Thomas Z. Ward [1], Gyula Eres [1] and Ho Nyung Lee [1,*]

[1] Oak Ridge National Laboratory, Oak Ridge, TN 37831, USA; lfan@zjut.edu.cn (L.F.);
g.xiang@outlook.com (X.G.); thomas.farmer27@hotmail.co.uk (T.O.F.); dongkyu@cec.sc.edu (D.L.);
ejguo@iphy.ac.cn (E.-J.G.); sai.mu1986321@gmail.com (S.M.); wangkai.mse@outlook.com (K.W.);
fitzsimmonsm@ornl.gov (M.R.F.); chisholmmf@ornl.gov (M.F.C.); wardtz@ornl.gov (T.Z.W.);
eresg@ornl.gov (G.E.)
[2] College of Mechanical Engineering, Zhejiang University of Technology, Hangzhou 310023, Zhejiang, China
[3] Department of Physics and Astronomy, University of Tennessee, Knoxville, TN 37996, USA
* Correspondence: hnlee@ornl.gov
† These authors contributed equally to this work.
‡ Current address: Center for High Pressure Science and Technology Advanced Research, Beijing
100094, China.

Received: 24 January 2020; Accepted: 29 February 2020; Published: 5 March 2020

Abstract: Micrometer-tall vertically aligned single-crystalline CoFe$_2$O$_4$ nanobrush architectures with extraordinarily large aspect ratio have been achieved by the precise control of a kinetic and thermodynamic non-equilibrium pulsed laser epitaxy process. Direct observations by scanning transmission electron microscopy reveal that the nanobrush crystal is mostly defect-free by nature, and epitaxially connected to the substrate through a continuous 2D interface layer. In contrast, periodic dislocations and lattice defects such as anti-phase boundaries and twin boundaries are frequently observed in the 2D interface layer, suggesting that interface misfit strain relaxation under a non-equilibrium growth condition plays a critical role in the self-assembly of such artificial architectures. Magnetic property measurements have found that the nanobrushes exhibit a saturation magnetization value of 6.16 μB/f.u., which is much higher than the bulk value. The discovery not only enables insights into an effective route for fabricating unconventional high-quality nanostructures, but also demonstrates a novel magnetic architecture with potential applications in nanomagnetic devices.

Keywords: cobalt ferrite; nanobrush; pulsed laser epitaxy; vertically aligned; single crystalline; magnetization; tailored anisotropy

---

1. Introduction

Metal oxides, particularly complex transition metal oxides, are highly desired for their wide range of novel properties and potential applications [1]. Exploration of new fabrication routes for unconventional metal oxide nanostructures has attracted increasing attention due to their novel size-dependent properties, which opens up new avenues of functionalities in diverse fields including electronics, photonics, sensors, catalysis, energy harvesting, and information storage [2–4]. In particular, vertically ordered single-crystalline nanostructures have significant advantages over conventional planar films due to the enhanced vertical surface area, control over the shape anisotropy, and efficient vertical transport of electrons and optical excitations, which are critical to the function and integration of components at the nanoscale [5–8]. The ability to fabricate such nanostructures is essential in modern science and technology; to whit, understanding their growth mechanisms at atomic level is imperative.

In our previous studies, we demonstrated the versatility of pulsed laser epitaxy (PLE) to tailor a variety of isolated vertically aligned single-crystalline nanostructures by balancing the kinetic and thermodynamic non-equilibrium PLE processes [9,10]. This balance provides a simple but intriguing strategy for constructing unconventional vertically aligned single-crystalline binary oxide nanostructures in terms of material diversity including CeO_2, Y_2O_3, and TiO_2 [9,10]. In this work, we demonstrate the application of the technique to the synthesis of a ternary ferrimagnetic spinel oxide, $CoFe_2O_4$ (CFO) nanobrush. In the past few decades, CFO has been intensively investigated due to intriguing potential applications in spintronics, for example, as one component in multi-ferroic heterstructure [11,12], or as a spin filter tunneling barrier [13,14], etc. The magnetic properties (saturation moment, magnetic ground states, anisotropy) of CFO is a strongly structure dependent, and thus can be manipulated by epitaxial strain through the magnetostriction effect [15–19], and the extent of the inversion of the spinel structure [20–22]. The fabrication of high-density vertically ordered arrays of single crystalline CFO nanomagnets is of fundamental importance to the emerging information technologies such as ultrahigh density storage devices, magnetic random access memory devices, and logic devices [23–26].

In this work, micrometer-tall vertically aligned single-crystalline CFO "nanobrushes" were fabricated within a small window of kinetic growth parameters. Diffusion limited aggregation (DLA) is a process whereby the incoming adatoms move in a random path to form clusters, the growth front roughness thus highly depends on the surface diffusion kinetics. The formation of CFO nanobrushes occurs due to a delicate balance between thermodynamic surface equilibration and kinetic DLA. It was found that the interface misfit strain relaxation plays a critical role in the construction of the unique nanobrush structure. The nanobrushes exhibit an abnormally high saturation magnetization value of 6.16 μB/f.u., associated with a thickness-dependent magnetic anisotropy.

2. Materials and Methods

CFO samples were epitaxially grown on (001) $SrTiO_3$ (STO) substrates by a home-made PLE system (Oak Ridge, TN, USA). The STO substrates were pretreated by buffered hydrofluoric acid (HF) for 30 s and thermally treated at 1100 °C in air for 1.5 h to ensure TiO_2-terminated surfaces. A KrF excimer laser (pulse duration: 25 ns, laser fluence: 1.2–2.1 J/cm^2, laser repetition rate: 15 Hz) was used to ablate a stoichiometric CFO ceramic target for deposition at a substrate temperature, T, in a range of 400–700 °C in an oxygen partial pressure, $p(O_2)$, with a range of 0.1–1 Torr. The typical deposition rate was 0.24 Å per pulse. 2D continuous CFO films were prepared at a $p(O_2)$ of 10 mTorr, a substrate temperature of 700 °C, and a laser fluence of 0.8 J/cm^2.

The structural quality of the CFO samples was examined by x-ray diffraction (XRD) using a four-circle high-resolution x-ray diffractometer (X'Pert Pro, PANalytical, Almelo, Netherlands; Cu K α_1 radiation). The surface morphology of the samples was characterized by scanning electron microscopy (SEM). The macroscopic magnetic properties were measured with magnetic field applied along both in-plane and out-of-plane directions using a superconducting quantum interference device (SQUID, Quantum Design, San Deigo, CA, USA).

Cross-sectional specimens oriented along the [110] STO direction for scanning transmission electron microscopy (STEM) analysis were prepared using ion milling after mechanical thinning and precision polishing (using water-free abrasive). High-angle annular dark-field STEM (HAADF-STEM) was carried out in a Nion UltraSTEM microscope operated at 100 keV (Nion Co., Kirkland, WA, USA). The microscope was equipped with a cold field-emission gun and a corrector of third- and fifth-order aberrations for sub-Ångstrom resolution. An inner detector semi-angle of ~78 mrad was used for HAADF imaging. The convergence semi-angle for the electron probe was set to 30 mrad.

3. Results

3.1. Self-Assembly of the CoFe₂O₄ Nanobrushes

As demonstrated in our previous work [9,10], self-assembly of highly epitaxial, vertically aligned, single-crystalline oxide nanostructures was achieved by controlling adatom surface equilibration and DLA far from thermodynamic equilibrium. In a similar manner, the development of the surface morphologies as a function of kinetic growth parameters (substrate temperature, reactive gas pressure, and laser fluence) was fully mapped out through the plan-view SEM images of the CFO film surfaces in Figure 1. As shown from the XRD θ–2θ scans in Figure S1, all films were epitaxially grown on STO (001) with only the (00*l*) series of the CFO peaks, except for the one prepared at 1 Torr that showed weak impurity peaks.

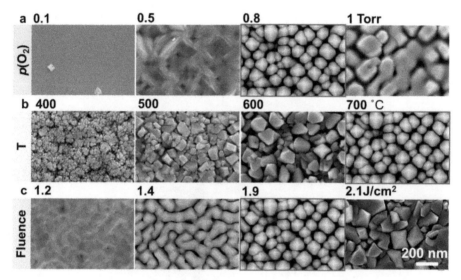

Figure 1. Plan-view SEM images of surface morphologies of CoFe₂O₄ samples prepared at different (**a**) oxygen pressure ($p(O_2)$), (**b**) substrate temperatures (T), and (**c**) laser fluences (J). Other growth conditions for the samples shown in (**a**) are $T = 700$ °C and $J = 1.9$ J/cm²; for (**b**), $p(O_2) = 0.8$ Torr and $J = 1.9$ J/cm²; for (**c**), $p(O_2) = 0.8$ Torr and $T = 700$ °C.

The CFO sample prepared at 700 °C, 0.8 Torr, and 1.9 J/cm² shows a distinct loosely packed surface morphology consisting of well-defined pyramid-like heads. It is noted that this unique porous structure formed at an extremely high $p(O_2)$ value, which is not normally used for conventional continuous 2D film epitaxy. The high reactive gas pressure is mainly responsible for the small kinetic energy of the arriving adatoms, giving rise to DLA and growth front roughness [5]. Furthermore, a high substrate temperature is essential to form the well-defined pyramids as opposed to the randomly shaped crystals observed at low temperatures. The large thermal energy provided by the hot substrate drives the adatoms to locally arrange into the lattice correctly, resulting in the growth of ordered epitaxial structures [27]. Relatively high laser fluence is also critical to the formation of this architecture, because it prevents adatoms from bouncing off and eliminating the long-distance diffusion of the adatoms by quickly covering them with the new incoming atoms [28–31]. This matches our previous observations with CeO₂, Y₂O₃, and TiO₂ systems [9,10], in which balancing the DLA and surface equilibrium allows spontaneous breakdown of the layer-by-layer growth while ensuring high crystallinity.

From the SEM image of the CFO sample (Figure 2a), the alignment of the pyramid-shaped heads was consistent along the <100> directions. Each individual pyramid has a size ranging from 80 to 150 nm in width. No impurity phases were detected in the XRD θ–2θ scan (Figure 2b) of the (001)-oriented CFO sample. Based on the Bragg diffraction theory, the CFO sample has a lattice constant of 8.386 ± 0.002 Å, which is located in between the reported values in JPCDS Card# 3-384 (8.377(7) Å) and the Card# 22-1086 (8.3919 Å) for stoichiometric CFO and comparable to the reported value of 8.39 Å in [32] for a cobalt ferrite bulk, which is slightly rich in Co with a Co:Fe ratio of 1:1.98. According to this, the CFO sample is near stoichiometric. The crystallinity of the CFO was evaluated by rocking curve measurements (Figure S2a). The typical full width at half-maximum (FWHM) of the 004 reflection was ~0.38°, revealing good epitaxial quality. The in-plane ϕ scans (Figure S2b) of the CFO 101 and STO 101 reflections at $\psi = 45°$ revealed a clear four-fold symmetry, demonstrating the cube-on-cube registration relationship between the CFO sample and the STO substrate. X-ray reciprocal space mapping (RSM) near the off-specular 114-reflection of the STO substrate was performed in order to check the strain state as displayed in Figure 2c. The CFO sample shows both in-plane 7.48% and out-of-plane 7.28% lattice strain relaxations. CFO and STO have in-plane lattice constants of 8.39 Å and 3.92 Å, respectively, leading to a lattice misfit value of 7.4%. The fact that the strain relaxation is close to the lattice mismatch between bulk CFO and STO suggests that both in-plane and out-of-plane lattice strains are fully relaxed in the CFO sample.

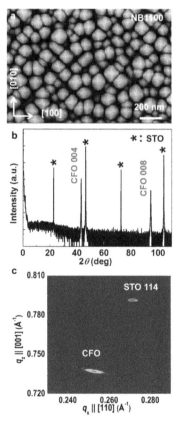

Figure 2. (**a**) A plan-view SEM image with pyramid-shaped tips clearly seen from a micrometer-tall CoFe$_2$O$_4$ nanobrush sample (NB1100), (**b**) a high-resolution XRD θ–2θ scan, and (**c**) an x-ray reciprocal space map around the 114-reflection of SrTiO$_3$ (STO) of sample NB1100 showing the biaxial strain of the nanobrush is highly relaxed. * denotes the STO {00l} substrate peaks.

3.2. Microstructure of the CoFe$_2$O$_4$ Nanobrushes

Cross-sectional STEM imaging showed the sample was composed of unique vertically aligned "nanobrushes" with a thickness of 1100 nm (Sample NB1100), as shown in Figure 3a. Sample NB1100 exhibits a two-layer structure: a dense thin interfacial layer connecting directly with the substrate, above which vertically aligned crystalline nanobrushes with a micrometer length were grown. The nanobrush diameter ranged from 80 to 120 nm and neighboring nanobrushes were separated by voids, forming a dense brush architecture.

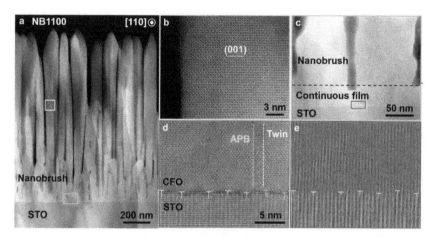

Figure 3. (**a**) A low-magnification cross-sectional high-angle annular dark-field (HAADF) image of Sample NB1100. High-resolution HAADF images of the regions marked by a yellow and a green rectangle in (**a**): (**b**) the nanobrush sidewall and (**c**) magnified interface microstructure. (**d**) High-resolution HAADF observation of a blue rectangle region in (**c**), showing the film/interface atom structure and (**e**) its fast-Fourier transformation (FFT) map.

To shed light on the formation of the nanobrushes, HAADF-STEM observation of Sample NB1100 along the CFO [110] direction was performed. The results showed that the single crystal nanobrushes were free from any obvious lattice defects (as shown in Figure 3b) and epitaxially grown. Typically, impurity phases were observed in epitaxial CFO films that lose coherence with the underlying substrate. We did not observe any impurity phase or even defects in the STEM nor XRD with the nanobrush structures that were not coherent with the substrate. Figure 3c clearly shows a continuous layer with an average thickness of 50 nm existing between the nanobrushes and the substrate. The interface displays a clear, albeit a less-abrupt interfacial region, between the dense interface layer and the substrate with slight intermix of the CFO lattice with the STO lattice no more than two unit cells (thin dark region of Figure 3d). The intermix can be attributed to damage caused by the incoming atoms present at the required high energy synthesis conditions.

As revealed by the fast-Fourier transformation (FFT) map in Figure 3e, periodic dislocation arrays were formed at the CFO/STO interface. The average periodicity was about 4.0 nm. Compared to the standard lattice misfit of 7.4% at the CFO/STO interface, the result suggests nearly full lattice misfit compensation by forming the interfacial dislocations. The results indicate that the thin CFO layer epitaxially grown on the substrate was initially under compressive stress due to the large lattice misfit, which favors the formation of interfacial dislocations to compensate for the misfit strain once the layer reaches the critical thickness [33]. The near-interface dense layer consists of nanodomains with anti-phase boundaries (APB) and twin boundaries formed in-between as shown in Figure 3d. The defects in the near-interface dense layer were not found in the nanobrush region. The areas between neighboring defects are energetically favorable for the epitaxial growth, leading to a rough growth front with the dimples corresponding to the defective area, whereas the peaks correspond to the defect-free areas [8]. A shadowing effect that reduces the probability of deposition in dimples will cause this growth front roughness to develop into extended nanobrushes [27].

3.3. Magnetization of the CoFe$_2$O$_4$ Nanobrushes

Figure 4a,b compare the in-plane and out-of-plane M–H hysteresis loops of the 1100 nm tall nanobrush Sample NB1100, a 70 nm tall nanobrush sample (NB70), and a 180 nm thick 2D continuous film (2DCF). The morphology and XRD characterization of these films are shown in Figure S3. The magnetic parameters extracted from the loops are listed in Table 1. As shown in Figure S3, NB1100 and NB70 exhibited mesoscale porous structures with a mean column width of 83.67 nm and 28.53 nm, respectively. Estimation of the volume fraction of CFO nanobrushes in the NB1100 is critical for accurate determination of its saturation magnetization value. The detailed estimation process of the volume fraction of CFO nanobrushes equal to 63 ± 10% is provided in the Supplementary Materials and Figure S4. Using this value for the volume fraction of nanobrushes, the NB1100 had an in-plane saturation magnetization value of M_s = 6.16 ± 0.7 μ_B/f.u. at 10 K. W. H. Wang reported the flux growth of bulk single crystal CFO with the 004 reflection FWHM of 0.15° and a saturation magnetization of 3.65 μ_B/f.u. [32]. The measured saturation magnetization of the NB1100 was almost three times that of the 2DCF and much higher than the reported bulk value [32]. The large saturation magnetization reduced slightly at 300 K. As shown in Figure S3, the NB70 and the 2DCF had much larger widths of the 004 Bragg reflections (e.g., full width at half maximum values 0.73° for the NB70 and 0.83° for the 2DCF) than the NB1100, 0.38°, suggesting lower crystalline quality (e.g., due to defects) compared to NB1100. The high saturation magnetization exhibited by NB1100 can be attributed to the high crystallinity of the nanobrushes and a possible extremely low inversion ratio, x, tending toward a normal spinel (all Fe^{3+} on octahedral sites and all Co^{2+} on tetrahedral sites). At thermodynamic equilibrium, CFO should possess an inverted spinel structure where half of the Fe^{+3} are on the tetrahedral sites and the remaining Fe^{3+} plus the Co^{2+} are on the octahedral sites. As the magnetic moments on the tetrahedral and octahedral lattices are in opposition, the CFO magnetization is strongly dependent on the degree of spinel inversion. Density functional theory calculations suggest the magnetization per formula unit of CFO varies from 3 μ_B in a fully inverted spinel structure where the magnetic moments of Fe^{3+} on the tetrahedral sites and the octahedral site are completed, compensated up to 7 μ_B as the inversion ratio decreases toward a normal spinal structure where all Fe^{3+} is on the octahedral sites and all Co^{2+} is on the tetrahedral sites [20–22]. The spinel inversion of CFO is highly sensitive to the growth conditions, where experimental values ranging from x = 0.62 to x = 0.93 have been reported [34,35]. W. H. Wang did not provide information on the inversion ratio of the bulk single crystal CFO, but indicated the CFO was not completely inverse [32]. The high saturation moment of 6.16 μ_B/f.u. may suggest that NB1100 has a partial inversion with a ratio below 0.5. Functionally, such a high magnetization observed in the CFO nanobrush architecture is highly desirable.

Table 1. Magnetic parameters (saturation magnetization M_s, coercivity H_c) of the CoFe$_2$O$_4$ samples.

T (K)	Sample	M_s_H//ab (μ_B/f.u.)	M_s_H//c (μ_B/f.u.)	H_c_H//ab (MA/m)	H_c_H//c (MA/m)
10	2DCF	2.27	2.19	0.99	0.65
	NB70	3.89	3.85	0.90	0.75
	NB1100	6.16 ± 0.71	6.00 ± 0.63	0.37	0.42
	Bulk [32]	3.65	-	0.10	-
300	2DCF	1.83	1.87	0.19	0.11
	NB70	3.13	3.17	0.15	0.10
	NB1100	5.90 ± 0.61	5.64 ± 0.59	0.05	0.07
	Bulk [32]	3.57	-	0.02	-

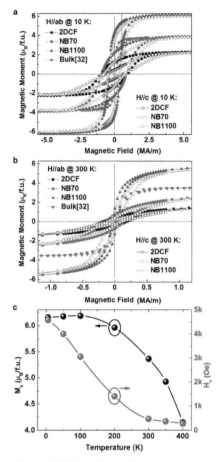

Figure 4. The M–H curves of Sample NB1100, a 70 nm tall nanobrush sample (NB70), and a 180 nm thick 2D continuous film (2DCF) under in-plane (H//ab) and out-of-plane (H//c) magnetic fields measured at temperatures of (**a**) 10 K and (**b**) 300 K, respectively. The M–H curve of the $CoFe_2O_4$ bulk was retrieved from [32] for comparison. (**c**) The coercive field (H_c) and the saturated magnetization (M_s) of the NB1100 plotted as a function of temperature.

The coercivity of different forms of CFO samples under both in-plane and out-of-plane magnetic field directions at 10 K and 300 K are listed in Table 1. Based on the Stoner–Wohlfarth model, coercivity comes from a combined effect of the anisotropy field($\sim K/M_s$, K is anisotropy constant) and the shape-anisotropy field ($\sim(1-3D)M_s$, D is a demagnetizing factor, $D = 0$ for long cylinder structures) [36]. A greater anisotropy gives rise to a larger coercivity. The H_c of bulk CFO are small at both 10 K and 300 K [32]. This could be attributed to the fact that cubic structure has weak anisotropy. As the NB1100 has a very large aspect ratio (length of each nanobrush divided by its width) of 9.2 and a higher M_s, we expected a more pronounced contribution to H_c from the shape-anisotropy field as compared to 2DCF and NB70. However, at 10 K, larger H_c was observed in 2DCF and NB70 than that in NB1100, which indicates a dominant contribution from the anisotropy field in 2DCF and NB70. In addition, we saw an in-plane H_c (H//ab) marginally larger than out-of-plane H_c (H//c) for the NB70 and 2DCF. However, the reverse was the case for the NB1100. The NB110 was fully relaxed as shown in the reciprocal space mapping analysis in Figure 2c and CFO had a large magnetostrictive constant, thus we attribute the difference to strain in the NB70 and 2DCF and lacking in the NB1100, and the ease with which magnetic response can be manipulated through symmetry change [37]. This is reasonable when considering the extremely large magnetostrictive constant of CFO [38]. From this, it is evident that control over the nanobrush height provides a mechanism for tailoring the magnetic anisotropy. A similar thickness dependent anisotropy has been observed in BFO/CFO nanocomposites [39].

Figure 4c shows the temperature-dependent $M_s(T)$ and $H_c(T)$ in the NB1100. While the $M_s(T)$ only reduces by 32.6% up to 300 K, the H_c is strongly temperature dependent and a pronounced reduction is observed. According to the Callen–Callen rule [40], the anisotropy constant is linked to the magnetization via power laws: $K(T) = K(0)\left(\frac{M_s(T)}{M_s(0)}\right)^n$, where the mth order anisotropy constant obeys $n = m(m + 1)/2$. The temperature dependent anisotropy of Fe and Co metals are well described by this model [36]. For a cubic structure as in CFO, a fourth order anisotropy is expected and therefore $n = 10$. However, from the scaling as shown in Figure S5, we observed a much faster H_c reduction than $n = 10$ in $M_s(T)^n$. The failure of the Callen–Callen model for this nanobrush system may be ascribed to the fact that the Callen–Callen model merely bridges the anisotropy to the net magnetization while neglecting the effect of different exchange fields on different sublattices. Since CFO is an uncompensated ferrimagnet [36], which involves multiple magnetic sublattices, the Callen–Callen model may be inadequate to describe the subtle temperature dependent exchange field on different sublattices, resulting in an error of the anisotropy at finite temperatures.

4. Discussion

A simple but novel synthesis strategy that can be applied extensively in diverse material systems for unconventional vertically aligned nanostructure fabrication was developed. Vertically aligned single-crystalline CFO nanobrushes were epitaxially grown on STO substrates by PLE. Thermodynamically and kinetically balancing DLA and surface equilibration through kinetic growth parameter control provides a route to design epitaxially grown self-assembled nanostructures. The magnetic properties are highly sensitive to both the morphology and brush length. It was found that the nanobrushes exhibited a very high saturation magnetization and a length-dependent magnetic anisotropy. This work provides new insight into the controllable synthesis of vertically aligned nanostructures and demonstrates a novel magnetic structure suitable for applications in nanomagnetic devices.

Supplementary Materials: The following are available online at http://www.mdpi.com/2079-4991/10/3/472/s1, Figure S1: XRD scans of the CFO samples prepared as functions of kinetic growth parameters; Figure S2: The crystallinity of Sample NB1100 evaluated by XRD characterization (rocking curve and in-plane ϕ scan); Figure S3: Morphology and structural characterization of NB1100, NB70, and a 2D continuous film; Figure S4: Estimation of the volume fraction of Sample NB1100; and Figure S5: Comparison of the normalized $H_c(T)$ curve with $M_s^{10}(T)$ curve according to the Callen–Callen rule.

Author Contributions: The individual contributions are as follows. Supervision and conceptualization: H.N.L.; Synthesis: L.F., D.L., T.Z.W., G.E., and H.N.L.; STEM characterization: X.G. and M.F.C.; Magnetization and other analysis: T.O.F., S.M., K.W., E.-J.G., M.R.F. and H.N.L. All authors have read and agreed to the published version of the manuscript.

Funding: This work was supported by the U.S. Department of Energy, Office of Science, Basic Energy Sciences, Materials Science and Engineering Division.

Acknowledgments: This work was supported by the U.S. Department of Energy (DOE), Office of Science, Basic Energy Sciences, Materials Science and Engineering Division. The STEM and SEM work used resources at the Center for Nanophase Materials Sciences, which is a U.S. DOE Office of Science User Facility operated by Oak Ridge National Laboratory. L. F. conducted this research while at ORNL and was in part supported by Zhejiang University of Technology for her efforts to write and revise the manuscript.

Conflicts of Interest: The authors declare no conflicts of interest.

References

1. Srivastava, A.K. *Oxide Nanostructures: Growth, Microstructures and Properties*; Chapter 1; Pan Stanford Pte. Ltd.: Boca Raton, FL, USA, 2014; pp. 1–77.

2. Xia, Y.; Yang, P.; Sun, Y.; Wu, Y.; Mayers, B.; Gates, B.; Yin, Y.; Kim, F.; Yan, H. One-Dimensional Nanostructures: Synthesis, Characterization, and Applications. *Adv. Mater.* **2003**, *15*, 353. [CrossRef]

3. Rorvik, P.M.; Grande, T.; Einarsrud, M.A. One-Dimensional Nanostructures of Ferroelectric Perovskites. *Adv. Mater.* **2011**, *23*, 4007. [CrossRef] [PubMed]

4. Xu, S.; Wang, Z.L. One-dimensional ZnO nanostructures: Solution growth and functional properties. *Nano Res.* **2011**, *4*, 1013–1098. [CrossRef]

5. Infortuna, A.; H arvey, A.S.; Gauckler, L.J. Microstructures of CGO and YSZ Thin Films by Pulsed Laser Deposition. *Adv. Funct. Mater.* **2008**, *18*, 127. [CrossRef]

6. Zhou, Y.; Park, C.S.; Wu, C.H.; Maurya, D.; Murayama, M.; Kumar, A.; Katiyar, R.S.; Priya, S. Microstructure and surface morphology evolution of pulsed laser deposited piezoelectric BaTiO3 films. *J. Mater. Chem. C* **2013**, *1*, 6308. [CrossRef]

7. Jiang, J.; Henry, L.L.; Gnansekar, K.I.; Chen, C.; Meletis, E.I. Self-Assembly of Highly Epitaxial (La,Sr)MnO3 Nanorods on (001) LaAlO3. *Nano Lett.* **2004**, *4*, 741. [CrossRef]

8. Zhang, K.; Dai, J.; Zhu, X.; Zhu, X.; Zou, X.; Zhang, P.; Hu, L.; Lu, W.; Song, W.; Sheng, Z.; et al. Vertical La$_{0.7}$Ca$_{0.3}$MnO3 nanorods tailored by high magnetic field assisted pulsed laser deposition. *Sci. Rep.* **2016**, *6*, 19483. [CrossRef]

9. Fan, L.; Gao, X.; Lee, D.K.; Guo, E.J.; Lee, S.B.; Snijders, P.C.; Ward, T.Z.; Chisholm, M.F.; Eres, G.; Lee, H.N. Kinetic controlled fabrication of single-crystalline TiO$_2$ nanobrush architecture. *Adv. Sci.* **2017**, *4*, 1700045. [CrossRef]

10. Lee, D.K.; Gao, X.; Fan, L.; Guo, E.J.; Farmer, T.O.; Heller, W.T.; Ward, T.Z.; Eres, G.; Fitzsimmons, M.R.; Chisholm, M.F.; et al. Non-equilibrium synthesis of highly porous single-crystalline oxide nanostructures. *Adv. Mater. Interfaces* **2017**, *4*, 1601034. [CrossRef]

11. Zheng, H.; Wang, J.; Lofland, S.E.; Ma, Z.; Mahaddes-Ardabili, L.; Zhao, T.; Salamanca-Riba, L.; Shinde, S.R.; Ogale, S.B.; Bai, F.; et al. Multiferroic BaTiO$_3$-CoFe$_2$O$_4$ nanostructures. *Science* **2004**, *303*, 661–663. [CrossRef]

12. Zavaliche, F.; Zheng, H.; Mohaddes-Ardabili, L.; Yang, S.Y.; Zhan, Q.; Shafer, P.; Reilly, E.; Chopdekar, R.; Jia, Y.; Wright, P.; et al. Electric Field-Induced Magnetization Switching in Epitaxial Columnar Nanostructures. *Nano Lett.* **2005**, *5*, 1793–1796. [CrossRef]

13. Chapline, M.G.; Wang, S.X. Room-temperature spin filerting in a CoFe$_2$O$_4$/MgAl$_2$O$_4$/Fe$_3$O$_4$ magnetic tunnel barrier. *Phys. Rev. B* **2006**, *74*, 014418. [CrossRef]

14. Ramos, A.V.; Santos, T.S.; Miao, G.X.; Guittet, M.-J.; Moussy, J.-B.; Moodera, J.S. Influence of oxidation on the spin-filtering properties of CoFe$_2$O$_4$ and the resultant spin polarization. *Phys. Rev. B* **2008**, *78*, 180402. [CrossRef]

15. Fritsch, D.; Ederer, C. Epitaxial strain effects in the spinel ferrites CoFe$_2$O$_4$ and NiFe$_2$O$_4$ from first principles. *Phys. Rev. B* **2010**, *82*, 104117. [CrossRef]

16. Zhou, S.; Potzger, K.; Xu, Q.; Keupper, K.; Talut, G.; Marko, D.; Mucklich, A.; Helm, M.; Fassbender, J.; Arenholz, E.; et al. Spinel ferrite nanocrystals embedded inside ZnO: Magnetic, electronic, and magnetotransport properties. *Phys. Rev. B* **2009**, *80*, 094409. [CrossRef]

17. Lisfi, A.; Williams, C.M. Magnetic anisotropy and domain structure in epitaxial CoFe$_2$O$_4$ thin films. *J. Appl. Phys.* **2003**, *93*, 8143–8145. [CrossRef]

18. Park, J.H.; Lee, J.H.; Kim, M.G.; Jeong, Y.K.; Oak, M.A.; Jang, H.M.; Choi, H.J.; Scott, J.F. In-plane strain control of the magnetic remanence and cation-charge redistribution in CoFe$_2$O$_4$ thin film grown on a piezoelectric substrate. *Phys. Rev. B* **2010**, *81*, 134401. [CrossRef]

19. Ma, J.X.; Mazumdar, D.; Kim, G.; Sato, H.; Bao, N.Z.; Gupta, A. A robust approach for the growth of epitaxial spinel ferrite films. *J. Appl. Phys.* **2010**, *108*, 063917. [CrossRef]

20. Walsh, A.; Wei, S.H.; Yan, Y.; Al-Jassim, M.M.; Turner, J.A. Structural, magnetic, and electronic properties of the Co-Fe-Al oxide spinel system: Density-functional theory calculations. *Phys. Rev. B* **2007**, *76*, 165119. [CrossRef]

21. Szotek, Z.; Temmerman, W.M.; Kodderitzsch, D.; Svane, A.; Petit, L.; Winter, H. Electronic structures of normal and inverse spinel ferrite from first principles. *Phys. Rev. B* **2006**, *74*, 174431. [CrossRef]

22. Hou, Y.H.; Zhao, Y.J.; Liu, Z.W.; Yu, H.Y.; Zhong, X.C.; Qiu, W.Q.; Zeng, D.C.; Wen, L.S. Structural, electronic and magnetic properties of partially inverse spinel CoFe$_2$O$_4$: A first principles study. *J. Phys. D Appl. Phys.* **2010**, *43*, 445003. [CrossRef]

23. Wang, Z.; Viswan, R.; Hu, B.; Harris, V.G.; Li, J.F.; Viehland, D. Tunable magnetic anisotropy of $CoFe_2O_4$ nanopillar arrays released from $BiFeO_3$ matrix. *Phys. Status Solidi RRL* **2012**, *6*, 92–94. [CrossRef]

24. Gao, X.; Liu, L.; Birajdar, B.; Ziese, M.; Lee, W.; Alexe, M.; Hesse, D. High-density periodically ordered magnetic cobalt ferrite nanodot arrays by template-assisted pulsed laser deposition. *Adv. Funct. Mater.* **2009**, *19*, 3450–3455. [CrossRef]

25. Guo, E.J.; Herklotz, A.; Kehlberger, A.; Cramer, J.; Jakob, G.; Klaui, M. Thermal generation of spin current in epitaxial $CoFe_2O_4$ thin films. *Appl. Phys. Lett.* **2016**, *108*, 022403. [CrossRef]

26. Mathew, D.S.; Juang, R.S. An overview of the structure and magnetism of spinel ferrite. *Chem. Eng. J.* **2007**, *129*, 51–65. [CrossRef]

27. Pelliccione, M.; Lu, T.M. *Evolution of Thin Film Morphology*; Chapter 9; Springer-Verlag: Berlin/Heidelber, Germany, 2008; pp. 121–136.

28. Thornton, J.A. High rate thick film growth. *Ann. Rev. Mater.* **1977**, *7*, 239–260. [CrossRef]

29. Muller, K.H. Dependence of thin-film microstructure on deposition rate by means of a computer simulation. *J. Appl. Phys.* **1985**, *58*, 2573. [CrossRef]

30. Yang, Y.G.; Johnson, R.A.; Wadley, H.N.G. A Monte Carlo simulation of the physical vapor deposition of nickel. *Acta Mater.* **1997**, *45*, 1455–1468. [CrossRef]

31. Lu, Y.; Wang, C.; Gao, Y.; Shi, R.; Liu, X.; Wang, Y. Microstructure map for self organized phase separation during film deposition. *Phys. Rev. Lett.* **2012**, *109*, 086101. [CrossRef]

32. Wang, W.H.; Ren, X. Flux growth of high-quality $CoFe_2O_4$ single crystals and their characterization. *J. Cryst. Growth* **2006**, *289*, 605–608. [CrossRef]

33. Moussy, J.B.; Gota, S.; Bataille, A.; Guittet, M.J.; Guatier-Soyer, M.; Delille, F.; Dieny, B.; Ott, F.; Doan, T.D.; Warin, P.; et al. Thickness dependence of anomalous magnetic behavior in epitaxial Fe_3O_4(111) thin films: Effect of density of antiphase boundaries. *Phys Rev. B* **2004**, *70*, 174448. [CrossRef]

34. Sawatzky, G.A.; Van Der Woude, F.; Morrish, A.H. Mossbauer study of several ferromagnetic spinel. *Phys. Rev.* **1969**, *187*, 747. [CrossRef]

35. Murray, P.J.; Linnett, J.W. Mössbauer studies in the spinel system $Co_xFe_{3-x}O_4$. *J. Phys. Chem. Solids* **1976**, *37*, 619. [CrossRef]

36. Skomski, R. *Simple Models for Magnetism*; Oxford university press: Oxford, United Kingdom, 2008.

37. Herklotz, A.; Gai, Z.; Sharma, Y.; Huon, A.; Rus, S.F.; Sun, L.; Shen, J.; Rack, P.D.; Ward, T.Z. Designing magnetic anisotropy through strain doping. *Adv. Sci.* **2018**, *5*, 1800356. [CrossRef]

38. Bozorth, R.M.; Walker, J.G. Magnetostriction of single crystals of cobalt and nickel ferrites. *Phys. Rev.* **1952**, *88*, 1209. [CrossRef]

39. Aimon, N.M.; Hun Kim, D.; Kyoon Choi, H.; Ross, C.A. Deposition of epitaxial $BiFeO_3/CoFe_2O_4$ nanocomposites on (001) $SrTiO_3$ by combinatorial pulsed laser deposition. *Appl. Phys. Lett.* **2012**, *100*, 092901. [CrossRef]

40. Callen, E.R.; Callen, H.B. Magnetoelastic Coupling in Cubic Crystal. *Phys. Rev.* **1963**, *129*, 578. [CrossRef]

nanomaterials

MDPI

Article

Ferroelectric Diode Effect with Temperature Stability of Double Perovskite Bi$_2$NiMnO$_6$ Thin Films

Wen-Min Zhong, Qiu-Xiang Liu, Xin-Gui Tang *, Yan-Ping Jiang, Wen-Hua Li, Wan-Peng Li and Tie-Dong Cheng

School of Physics Optoelectric Engineering, Guangdong University of Technology, Guangzhou Higher Education Mega Center, Guangzhou 510006, China; zhongwen_min@163.com (W.-M.Z.); liuqx@gdut.edu.cn (Q.-X.L.); ypjiang@gdut.edu.cn (Y.-P.J.); liwenhuat@gdut.edu.cn (W.-H.L.); liwanpeng361@163.com (W.-P.L.); chengtiedong@126.com (T.-D.C.)
* Correspondence: xgtang@gdut.edu.cn; Tel./Fax: +86-20-3932-2265

Received: 20 November 2019; Accepted: 10 December 2019; Published: 15 December 2019

Abstract: Double perovskite Bi$_2$NiMnO$_6$ (BNMO) thin films grown on p-Si (100) substrates with LaNiO$_3$ (LNO) buffer layers were fabricated using chemical solution deposition. The crystal structure, surface topography, surface chemical state, ferroelectric, and current-voltage characteristics of BNMO thin films were investigated. The results show that the nanocrystalline BNMO thin films on p-Si substrates without and with LNO buffer layer are monoclinic phase, which have antiferroelectric-like properties. The composition and chemical state of BNMO thin films were characterized by X-ray photoelectron spectroscopy. In the whole electrical property testing process, when the BNMO/p-Si heterojunction changed into a BNMO/LNO/p-Si heterojunction, the diode behavior of a single diode changing into two tail to tail diodes was observed. The conduction mechanism and temperature stability were also discussed.

Keywords: Bi$_2$NiMnO$_6$; thin films; diode effect; oxygen defect; conduction mechanism

---

1. Introduction

In the past few decades, electronic devices prepared using a semiconductor have become an important research project in the field of materials science [1–4]. These devices have garnered attention for their practical applications, such as magnetoresistance, photodetectors, p-n diodes and thin film transistors [5,6]. Bi$_2$NiMnO$_6$ has been widely studied as a multiferroic material. The ferromagnetic and ferroelectric Bi$_2$NiMnO$_6$ was successfully prepared at 6 GPa as reported by Azuma et al. [7]. Low temperature (about 100 K) ferroelectric properties in pulsed laser-deposition drive Bi$_2$NiMnO$_6$ thin films on (001)-oriented SrTiO$_3$ single crystal substrates were reported by Sakai et al. [8]. The phase transition temperature of epitaxial Bi$_2$NiMnO$_6$ thin films affected by single crystal substrates was studied using Raman spectroscopy [9]. The magnetodielectric effect was obtained in single-phase and epitaxial thin film of multiferroic Bi$_2$NiMnO$_6$, as reported by Padhan et al. and Rathi et al. [10–12]. The ferroelectric behavior and magnetic exchange interaction effect of Bi$_2$NiMnO$_6$ with the electric polarization 19.01 μC/cm^2 was reported by Zhao et al. [13]. Theoretical and experimental results confirm that Bi$_2$NiMnO$_6$ thin films are multiferroic materials [8,14–18]. The ferroelectric and current leakage characteristics of La-doped Bi$_2$NiMnO$_6$ and Bi$_2$NiMnO$_6$ thin film was reported by Li et al. [19].

However, the ferroelectric diode effect and temperature stability of Bi$_2$NiMnO$_6$ thin film has never been reported. Therefore, in this work, a thin film of Bi$_2$NiMnO$_6$ was growth on p-Si and LaNiO$_3$/p-Si substrates using chemical solution deposition technology, the ferroelectric diode effect and temperature stability of Bi$_2$NiMnO$_6$ thin film was first investigated, as was the conduction mechanism.

2. Materials and Methods

The Bi$_2$NiMnO$_6$ (abbreviated as BNMO) precursor was prepared by dissolving nitrogen salt, bismuth, manganese acetate, and nickel acetate in a ratio of 2.2:1:1 in ethylene glycol solution [16]. The excess of 10% bismuth was to prevent evaporation of the film during the drying and annealing process. Three milliliters of acetic acid was added to the solution to prevent precipitation. The precursor BNMO concentration is 0.2 M. The LaNiO$_3$ (LNO) precursor was prepared by dissolving nickel acetate and lanthanum nitrate in a ratio of 1:1 in ethylene glycol solution and the resulting concentration was 0.3 M. The 5 mL acetone acetate stabilizer was also added to the precursor. The precursors were then aged for 3 days. The wet LNO/p-Si thin film was synthesized by a spin-coating process at a rate of 2500 rpm for 15 s. The LNO/p-Si substrate was made using a drying process at 573 K for 5 min and an annealing process at 973 k for 30 min. The BNMO/p-Si and BNMO/LNO/p-Si heterojunctions are prepared using the spin coating process at 3000 rpm for 15 s, the dry process at 573 K for 5 min and annealing process at 973 K for 10 min by rapid thermal annealing (RTA) in air atmosphere. A gold electrode with a diameter of 0.3 mm was plated on the surface of the film sample by a small high vacuum coater to form a film capacitor structure.

The crystal structure analysis was measured using XRD (Bruker D8 Advance, AXS, Germany) and the chemical states were examined using X-ray photoelectron spectroscopy (XPS, Escalab 250Xi, Sussex, UK). The surface topography and elemental analysis were performed with the field emission scanning electron microscope (FE-SEM, SU8220, Hitachi, Japan). The ferroelectric properties of the BNMO thin films were measured with a ferroelectric test system (Radiant Technologies Precision Workstation, Albuquerque, NM, USA). The current-voltage characteristic was measured with a Keithley 2400 system.

3. Results and Discussion

The XRD patterns are shown in Figure 1. From Figure 1, there are six peaks at (100), (110), (111), (200), (210) and (211), the crystal structure of LNO grown on p-Si can be judged by PDF card of 33-0710 as a cubic phase. The crystal structure of BNMO film grown on p-Si substrate without and with LNO can be determined by diffraction angles of 23.62° and 31.18°. It has the monoclinic structure reported by Azuma et al. [7]. The unit cell of BNMO was considered to be similar to BiMnO$_3$. The three possible transition metal sites of M1, M2, and M3 were filled with Bi^{3+}, Ni^{2+} and Mn^{4+} cations and the Mn^{4+}–O^{2-}–Ni^{2+} chemical links and Bi^{3+}–O^{2-} links were the main chain segment. The diffraction angle of 27.58°and 29.22° matches the results of Li et al. [19], and it is a monoclinic structure with C2 space group [20]. In theory, The NiO$_6$ and MnO$_6$ octahedral of BNMO are isotropic and do not cause distortion.

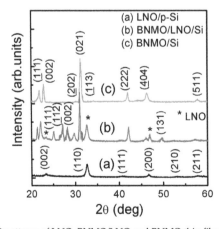

Figure 1. The XRD patterns of LNO, BNMO/LNO and BNMO thin films on p-Si substrates.

Nanomaterials **2019**, *9*, 1783

Figure 2 show the surface and cross-section topographies of BNMO thin film and BNMO/LNO thin film grown p-Si substrates, respectively. From Figure 2a,b it can be clearly observed from the surface topography that the grain size of BNMO/p-Si thin film is nearly 15 nm, but the BNMO/LNO/p-Si thin film is nearly 10 nm. The BNMO thin film was formed by the Mn^{4+} cations and Ni^{2+} cations, and the thin films was annealed at air atmosphere for only 10 min. Therefore, the growth of crystal grains is relatively difficult, resulting in a small grain size. Observed from the cross-section images, it can be obtained that the thickness of the BNMO layer growth on the p-Si substrata is nearly 100 nm (see Figure 2c), the BNMO layer on LNO/p-Si substrate is 140 nm (see Figure 2d). The different film thickness of the BNMO layers may be caused by the different adhesion of LNO and Si to the solution and the first layer, respectively.

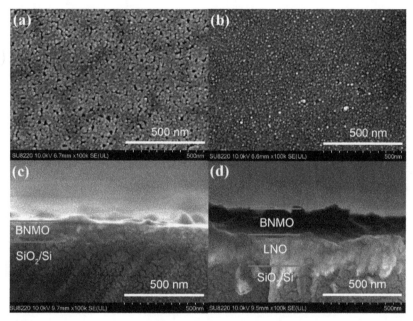

Figure 2. The surface topography and cross-section images: (**a,c**) for BNMO/p-Si, (**b,d**) for BNMO/LNO/p-Si.

The XPS spectra of Ni 2p and Mn 2p are shown in Figure 3. The binding energy of Mn $2p_{3/2}$ and Mn $2p_{1/2}$ of the BNMO/p-Si heterojunctions was 641.35 eV and 653.2 eV [21]. The binding energy of Mn $2p_{3/2}$ and Mn $2p_{1/2}$ of the BNMO/LNO heterojunction was 641.6 eV and 653.45 eV. The binding energy of Ni $2p_{3/2}$ and Ni $2p_{1/2}$ of the BNMO/p-Si heterojunction was 872.45 eV and 861.3 eV [22]. The binding energy of Ni $2p_{3/2}$ and Ni $2p_{1/2}$ of the BNMO/LNO/p-Si heterojunction was 855.1 eV and 872.55 eV.

The XPS spectrum is subject to peak processing. The binding energy of 638.85 eV, 641.65 eV, and 644.25 eV of the BNMO/p-Si heterojunction and 638.25 eV, 641.5 eV, and 643.95 eV of the BNMO/LNO/p-Si heterojunction indicates the Mn^{2+}, Mn^{4+} and Mn^{6+} cation. The binding energy of 855 eV and 857.65 eV of the BNMO/p-Si heterojunction and 855 eV and 857.2 eV of the BNMO/LNO/p-Si heterojunction shows the Ni^{2+} and Ni^{3+} cation. The ion ratio of Mn^{2+}:Mn^{4+}:Mn^{6+} on BNMO/p-Si thin film is 0.15:1:0.3 and for BNMO/LNO/p-Si thin film is 0.16:1:0.27. The ratio of Ni^{2+}:Ni^{3+} on the BNMO/p-Si heterojunction and on the BNMO/LNO heterojunction is 1:0.22 and 1:0.24, respectively. By analyzing the XPS spectrum, a variety of Ni, Mn ions are found in the BNMO heterojunction device. The cubic crystal structure of NiO and $BiMnO_3$ octahedral crystallites interferes with the formation of pure phase monoclinic crystals and causes crystal defects.

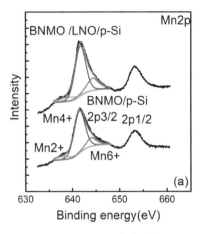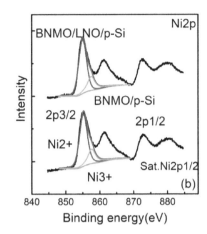

Figure 3. The fitted narrow-scan spectra for (**a**) Mn 2p and (**b**) Ni 2p.

The room temperature polarization-electric field (*P-E*) properties are shown in Figure 4. The results show that the two films have antiferroelectric-like properties. The saturated polarization ($2P_s$), remnant polarization ($2P_r$), and coercive field (E_c) of BNMO/LNO/Si thin film were 0.875 µC/cm², 0.150 µC/cm², and 40.0 kV/cm, and 1.03 µC/cm², 0.202 µC/cm², and 38.4 kV/cm, respectively for BNMO/Si and BNMO/LNO/Si thin films at 500 Hz. The ferroelectric polarization was enhanced by using LNO as a buffer layer. The room temperature ferroelectric polarization phenomenon could be due to the incompletely symmetric monoclinic structure preventing the ferroelectric domain from flipping. It is a pinning effect caused by the interaction of defect dipoles in the BNMO layer. The BNMO growth on the LNO/p-Si substrate is a nanocrystalline state, resulting in more lattice defects, preventing the deformation of the crystal [7].

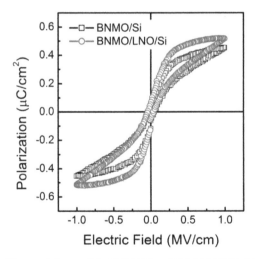

Figure 4. The typical hysteresis loops of the BNMO/Si and BNMO/LNO/Si thin films measured at 500 Hz.

The current-voltage characteristics of BNMO/Si and BNMO/LNO/Si thin films are shown in Figure 5. From Figure 5a, we know that from 0 to −1.0 V and from 0 to 0.25 V, the current hardly changed with the increase of absolute voltage value, when the voltage increases from −1.0 to −1.5 V, the current increased, and when the voltage increased from 0.65 to 1.5 V, the current increased rapidly. The results show typical diode characteristics.

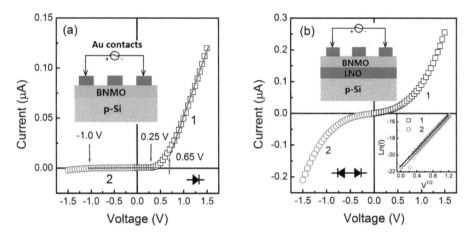

Figure 5. I-V characteristics of BNMO thin films without (**a**) and with (**b**) a LNO buffer layer on p-Si substrates. Insets show the schematics of the diode cell used for measurement.

The BNMO/p-Si heterojunction exhibited rectification effect behavior of p-n junction. It is a forward-conducting heterojunction device with an ON/OFF ratio ($R = I_{on}/I_{off}$) of 65. The n-type semiconductor properties of BNMO have not been reported. It can be inferred from the results, and the process of converting Mn^{2+} ions into Mn^{4+} ions can release electrons.

As the bottom electrodes of the LNO grew on the p-Si substrate, the rectification effect of the forward conduction was forcibly converted into a reverse conduction rectification effect, and the ON/OFF ratio is increased to 1.4. The result shows two tail to tail diodes. The cubic phase of LNO acted as an n-type semiconductor to release electrons, while the nano-crystallinity of the BNMO layer defects acted as a hole-absorbing electron [23–25].The Schottky emission mechanism is determined by the linear relationship of Ln(I) versus $V^{1/2}$ [26,27]. If the relationship is linear, this is due to the thermionic emission by holes, vacancies and defects [28–32], respectively. The restricted behavior of the interface and the hole trapping behavior are considered to be a case of the Schottky emission mechanism, as expected with linear relationship of Ln(I) versus $V^{1/2}$ for the BNMO/LNO/p-Si heterojunction (see the inset of Figure 5b). Therefore, the large ON/OFF ratio of the BNMO/LNO/p-Si heterojunction is caused by the hole of the BNMO layer being filled.

The voltage-current characteristics measured at different temperatures were shown in Figure 6. The conductivity of the BNMO/p-Si heterojunction and the BNMO/LNO/p-Si heterojunction was increased due to increased test temperature. The ON/OFF ratio of the BNMO/p-Si heterojunction is decreased from 65 to 3.8, and the ON/OFF ratio of the BNMO/LNO/p-Si heterojunction is increased from 1.4 to 13.4. The result can be interpreted with the thermion emission effect of Schottky diodes [33–36]. The electrons are excited by the lattice defects caused by the heat. Therefore, the conductivity of the film increases. At the same time, the space charge accumulated at the interface of the heterojunction is also subject to thermal radiation. The BNMO/p-Si heterojunction is excited by thermal radiation, causing the electrons of the defect to be excited, forming a space charge region at the interface of the heterojunction, and finally the rectification effect is improved. The BNMO/LNO/p-Si heterojunction is thermally radiated, and the electron trapping ability of the hole is weakened, resulting in the rectification effect being weakened.

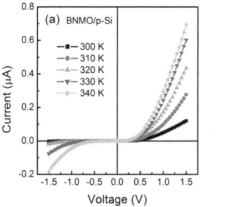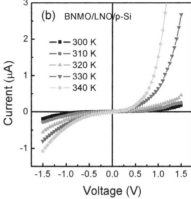

Figure 6. The I-V characteristics of BNMO thin films (**a**) without and (**b**) with a LNO buffer layer on p-Si substrates measured at different temperatures.

4. Conclusions

In conclusion, the BNMO thin films on Si and LNO/Si substrates show a monoclinic phase with C2 space group and the porosity of BNMO thin film with LNO layer is smaller than without, resulting in a larger ferroelectric polarization. The conduction mechanism of the BNMO/Si and BNMO/LNO/Si heterojunctions were dominated by Ohmic conduction and Schottky emission mechanisms, respectively. In the case of temperature rise, the rectification effect of BNMO/Si will decrease due to the energy of the hole trapping electrons being weakened and the rectification effect of the BNMO/LNO/p-Si heterojunction will increase due to the charge accumulation, respectively. XPS testing shows that BNMO synthesized under normal pressure is a material with the coexistence of different valence cations.

Author Contributions: Author X.-G.T. and Q.-X.L., conceptualized the idea. The test was helped by the W.-M.Z., W.-P.L. and T.-D.C. The draft of the manuscript was reviewed and revised by W.-M.Z. and X.-G.T. The work was supported and supervised by Q.-X.L., Y.-P.J. and W.-H.L. All authors read and approved the final manuscript.

Funding: This research was funded by "the National Natural Science Foundation of China (Grant Nos. 11574057 and 51604087)", "the Guangdong Provincial Natural Science Foundation of China (Grant No. 2016A030313718)", and "the Science and Technology Program of Guangdong Province of China (Grant Nos. 2016A010104018 and 2017A010104022)".

Conflicts of Interest: The authors declare no conflict of interest.

References

1. Park, S.; Lee, B.; Bae, B.; Chai, J.; Lee, S.; Kim, C. Ambipolar thin-film transistors based on organic semiconductor blend. *Synth. Met.* **2019**, *253*, 40–47. [CrossRef]
2. Shiryaev, A.O.; Rozanov, K.N.; Vyzulin, S.A.; Kevraletin, A.L.; Syr'ev, N.E.; Vyzulin, E.S.; Lahderanta, E.; Maklakov, S.A.; Granovsky, A.B. Magnetic resonances and microwave permeability in thin Fe films on flexible polymer substrates. *J. Magn. Magn. Mater.* **2018**, *461*, 76–81. [CrossRef]
3. Makhlouf, M.M. Preparation and optical characterization of β-MnO$_2$ nano thin films for application in heterojunction photodiodes. *Sens. Actuators A Phys.* **2018**, *279*, 145–156. [CrossRef]
4. Munshi, A.H.; Sasidharan, N.; Pinkayan, S.; Barth, K.L.; Sampath, W.S.; Ongsakul, W. Thin-film CdTe photovoltaics—The technology for utility scale sustainable energy generation. *Sol. Energy* **2018**, *173*, 511–516. [CrossRef]
5. Yin, L.; Wang, C.; Li, L.; Shen, Q.; Zhang, L. Large room temperature magnetoresistance in La$_{0.9}$Sr$_{0.1}$MnO$_3$ thin films. *J. Alloys Compd.* **2018**, *730*, 327–332. [CrossRef]

6. Mitta, S.B.; Reddeppa, M.; Vellampatti, S.; Dugasani, S.R.; Yoo, S.; Lee, S.; Kim, M.D.; Park, S.H. Gold nanoparticle-embedded DNA thin films for ultraviolet photodetectors. *Sens. Actuators B Chem.* **2018**, *275*, 137–144. [CrossRef]

7. Azuma, M.; Takata, K.; Saito, T.; Ishiwata, S.; Shimakawa, Y.; Takano, M. Designed Ferromagnetic, Ferroelectric Bi_2NiMnO_6. *J. Am. Chem. Soc.* **2005**, *127*, 8889–8892. [CrossRef]

8. Sakai, M.; Masuno, A.; Kan, D.; Hashisaka, M.; Takata, K.; Azuma, M.; Takano, M.; Shimakawa, Y. Multiferroic thin film of Bi_2NiMnO_6 with ordered double-perovskite structure. *Appl. Phys. Lett.* **2007**, *90*, 072903. [CrossRef]

9. Iliev, M.N.; Padhan, P.; Gupta, A. Temperature-dependent Raman study of multiferroic Bi_2NiMnO_6 thin films. *Phys. Rev. B* **2008**, *77*, 172303. [CrossRef]

10. Padhan, P.; LeClair, P.; Gupta, A.; Srinivasan, G. Magnetodielectric response in epitaxial thin films of multiferroic Bi_2NiMnO_6. *J. Phys. Condens. Matter* **2008**, *20*, 355003. [CrossRef]

11. Padhan, P.; LeClair, P.; Gupta, A.; Subramanian, M.A.; Srinivasan, G. Magnetodielectric effect in Bi_2NiMnO_6-La_2NiMnO_6 superlattices. *J. Phys. Condens. Matter* **2009**, *21*, 306004. [CrossRef] [PubMed]

12. Rathi, A.; Anshul, A.; Gupta, A.; Rout, P.K.; Maurya, K.K.; Kotnala, R.K.; Pant, R.P.; Basheed, G.A. Large low-field magnetodielectric response in multiferroic Bi_2NiMnO_6 thin film. *J. Phys. D Appl. Phys.* **2017**, *50*, 135006. [CrossRef]

13. Zhao, H.J.; Chen, X.M. First-principles study on the differences of possible ferroelectric behavior and magnetic exchange interaction between Bi_2NiMnO_6 and La_2NiMnO_6. *AIP Adv.* **2012**, *2*, 042143. [CrossRef]

14. Ciucivara, A.; Sahu, B.; Kleinman, L. Density functional study of multiferroic Bi_2NiMnO_6. *Phys. Rev. B* **2007**, *76*, 064412. [CrossRef]

15. Shimakawa, Y.; Kan, D.; Kawai, M.; Sakai, M.; Inoue, S.; Azuma, M.; Kimura, S.; Sakata, O. Direct observation of B-site ordering in multiferroic Bi_2NiMnO_6 thin film. *Jpn. J. Appl. Phys.* **2007**, *46*, L845–L847. [CrossRef]

16. Bahoosh, S.G.; Wesselinowa, J.M.; Trimper, S. The magnetoelectric effect and double-perovskite structure. *Phys. Status Solidi B* **2012**, *249*, 1602–1606. [CrossRef]

17. Zhao, H.J.; Liu, X.Q.; Chen, X.M. Density functional investigations on electronic structures, magnetic ordering and ferroelectric phase transition in multiferroic Bi_2NiMnO_6. *AIP Adv.* **2012**, *2*, 022115. [CrossRef]

18. Dieguez, D.; Iniguez, J. Multiferroic Bi_2NiMnO_6 thin films: A computational prediction. *Phys. Rev. B* **2017**, *95*, 085129. [CrossRef]

19. Li, W.P.; Liu, Q.X.; Tang, X.G.; Lai, J.L.; Jiang, Y.P. Low leakage current in $(Bi_{0.95}La_{0.05})_2NiMnO_6$ double-perovskite thin films prepared by chemical solution deposition. *Mater Lett.* **2014**, *120*, 23–25. [CrossRef]

20. Harijan, P.K.; Singh, A.; Upadhyay, C.; Pandey, D. Néel transition in the multiferroic $BiFeO_3$-$0.25PbTiO_3$ nanoparticles with anomalous size effect. *J. Appl. Phys.* **2019**, *125*, 024102. [CrossRef]

21. Cerrato, J.M.; Hochella, M.F.; Knocke, W.R.; Dietrich, A.M.; Cromer, T.F. Use of XPS to Identify the Oxidation State of Mn in Solid Surfaces of Filtration Media Oxide Samples from Drinking Water Treatment Plants. *Environ. Sci. Technol.* **2010**, *44*, 5881–5886. [CrossRef] [PubMed]

22. Amaya, Á.A.; González, C.A.; Niño-Gómez, M.E.; Martínez, O.F. XPS fitting model proposed to the study of Ni and La in deactivated FCC catalysts. *J. Electron. Spectrosc.* **2019**, *233*, 5–10. [CrossRef]

23. Afroz, K.; Moniruddin, M.; Bakranov, N.; Kudaibergenov, S.; Nuraje, N.A. Heterojunction strategy to improve the visible light sensitive water splitting performance of photocatalytic materials. *J. Mater. Chem. A* **2018**, *6*, 21696–21718. [CrossRef]

24. Chang, C.M.; Hsu, C.H.; Liu, Y.W.; Chien, T.C.; Sung, C.H.; Yeh, P.H. Interface engineering: Broadband light and low temperature gas detection abilities using a nano-heterojunction device. *Nanoscale* **2015**, *7*, 20126–20131. [CrossRef] [PubMed]

25. Kumar, S.G.; Rao, K.S.R.K. Physics and chemistry of CdTe/CdS thin film heterojunction photovoltaic devices: Fundamental and critical aspects. *Energy Environ. Sci.* **2014**, *7*, 45–102. [CrossRef]

26. Dietz, G.W.; Waser, R. Charge injection in $SrTiO_3$ thin films. *Thin Solid Film.* **1997**, *299*, 53–58. [CrossRef]

27. Wang, Y.G.; Tang, X.G.; Liu, Q.X.; Jiang, Y.P.; Jiang, L.L. Room temperature tunable multiferroic properties in sol-gel derived nanocrystalline $Sr(Ti_{1-x}Fe_x)O_{3-\delta}$ thin films grown on $LaNiO_3$ buffered silicon substrates. *Nanomaterials* **2017**, *7*, 264. [CrossRef]

28. Barnett, C.J.; Mourgelas, V.; McGettrick, J.D.; Maffeis, T.G.G.; Barron, A.R.; Cobley, R.J. The effects of vacuum annealing on the conduction characteristics of ZnO nanorods. *Mater. Lett.* **2019**, *243*, 144–147. [CrossRef]

29. Lee, H.S. Electrokinetic analyses in biofilm anodes: Ohmic conduction of extracellular electron transfer. *Bioresour. Technol.* **2018**, *256*, 509–514. [CrossRef]

30. Abbaszadeh, D.; Blom, P.W.M. Efficient Blue Polymer Light-Emitting Diodes with Electron-Dominated Transport Due to Trap Dilution. *Adv. Electron. Mater.* **2016**, *2*, 1500406. [CrossRef]

31. Rani, V.; Sharma, A.; Kumar, P.; Singh, B.; Ghosh, S. Charge transport mechanism in copper phthalocyanine thin films with and without traps. *RSC Adv.* **2017**, *7*, 54911–54919. [CrossRef]

32. Shah, S.S.; Hayat, K.; Ali, S.; Rasool, K.; Iqbal, Y. Conduction mechanisms in lanthanum manganite nanofibers. *Mat. Sci. Semicon. Proc.* **2019**, *90*, 65–71. [CrossRef]

33. Perello, D.J.; Lim, S.C.; Chae, S.J.; Lee, I.; Kim, M.J.; Lee, Y.H.; Yun, M. Thermionic Field Emission Transport in Carbon Nanotube Transistors. *ACS Nano* **2011**, *5*, 1756–1760. [CrossRef] [PubMed]

34. Rodriguez-Nieva, J.F.; Dresselhaus, M.S.; Levitov, L.S. Thermionic Emission and Negative dI/dV in Photoactive Graphene Heterostructures. *Nano Lett.* **2015**, *15*, 1451–1456. [CrossRef]

35. Chen, C.C.; Aykol, M.; Chang, C.C.; Levi, A.F.J.; Cronin, S.B. Graphene-Silicon Schottky Diodes. *Nano Lett.* **2011**, *11*, 1863–1867. [CrossRef]

36. Lao, C.S.; Liu, J.; Gao, P.; Zhang, L.; Davidovic, D.; Tummala, R.; Wang, Z.L. ZnO Nanobelt/Nanowire Schottky Diodes Formed by Dielectrophoresis Alignment across Au Electrodes. *Nano Lett.* **2006**, *6*, 263–266. [CrossRef]

Article

Bi$_{1-x}$Eu$_x$FeO$_3$ Powders: Synthesis, Characterization, Magnetic and Photoluminescence Properties

Vasile-Adrian Surdu [1], Roxana Doina Truşcă [1], Bogdan Ştefan Vasile [1,*], Ovidiu Cristian Oprea [2], Eugenia Tanasă [1], Lucian Diamandescu [3], Ecaterina Andronescu [1] and Adelina Carmen Ianculescu [1,*]

[1] Department of Science and Engineering of Oxide Materials and Nanomaterials, Faculty of Applied Chemistry and Materials Science, "Politehnica" University of Bucharest, Gh. Polizu Street no. 1-7, 011061 Bucharest, Romania; adrian.surdu@upb.ro (V.-A.S.); truscaroxana@yahoo.com (R.D.T.); eugenia.vasile27@gmail.com (E.T.); ecaterina.andronescu@upb.ro (E.A.)

[2] Department of Inorganic Chemistry, Physical Chemistry and Electrochemistry, Faculty of Applied Chemistry and Materials Science, "Politehnica" University of Bucharest, Gh. Polizu Street no. 1-7, 011061 Bucharest, Romania; ovidiu73@yahoo.com

[3] National Institute of Materials Physics, 077125 Bucharest-Măgurele, Romania; diamand@infim.ro

* Correspondence: bogdan.vasile@upb.ro (B.Ş.V.); a_ianculescu@yahoo.com (A.C.I.)

Received: 23 September 2019; Accepted: 12 October 2019; Published: 16 October 2019

Abstract: Europium substituted bismuth ferrite powders were synthesized by the sol-gel technique. The precursor xerogel was characterized by thermal analysis. Bi$_{1-x}$Eu$_x$FeO$_3$ (x = 0–0.20) powders obtained after thermal treatment of the xerogel at 600 °C for 30 min were investigated by X-ray diffraction (XRD), scanning electron microscopy (FE-SEM), transmission electron microscopy (TEM), Raman spectroscopy, and Mössbauer spectroscopy. Magnetic behavior at room temperature was tested using vibrating sample magnetometry. The comparative results showed that europium has a beneficial effect on the stabilization of the perovskite structure and induced a weak ferromagnetism. The particle size decreases after the introduction of Eu^{3+} from 167 nm for x = 0 to 51 nm for x = 0.20. Photoluminescence spectroscopy showed the enhancement of the characteristic emission peaks intensity with the increase of Eu^{3+} concentration.

Keywords: bismuth ferrite; sol-gel process; magnetic properties; photoluminescence properties

1. Introduction

Among various multiferroic compounds, bismuth ferrite (BiFeO$_3$) stands out because it is one of the few magnetic ferroelectrics at room temperature. Therefore, there has been intensive research in the past decades to make it useful in practical applications. There are certain issues that are still open in what concerns voltage-induced changes, the possibility of reading magnetic data or the mechanism of magnetoelectric coupling, and whether it may be controlled [1]. Besides, extensive studies search for the possibility of using BiFeO$_3$ based materials for applications such as actuators, transducers, magnetic field sensors, information storage devices, optical imaging, photocatalysis, or gas sensors [2–6]. Recently, BiFeO$_3$ nanopowders were found to exhibit catalytic activity for doxorubicine degradation [7].

The antiferromagnetic structure in BiFeO$_3$ is quite complex, usually being considered as a G-type with a spiral spin arrangement (about 62 nm wavelength), due to the interplay between exchange and spin-orbit coupling interactions involving Fe ions. There are several strategies to enhance its magnetic properties, including chemical modifications, or control of morphology and structure. In terms of morphologies, BiFeO$_3$-based nanostructures exhibit increased magnetization than the corresponding bulks, due to the perturbation of the helimagnetic order by structural peculiarities (e.g., local defects) or the specific size of nanoparticles [8–12].

Another way to modify the magnetic structure consists in replacing Bi by rare earth ions, based on the fact that in the perovskite-like structure, the superexchange interaction between the localized RE$4f$ and Fe$3d$ electrons may play an important role. The effect of several rare-earth dopants/solutes, as Ho [13,14], Sm [15,16], La [17,18], Dy [19], Gd [20], or Nd [21] on the properties of bismuth ferrite have been investigated. There are some works which described the effect of Eu^{3+} used as *A* site solute on the characteristics of bismuth ferrite powders prepared by various non-conventional techniques, such as hydrothermal process [22], ball milling [23], or different variants of the sol-gel methods [24–26], etc. Even if the magnetic behavior of these powders was extensively analyzed, however no data regarding other properties as photoluminescence were reported.

The aim of this work is to study the influence of europium addition on the phase purity, crystal structure, morphology, magnetic behavior, and optical properties of $Bi_{1-x}Eu_xFeO_3$ powders (x = 0; 0.05; 0.10; 0.15; 0.20) prepared by the sol-gel route. In order to be able to assess only the contribution of Eu substitution on the *A*-site of the perovskite structure, all the processing parameters were constantly maintained.

2. Materials and Methods

Synthesis of $Bi_{1-x}Eu_xFeO_3$ powders (x = 0; 0.05; 0.10; 0.15; 0.20) was carried out through sol-gel route. All the solvents and chemicals were of analytical grade and used without further purification. The precursor solution was prepared by dissolution of $Bi(NO_3)_3 \cdot 5H_2O$ (Sigma Aldrich, St. Louis, MO, USA ≥98%), $Eu(NO_3)_3 \cdot 5H_2O$ (Sigma Aldrich, 99.9%) and $Fe(NO_3)_3 \cdot 9H_2O$ (Aldrich, 99.99%) in stoichiometric ratios in acetic acid solution (Honeywell Fluka, Wabash, IN, USA ACS Reagent, ≥99.7%). A transparent brownish red sol resulted after the complete dissolution (≈ 1 h) of the nitrates. The sol was stabilized with 2-methoxyetanol which was added in a 1:1 volume ratio with respect to acetic acid. The amounts of precursors are summarized in Table 1. After 1 h mixing at 400 rpm, the temperature was set to 80 °C and the sol was kept under magnetic stirring at this temperature for 12 h until a gel was obtained. Gel drying was carried out in a forced convection oven (Memmert Universal Oven U, Schwabach, Germany) in air at 120 °C for 12 h to obtain the xerogels. The precursor powders were heat treated in air at 600 °C with a soaking time of 30 min, a heating rate of 5 °C/min and then were slowly cooled at the normal rate of the oven (CWF 1200, Carbolite Gero, Hope Valley, England).

Table 1. Amounts of precursors for $Bi_{1-x}Eu_xFeO_3$ sol-gel synthesis.

	x = 0	x = 0.05	x = 0.10	x = 0.15	x = 0.20
$Bi(NO_3)_3 \cdot 5H_2O$	1.4554 g	1.3826 g	1.3098 g	1.2371 g	1.1643 g
$Eu(NO_3)_3 \cdot 5H_2O$	0 g	0.0643 g	0.1285 g	0.1928 g	0.2571 g
$Fe(NO_3)_3 \cdot 9H_2O$	1.2121 g	1.2121 g	1.2121 g	1.2121 g	1.2121 g
2-methoxyetanol	125 mL	125 mL	125 mL	125 mL	125 mL
Acetic acid	125 mL	125 mL	125 mL	125 mL	125 mL

Thermal behavior of the precursor powders was investigated by differential scanning calorimetry–thermogravimetry (DSC-TG) analyses carried out with a TG 449C STA Jupiter (Netzsch, Selb, Germany) thermal analyzer. Samples were placed in alumina crucible and heated with 10 °C/min from room temperature to 900°, under dried air flow of 20 mL/min.

Room temperature X-ray diffraction (XRD) measurements were performed to investigate the phase purity and structure of the (Bi,Eu)FeO$_3$ powders. For this purpose, an Empyrean diffractometer (PANalytical, Almelo, The Netherlands), using Ni-filtered Cu-Kα radiation (λ = 1.5418 Å) with a scan step increment of 0.02° and a counting time of 255 s/step, for 2θ ranged between 20–80°was used. Lattice parameters were refined by the Rietveld method [27], using the HighScore Plus 3.0e software (PANalytical, Almelo, The Netherlands). After removing the instrumental contribution, the full-width at half-maximum (FWHM) of the diffraction peaks can be interpreted in terms of crystallite size and lattice strain. A pseudo-Voigt function was used to refine the shapes of the $BiFeO_3$ peaks.

The local order and the cation coordination in the calcined powders were studied by Raman spectroscopy carried out at room temperature, using a LabRAm HR Evolution spectrometer (Horiba, Kyoto, Japan). Raman spectra were recorded using the 514 nm line of an argon ion laser, by focusing a 125 mW beam of a few micrometer sized spots on the samples under investigation.

Mössbauer spectroscopy ICE Oxford Mössbauer cryomagnetic system (WissEL, Mömbris, Germany) was used to analyze the state of iron ions in the perovskite lattice. The system was equipped with a 10 mCi $^{57}Co(Rh)$ source and the velocity was calibrated using a α-Fe standard foil.

Morphology and crystallinity degree of the $(Bi,Eu)FeO_3$ particles were investigated by scanning electron microscopy operated at 30 kV (Inspect F50, FEI, Hillsboro, OR, USA) and transmission electron microscopy operated at 300 kV (Tecnai^{TM} G2 F30 S-TWIN, FEI, Hillsboro, OR, USA). The average particle size of the $(Bi,Eu)FeO_3$ powders was estimated from the particle size distributions, which were determined using the OriginPro 9.0 software (OriginLab, Northampton, MA, USA) by taking into account size measurements on ~100 particles performed by means of the software of the electron microscopes (ImageJ 1.50b, National Institutes of Health and the Laboratory for Optical and Computational Instrumentation, Madison, WI, USA) in the case of SEM, and Digital Micrograph 1.8.0 (Gatan, Sarasota, FL, USA) in the case of transmission electron microscopy (TEM).

Vibrating sample magnetometry (7404-s VSM, LakeShore, Westerville, OH, USA) was used in order to investigate the magnetic behavior of the processed powders. Hysteresis loops were recorded at room temperature with an applied field up to 15 kOe, with increments of 200 Oe and a ramp rate of 20 Oe/s.

The fluorescence spectra were recorded with a LS 55 spectrometer (Perkin Elmer, Waltham, MA, USA) using an Xe lamp as a UV light source, at ambient temperature, in the range 350–650 nm, with all the samples in solid state. The measurements were made with a scan speed of 200 nm/min, excitation and emission slits of 10 nm, and a cut-off filter of 350 nm. An excitation wavelength of 320 nm was used.

3. Results

3.1. Thermal Behavior of the Precursor Powders

The TG-DSC curves of the $Bi_{1-x}Eu_xFeO_3$ xerogels are shown in Figure 1. The peaks corresponding to the exothermic effects and associated mass loss are illustrated in Table 2.

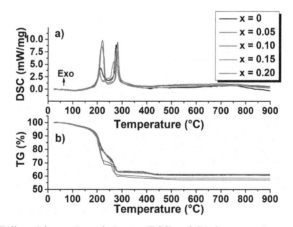

Figure 1. (a) Differential scanning calorimetry (DSC) and (b) thermogravimetry (TG) curves of $Bi_{1-x}Eu_xFeO_3$ xerogels.

Table 2. TG-DSC effects corresponding to $Bi_{1-x}Eu_xFeO_3$ xerogels.

	x = 0		x = 0.05		x = 0.10		x = 0.15		x = 0.20
T (°C)	Mass Loss (%)	T (°C)	Mass Loss (%)	T (°C)	Mass Loss (%)	T (°C)	Mass Loss (%)	T (°C)	Mass Loss (%)
103.3	−4	103	−4.4	105.1	−3.7	108.3	−4.7	104.5	−4.7
214.1	−18.2	218.4	−26.5	221.6	−24.6	206.7	−12.4	208.7	−13.4
275.9	−14	278.0	−5.5	282.1	−8.7	239.2	−11.4	240.3	−11.1
413.5	−2.4	419.9	−2	422.2	−2.1	269.9	−10.8	277.9	−10.8
						427.8	−2.2	391.9	−2.7

Thermal analysis reveals four step decomposition in the case of (Bi,Eu)FeO₃ powders with x = 0, x = 0.05, x = 0.10, and five step decomposition for the samples with x = 0.15 and x = 0.20. The first step decomposition at 103–108° was attributed to dehydration of the xerogels.

The second decomposition step (110–230 °C) associated with exothermic reactions with the highest mass loss, between 18.2% and 26.5% for the selected compositions, correspond to decarboxylation of acetic acid and decomposition of small groups such as NO_3^-. For the powders with x = 0.15 and x = 0.20, this reaction takes place in two steps, one at 206.7 °C and 208.7 °C, respectively, and the other at 239.2 °C and 240.3 °C, respectively [28].

The exothermic effect at 270–282 °C could be assigned to the collapse of the gel network and combustion of most organic materials. A small weight loss ≈2.5% occurring up to 430 °C corresponds to the end of CO_2 release [28].

3.2. Phase Composition and Structure of the (Bi,Eu)FeO₃ Powders

The room-temperature XRD patterns of $Bi_{1-x}Eu_xFeO_3$ powders are illustrated in Figure 2. The profiles of the peaks indicate a high crystallinity. A rhombohedral perovskite structure with space group R3c was indexed for the powders with x ≤ 0.10 [23]. A small amount of $Bi_{25}FeO_{40}$ sillenite phase is also detected for these compositions. Upon increasing the substitution ratio, the secondary phase diminishes until vanishing, which proves the beneficial effect of Eu^{3+} in what concerns the stabilization of the perovskite phase. All the reflections corresponding to the major perovskite phase are shifted to higher values of the diffraction angle when x is increased. Besides, in the case of the compositions with x ≥ 0.15 it may be observed that the (012) peak is split and a supplementary interference occurs at 2θ ≈ 34°. These are arguments that suggest Eu^{3+} ions have substituted Bi^{3+} in the BiFeO₃ lattice and that a phase transition from rhombohedral R3c (α phase) to orthorhombic Pnma (β phase) crystal symmetry has occurred [29,30].

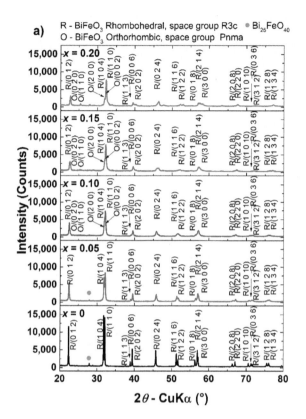

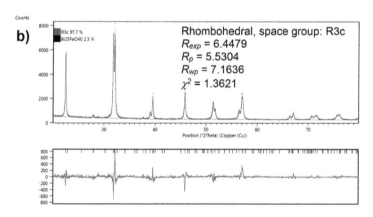

Figure 2. *Cont.*

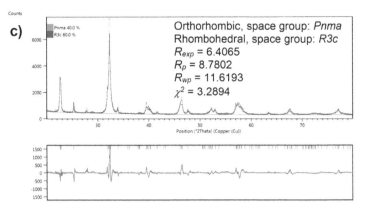

Figure 2. (a) X-ray diffraction (XRD) patterns of $Bi_{1-x}Eu_xFeO_3$ calcined powders, (b,c) Rietveld refined patterns for x = 0.05 and x = 0.20.

Rietveld refinement was performed in order to accurately determine the phase composition and structure of the powders. For the specimens with x ≥ 0.15, the best fit to data was obtained when using a mixture of rhombohedral R3c and orthorhombic Pnma polymorphs. The quality of the fits is indicated by the agreement indices obtained from Rietveld refinement (Table 3).

Table 3. Agreement indices from Rietveld refinement for $Bi_{1-x}Eu_xFeO_3$ calcined powders.

Agreement Indices	x = 0	x = 0.05	x = 0.10	x = 0.15	x = 0.20
R_{exp}	6.1381	6.4479	6.4451	6.3596	6.4065
R_p	5.5304	5.0238	6.7391	8.6136	8.7802
R_{wp}	7.1636	6.4951	8.9802	11.6540	11.6193
χ^2	1.3621	1.0147	1.9414	3.3581	3.2894

The phase composition evolution versus Eu^{3+} substitution degree is shown in Figure 3. For x ≥ 0.15, the $Bi_{25}FeO_{40}$ secondary phase vanishes in the limit of detection of X-ray diffraction. Stabilization of the perovskite phase is also accompanied by rapid polymorph transition. When increasing x from 0.10 to 0.15, phase composition changes from 97.4% R3c bismuth ferrite and 2.6% sillenite in the secondary phase to 62% R3c bismuth ferrite polymorph and 40% Pnma bismuth ferrite polymorph, respectively. These results are in good agreement with those reported by Iorgu et al. [31] and Khomchenko et al. [32] who also found a second orthorhombic polymorph in their Eu-substituted bismuth ferrite, with x ≥ 0.10 obtained by combustion method and solid state reaction, respectively.

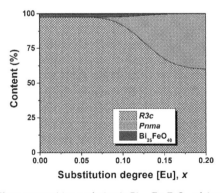

Figure 3. Phase composition evolution in $Bi_{1-x}Eu_xFeO_3$ calcined powders.

Unit cell parameters and cell volume (Figure 4) decrease with the increasing amount of Eu solute. This, together with the phase transition, is supported most likely by the smaller ionic radius of Eu^{3+} (1.07 Å) than that of Bi^{3+} (1.17 Å) [33].

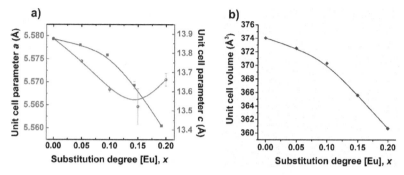

Figure 4. (**a**) Unit cell parameters corresponding to R3c polymorph, and (**b**) unit cell volume for $Bi_{1-x}Eu_xFeO_3$ calcined powders.

As expected, the formation of $(Bi,Eu)FeO_3$ solid solutions drives to the decrease of the crystallite size and the concurrent increase of the internal microstrains (Figure 5).

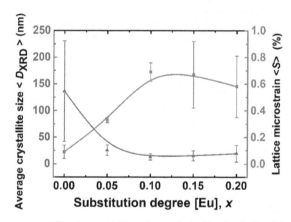

Figure 5. Average crystallite size and lattice microstrain for $Bi_{1-x}Eu_xFeO_3$ calcined powders.

Raman spectroscopy is a powerful technique, which is sensitive to structural phase transitions and it has been carried out to further support the Rietveld analysis of the XRD patterns. The active Raman modes of the $BiFeO_3$ solid solutions with rhombohedral R3c structure may be summarized using the irreducible representation of $\Gamma_{Raman, R3c} = 4A_1 + 9E$ [34–36].

In the present study, for the powders with lower Eu content (x ≤ 0.10), the modes A_1-1 and A_1-2, attributed to Bi-O bonds shift to higher-frequency region. This may be explained by the partial substitution of Bi^{3+} with Eu^{3+} because the frequency of the mode is inversely proportional to the mass, M, at A-site. Since the mass of Eu is about 27% lower than the mass of Bi, substitution will induce the shift in the frequency of vibration of the modes, which is consistent to the data presented in Figure 6. When x increases from 0.10 to 0.15, the most significant feature in the Raman spectra is that A_1-1 and A_1-2 modes almost vanish and severely broaden, while the E mode at ≈290 cm^{-1} shifts to a higher frequency and increases in intensity. Such peak has been reported for orthorhombic rare earth ferrites and can be assigned to A_g mode [37]. The further increase of x from 0.15 to 0.20 indicate a visible distortion of FeO_6 octahedra, which is evidenced by the increase of intensity of the 500 and 600 cm^{-1}

modes [38]. All the discussed features are arguments that Eu^{3+} is incorporated on the Bi site of the perovskite lattice of $BiFeO_3$ forming solid solutions, and that when the substitution degree exceeds the value of 0.15, using the processing parameters in the present work, it induces a structural phase transition from rhombohedral to orthorhombic symmetry.

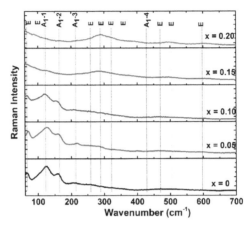

Figure 6. Typical Raman scattering spectra of $Bi_{1-x}Eu_xFeO_3$ calcined powders.

The ^{57}Fe Mössbauer spectra for the selected compositions with x = 0 and x = 0.20 were recorded at room temperature. Results show that the spectra corresponding to the investigated samples present hyperfine magnetic sextet (Figure 7).

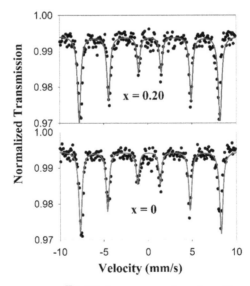

Figure 7. Room temperature ^{57}Fe Mössbauer spectra of powders with x = 0 and x = 0.20.

The refining of the spectra under the assumption of the Lorentzian shape of the Mössbauer line allowed obtaining of the characteristic parameters: isomeric shift (IS), quadrupole splitting (ΔEq) and hyperfine field (H_{hf}), which are presented in Table 4.

Table 4. Mössbauer parameters for x = 0 and x = 0.20 samples.

	IS (mm/s)	ΔEq (mm/s)	H_{hf} (T)
x = 0	0.335	0.179	48.90
x = 0.20	0.335	−0.054	48.89
SE	±0.001	±0.002	±0.02

The values of IS and ΔEq prove that Fe occupies only the B-site in the perovskite structure and correspond to high-spin Fe^{3+} ions.

Upon introduction of Eu^{3+} in the lattice, ΔEq switches from positive values (0.179 mm/s) in the case of x = 0 to negative values (−0.054 mm/s) in the case of x = 0.20. This means that the electric field gradient is drastically changed by the substitution and may be assigned to structural phase transition from rhombohedral to orthorhombic symmetry, as seen for $Bi_{1-x}Dy_xFeO_3$ nanoparticles obtained by Qian et al. [39]

In what concerns the H_{hf}, the substitution of Bi^{3+} with Eu^{3+} does not affect the obtained values nor the charge density reflected in the IS parameter which remains constant. All Mössbauer parameters are in good agreement with those obtained by Prado-Gonjal et al. for microwave-assisted hydrothermal processed $BiFeO_3$ powders [40].

3.3. Morphology

Scanning electron microscopy (FE-SEM) images depicting the morphology and the particle size distribution of the calcined (Bi,Eu)FeO$_3$ powders are shown in Figure 8. A general view of two selected compositions, x = 0 (Figure 8a) and x = 0.10 (Figure 8b), illustrate porous networks with pores in the micrometer and submicrometer range, which were formed after heat treatment of the precursor gels. The walls of the pores are dense and consist of agglomerated particles as it may be seen in the detail in the images of Figure 8c,e,g,i,k. In each case, the particles exhibit polyhedral shapes and as x increases the particles tend to have a more rounded aspect. Moreover, for ternary compositions, a tendency toward coarsening was observed. In what concerns the particle size, one can see a decrease after the introduction of Eu^{3+} as a substituent in the perovskite lattice from 167 nm for x = 0 to 85 nm for x = 0.05, which becomes even more evident for the compositions where the polymorphic transformation occurs (x = 0.15 and x = 0.20). In the latter case, the particle size decreases from 78 nm for x = 0.10 to 56 nm for x = 0.15. This kind of effect is consistent with other studies regarding substituted $BiFeO_3$ particles prepared by various techniques [24,26,31,41]. Moreover, Dai et al. explained this in the case of (Eu, Ti) co-substituted ceramics as a result of suppression of oxygen vacancies by the solutes, which slows oxygen ion motion and, consequently, grain growth rate [42]. The particle size distribution is unimodal (Figure 8d,f,h,j,l) and becomes narrower as the solute concentration increases. Thus, in the $BiFeO_3$ sample, the unimodal distribution show 20%–25% of nanoparticles in the size range of 140–180 nm. Besides, the influence of the addition of Eu^{3+} on the size and particle size distribution should be noted. The introduction of 5% Eu^{3+} results in the particle size distribution shown in Figure 8f. The entire particle size distribution is between 50 and 120 nm, with a maxima at 80–90 nm, which represents a proportion of 35%. Actually, all the nanoparticles present sizes below those characteristic to $BiFeO_3$ (80–280 nm). The slowing particle growth effect of europium is better observed when its concentration in the perovskite solid solution increases. Thus, for x = 0.10, even if 30% of the nanoparticles correspond to the size range of 80–90 nm, the unimodal distribution is asymmetric due to the increase of the ratio of nanoparticles in the range size of 50–80 nm. More obvious contribution of the solute is shown in the case of x = 0.15 and x = 0.20, where one can observe that 35%–40% of the nanoparticles are in the size range of 50–60 nm and, respectively, 40–60 nm. The measurements of the sizes and the corresponding distributions from FE-SEM data illustrate a clear influence of the Eu^{3+} solute on the reduction of the particles size, as well as on the narrowing of the particle size distribution with the increase of the substitution rate.

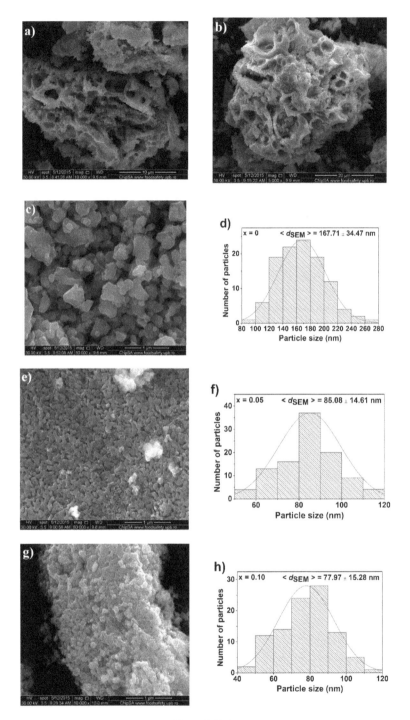

Figure 8. *Cont.*

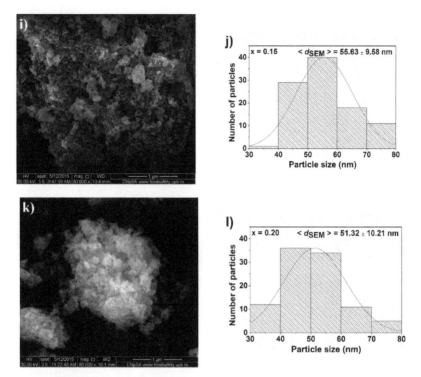

Figure 8. FE-SEM images showing the morphology and corresponding histograms for particle size distribution of $Bi_{1-x}Eu_xFeO_3$ powders: (**a,b**) General view for x = 0 and x = 0.10, (**c,d**) x = 0: (**c**) detail, (**d**) particle size distribution, (**e,f**) x = 0.05: (**e**) detail, (**f**) particle size distribution, (**g,h**) x = 0.10: (**g**) detail, (**h**) particle size distribution, (**i,j**) x = 0.15: (**i**) detail, (**j**) particle size distribution, (**k,l**) x = 0.20: (**k**) detail, (**l**) particle size distribution.

TEM investigations sustain FE-SEM observations. The coarsening of the particles is observed better in Bright-field TEM images (Figure 9a,e,i,m,q) by means of necks at the particles limits. Particle size distributions (Figure 9b,f,j,n,r) are similar to those measured from FE-SEM images, as the small differences are in the limits of the standard deviation. Morphology evolution with increasing Eu^{3+} solute degree is similar to that reported by Bahraoui et al. who synthetized $Bi_{1-x}Eu_xFeO_3$ powders by the sol-gel method with calcination treatment at 500 °C for 24 h, but the average particle size is almost four times higher [26]. This shows that although the time of heat treatment at 600 °C was relatively short (30 min), the temperature has a stronger influence on the particle size growth.

The powders show a high crystallinity degree as assessed from the selected area electron diffraction (SAED) patterns (Figure 9c,g,k,o,s), which consist of well-defined diffraction spots arranged in concentric diffraction rings. For the pure $BiFeO_3$ powder (x = 0), the diffraction rings are less visible due to the fact that both crystallite size and particle size are situated in the submicrometer scale and because the coarsening process may induce some preferential orientations of the aggregated particles. In the case of the samples with higher Eu^{3+} content (x = 0.15 and x = 0.20), the patterns are more complicated because of the coexistence of rhombohedral and orthorhombic polymorphs which are homogeneously distributed.

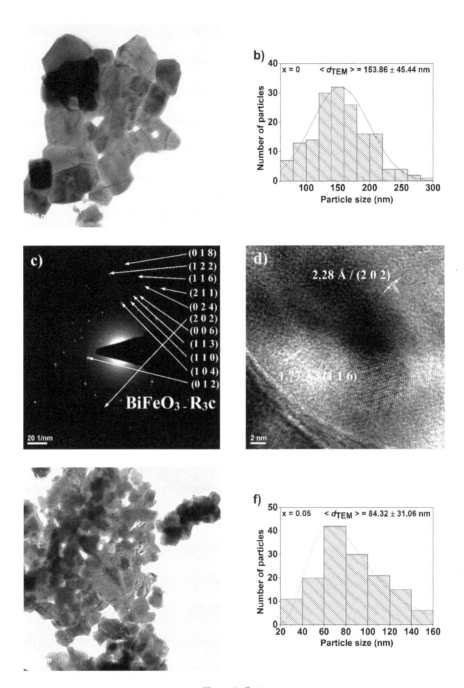

Figure 9. *Cont.*

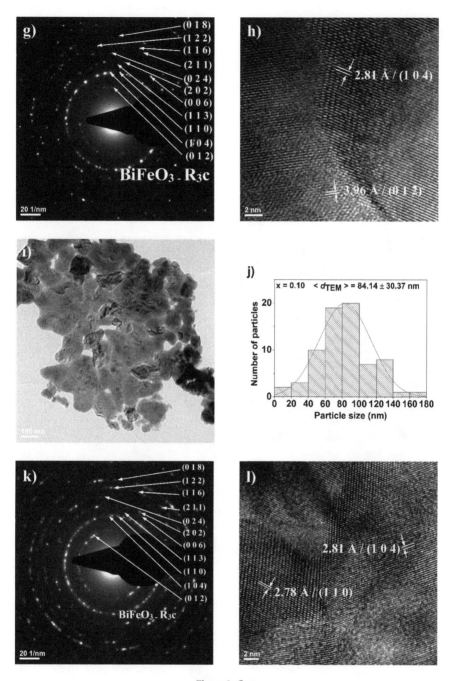

Figure 9. *Cont.*

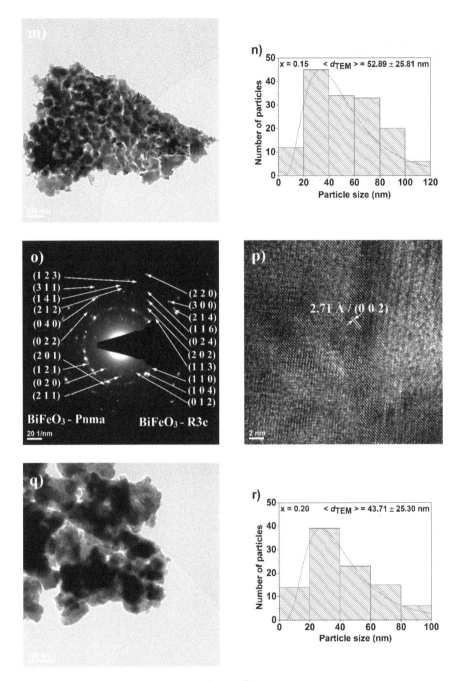

Figure 9. *Cont.*

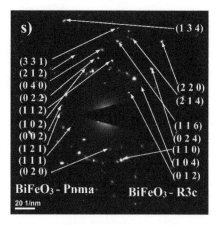
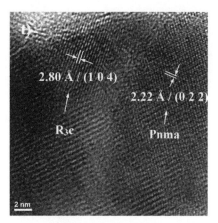

Figure 9. (**a,e,i,m,q**) Bright field TEM images, (**b,f,j,n,r**) particle size distributions, (**c,g,k,o,s**) Selected area electron diffraction patterns, and (**d,h,l,p,t**) High resolution TEM images corresponding to $Bi_{1-x}Eu_xFeO_3$ powders for: x = 0 (**a–d**), 0.05 (**e–h**), 0.10 (**i–l**), 0.15 (**m–p**) and 0.20 (**q–t**), respectively.

High resolution transmission electron microscopy (HR-TEM) investigations reveal long-range highly ordered fringes with spacing at 2.28 Å and 1.77 Å corresponding to the (2 0 2) and (1 1 6) crystalline planes of the rhombohedrally-distorted perovskite structure in the case of x = 0. For x = 0.05 and x = 0.10, there were also identified the crystallographic planes specific to rhombohedral polymorphs (Figure 9f,h). In the case of x = 0.20, both polymorphs were identified in the same particles consisting of multiple crystallites. It is worth mentioning that the substitution also induces the forming of polycrystalline particles, which is also evidenced in the HR-TEM images.

In order to have a better understanding of the nature of the particles, in Figure 10 a comparison between average crystallite size determined from XRD data and average particle size measured on SEM and TEM images was depicted. In the case of unsubstituted $BiFeO_3$ particles, the three values are almost equal. Slightly differences that occur are in the range of standard deviation. This means that in this case, the particles are single crystals. Interestingly, when comparing the values obtained for Eu-substituted $BiFeO_3$ compositions, one can observe that the values for the average nanoparticle size determined from SEM and TEM investigations are very close, whereas the average crystallite size presents at most a half value of the average particle size, proving that for x ≥ 0.05, the particles are polycrystalline and consist of two or more crystallites, which sustains the HR-TEM observations.

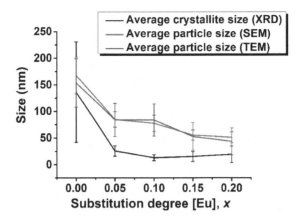

Figure 10. Comparison between average crystallite size determined from XRD data and average particle size measured from SEM and TEM images.

3.4. Magnetic Behavior

Figure 11 and Table 5 show the room-temperature M = f(H) hysteresis loops up to 15,000 Oe, and M_s, M_r, and H_c parameters of the $Bi_{1-x}Eu_xFeO_3$ powders.

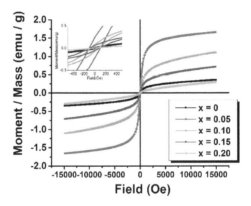

Figure 11. M-H hysteresis loops for $Bi_{1-x}Eu_xFeO_3$ powders: Inset showing the low-field M = f(H) dependence.

Table 5. M_s, M_r and H_c for $Bi_{1-x}Eu_xFeO_3$ powders.

	x = 0	x = 0.05	x = 0.10	x = 0.15	x = 0.20
M_s (emu/g)	0.3529	1.6570	1.1089	0.7113	0.2968
M_r (emu/g)	0.0128	0.2287	0.0734	0.0551	0.0457
H_c (Oe)	51.290	101.160	69.883	91.156	36.815

It can be observed that the sample with x = 0 shows a continuous linear increase of magnetization versus magnetic field which suggests the presence of the antiferromagnetic phase, involving relative low exchange integrals in order to progressively reorient the spins along the field direction.

However, at very low fields, there is a much faster variation of saturation magnetization of 0.3529 emu/g and a coercive field of 51.290 Oe. Unlike this sample, in the case of the samples with $0.05 \leq x \leq 0.15$, a weak ferromagnetic behavior, with a saturation magnetization of 1.6570 emu/g for x = 0.05,

1.1089 emu/g for x = 0.10 and 0.7113 emu/g for x = 0.15 is present. A possible conclusion of these aspects is that Eu content influences the spin spiral structure, most likely by a perturbation of the superexchange interactions between localized Eu4*f* and Fe3*d* electrons.

At the maximum substitution degree studied in this work, the magnetic behavior shows a decrease in M_s, M_r, and H_c compared to pristine $BiFeO_3$ particles, suggesting that increasing Eu content above x = 0.15 does not improve the magnetic behavior of the particles. This suggests that the presence of the orthorhombic Pnma polymorph affects the magnetic order.

3.5. Photoluminescence Properties

$BiFeO_3$ is an interesting optical material, which shows promising applications in photocatalysts and photoconductive devices. Thus, photoluminescence spectroscopy was used to study the optical property of $Bi_{1-x}Eu_xFeO_3$ nanoparticles.

During synthesis there are generated several deep and shallow oxygen vacancies and surface defects that introduce localized electronic levels in-band [43]. Therefore, the PL spectra (Figure 12) are complex and present more than a single peak from band-to-band transition.

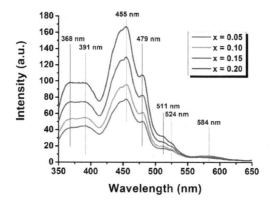

Figure 12. Fluorescence spectra for $Bi_{1-x}Eu_xFeO_3$ powders.

The most intense blue emission peak at the wavelength of 455 nm (2.72 eV) originates from self-activated centers in the synthesized nanoparticles [44,45]. The emission peak is broad and asymmetric, with a clear overlap with the peak from 479 nm (2.58 eV). This indicate the existence of another transition below the conduction band, due to the presence of defects in grain boundaries or oxygen vacancies, usually referred as near-band edge (NBE) transition [45–47].

In the blue-green region there are further shoulders at the 511–524 nm range which can be attributed to oxygen vacancies and a small, but broad peak in the range of 569–584 nm which has an unknown origin [48]. These peaks are usually referred as defect-level emissions (DLE).

The intensity of emission peaks increases with the increase of the solute content from the Eu^{3+}-doped $BiFeO_3$ with x = 0.05 to x = 0.20. This behavior cannot be explained in terms of the difference in nanoparticles dimensions, taking into account that the 5% and 10%-doped samples, as well as the 15% and 20%-doped powders exhibit roughly similar sizes.

In the first instance, for 5% Eu^{3+}-doped sample, the photo-generated electron-hole pairs present a lower recombination rate, which leads to lower intensity of emission peaks. For the next three samples the increase of luminescent emission with the europium amount could be related to a higher concentration of surface defects as new crystalline phase is formed. These defects can contribute to the capture of photo-generated electrons, to produce excitons, which will enhance the emission intensity. A similar behavior was reported for Sn^{4+}/Gd^{3+} or Mn^{2+}-doped $BiFeO_3$ samples [49,50].

In the 550–650 nm range there are no peaks that can be assigned to Eu^{3+} ions emission spectrum, indicating either a masking effect from $BiFeO_3$ luminescence or, simply, a quenching of europium fluorescence. This effect was also observed for other rare earth ions used as soluted for bismuth ferrite [49,51,52].

4. Conclusions

$Bi_{1-x}Eu_xFeO_3$ powders were prepared by the sol-gel method. XRD and Raman spectroscopy investigations indicated phase-pure particles and a structural phase transition for $x \geq 0.15$ when using the processing parameters presented in the present work. Mössbauer spectroscopy showed only the presence of Fe^{3+} and a hyperfine magnetic sextet. FE-SEM and TEM analysis evidenced obtaining submicron-sized single-crystal particles for pure $BiFeO_3$ composition, and polycrystalline nanoparticles in the case of Eu^{3+}-substituted powders. The most pronounced ferromagnetic behavior was observed for $Bi_{0.95}Eu_{0.05}FeO_3$ composition, which exhibited a saturation magnetization of 1.65 emu/g and a coercive field of 100 Oe, which occurs, most likely, by a perturbation of the superexchange interactions between localized Eu*4f* and Fe*3d* electrons. This work shows a possibility to tailor magnetic behavior of bismuth ferrite using rare earth metal solute on the *A*-site of the perovskite structure. The luminescence emission increases with the increase of the Eu^{3+} content, but the quenching of the fluorescence specific to europium ions seems to be induced by a masking effect of $BiFeO_3$, as in other rare-earth doped bismuth ferrite systems.

Author Contributions: The authors contributions are as follows: methodology, V.-A.S. and A.C.I.; investigation, O.C.O., E.T., and L.D.; data curation, V.-A.S., R.D.T., and B.S.V.; formal analysis, A.C.I.; writing—original draft preparation V. -A.S.; writing—review and editing, E.A. and A.C.I.; visualization, B.Ș.V. and A.C.I.

Funding: This research was funded by Romanian National Authority for Scientific Research, CNCS-UEFISCDI, Project No. PN-III-P4-ID-PCE-2016-0072.

Conflicts of Interest: The authors declare no conflict of interest. The funders had no role in the design of the study; in the collection, analyses, or interpretation of data; in the writing of the manuscript; or in the decision to publish the results.

References

1. Catalan, G.; Scott, J.F. Physics and applications of bismuth ferrite. *Adv. Mater.* **2009**, *21*, 2463–2485. [CrossRef]
2. Clarke, G.; Rogov, A.; McCarthy, S.; Bonacina, L.; Gun'Ko, Y.; Galez, C.; le Dantec, R.; Volkov, Y.; Mugnier, Y.; Prina-Mello, A. Preparation from a revisited wet chemical route of phase-pure, monocrystalline and SHG-efficient BiFeO₃ nanoparticles for harmonic bio-imaging. *Sci. Rep.* **2018**, *8*, 10473. [CrossRef] [PubMed]
3. Haruna, A.; Abdulkadir, I.; Idris, S.O. Synthesis, characterization and photocatalytic properties of Bi₀.₈₅₋ₓMₓBa₀.₁₅FeO₃ (M = Na and K, X = 0, 0.1) perovskite-like nanoparticles using the sol-gel method. *J. King Saud Univ. Sci.* **2019**. [CrossRef]
4. Ramos-Gomes, F.; Möbius, W.; Bonacina, L.; Alves, F.; Markus, M.A. Bismuth ferrite second harmonic nanoparticles for pulmonary macrophage tracking. *Small* **2019**, *15*, 1803776. [CrossRef]
5. Chakraborty, S.; Pal, M. Highly selective and stable acetone sensor based on chemically prepared bismuth ferrite nanoparticles. *J. Alloys Compd.* **2019**, *787*, 1204–1211. [CrossRef]
6. Sinha, A.K.; Bhushan, B.; Mishra, S.K.; Sharma, R.K.; Sen, S.; Mandal, B.P.; Meena, S.S.; Bhatt, P.; Prajapat, C.L.; Priyam, A.; et al. Enhanced dielectric, magnetic and optical properties of Cr-doped BiFeO₃ multiferroic nanoparticles synthesized by sol-gel route. *Results Phys.* **2019**, *13*, 102299. [CrossRef]
7. Dumitru, R.; Ianculescu, A.; Păcurariu, C.; Lupa, L.; Pop, A.; Vasile, B.; Surdu, A.; Manea, F. BiFeO₃-synthesis, characterization and its photocatalytic activity towards doxorubicin degradation from water. *Ceram. Int.* **2019**, *45*, 2789–2802. [CrossRef]
8. Philip, G.G.; Senthamizhan, A.; Natarajan, T.S.; Chandrasekaran, G.; Therese, H.A. The effect of gadolinium doping on the structural, magnetic and photoluminescence properties of electrospun bismuth ferrite nanofibers. *Ceram. Int.* **2015**, *41*, 13361–13365. [CrossRef]
9. Godara, S.; Sinha, N.; Kumar, B. Enhanced electric and magnetic properties in Ce-Cr co-doped bismuth ferrite nanostructure. *Mater. Lett.* **2014**, *136*, 441–444. [CrossRef]

10. Dutta, D.P.; Mandal, B.P.; Mukadam, M.D.; Yusuf, S.M.; Tyagi, A.K. Improved magnetic and ferroelectric properties of Sc and Ti codoped multiferroic nano $BiFeO_3$ prepared via sonochemical synthesis. *Dalt. Trans.* **2014**, *43*, 7838–7846. [CrossRef]

11. Chaudhuri, A.; Mandal, K. Study of structural, ferromagnetic and ferroelectric properties of nanostructured barium doped Bismuth Ferrite. *J. Magn. Magn. Mater.* **2014**, *353*, 57–64. [CrossRef]

12. Huang, F.; Wang, Z.; Lu, X.; Zhang, J.; Min, K.; Lin, W.; Ti, R.; Xu, T.; He, J.; Yue, C.; et al. Peculiar magnetism of $BiFeO_3$ nanoparticles with size approaching the period of the spiral spin structure. *Sci. Rep.* **2013**, *3*, 2907. [CrossRef] [PubMed]

13. Zhang, Y.; Wang, Y.; Qi, J.; Tian, Y.; Zhang, J.; Wei, M.; Liu, Y.; Yang, J. Structural, magnetic and impedance spectroscopy properties of Ho^{3+} modified $BiFeO_3$ multiferroic thin film. *J. Mater. Sci. Mater. Electron.* **2019**, *30*, 2942–2952. [CrossRef]

14. Muneeswaran, M.; Lee, S.H.; Kim, D.H.; Jung, B.S.; Chang, S.H.; Jang, J.W.; Choi, B.C.; Jeong, J.H.; Giridharan, N.V.; Venkateswaran, C. Structural, vibrational, and enhanced magneto-electric coupling in Ho-substituted $BiFeO_3$. *J. Alloys Compd.* **2018**, *750*, 276–285. [CrossRef]

15. Xu, X.; Guoqiang, T.; Huijun, R.; Ao, X. Structural, electric and multiferroic properties of Sm-doped $BiFeO_3$ thin films prepared by the sol-gelprocess. *Ceram. Int.* **2013**, *39*, 6223–6228. [CrossRef]

16. Yotburut, B.; Thongbai, P.; Yamwong, T.; Maensiri, S. Synthesis and characterization of multiferroic Sm-doped $BiFeO_3$ nanopowders and their bulk dielectric properties. *J. Magn. Magn. Mater.* **2017**, *437*, 51–61. [CrossRef]

17. Reddy, B.P.; Sekhar, M.C.; Prakash, B.P.; Suh, Y.; Park, S.H. Photocatalytic, magnetic, and electrochemical properties of La doped $BiFeO_3$ nanoparticles. *Ceram. Int.* **2018**, *44*, 19512–19521. [CrossRef]

18. Castillo, M.E.; Shvartsman, V.V.; Gobeljic, D.; Gao, Y.; Landers, J.; Wende, H.; Lupascu, D.C. Effect of particle size on ferroelectric and magnetic properties of $BiFeO_3$ nanopowders. *Nanotechnology* **2013**, *24*, 355701. [CrossRef]

19. Dhir, G.; Uniyal, P.; Verma, N.K. Effect of particle size on magnetic and dielectric properties of nanoscale Dy-doped $BiFeO_3$. *J. Supercond. Nov. Magn.* **2014**, *27*, 1569–1577. [CrossRef]

20. Vijayasundaram, S.V.; Suresh, G.; Mondal, R.A.; Kanagadurai, R. Composition-driven enhanced magnetic properties and magnetoelectric coupling in Gd substituted $BiFeO_3$ nanoparticles. *J. Magn. Magn. Mater.* **2016**, *418*, 30–36. [CrossRef]

21. Zhang, H.; Liu, W.F.; Wu, P.; Hai, X.; Wang, S.Y.; Liu, G.Y.; Rao, G.H. Unusual magnetic behaviors and electrical properties of Nd-doped $BiFeO_3$ nanoparticles calcined at different temperatures. *J. Nanopart. Res.* **2014**, *16*, 2205. [CrossRef]

22. Li, X.; Zhu, Z.; Yin, X.; Wang, F.; Gu, W.; Fu, Z.; Lu, Y. Enhanced magnetism and light absorption of Eu-doped $BiFeO_3$. *J. Mater. Sci. Mater. Electron.* **2016**, *27*, 7079–7083. [CrossRef]

23. Freitas, V.F.; Grande, H.L.C.; de Medeiros, S.N.; Santos, I.A.; Cótica, L.F.; Coelho, A.A. Structural, microstructural and magnetic investigations in high-energy ball milled $BiFeO_3$ and $Bi_{0.95}Eu_{0.05}FeO_3$ powders. *J. Alloys Compd.* **2008**, *461*, 48–52. [CrossRef]

24. Reshak, A.H.; Tlemçani, T.S.; Bahraoui, T.E.; Taibi, M.; Plucinski, K.J.; Belayachi, A.; Abd-Lefdil, M.; Lis, M.; Alahmed, Z.A.; Kamarudin, H.; et al. Characterization of multiferroic $Bi_{0.8}RE_{0.2}FeO_3$ powders (RE=Nd^{3+}, Eu^{3+}) grown by the sol-gel method. *Mater. Lett.* **2015**, *139*, 104–107. [CrossRef]

25. Liu, J.; Fang, L.; Zheng, F.; Ju, S.; Shen, M. Enhancement of magnetization in Eu doped $BiFeO_3$ nanoparticles. *Appl. Phys. Lett.* **2009**, *95*, 022511. [CrossRef]

26. Bahraoui, T.E.; Tlemçani, T.S.; Taibi, M.; Zaarour, H.; Bey, A.E.; Belayachi, A.; Silver, A.T.; Schmerber, G.; Naggar, A.M.E.; Albassam, A.A.; et al. Characterization of multiferroic $Bi_{1-x}Eu_xFeO_3$ powders prepared by sol-gel method. *Mater. Lett.* **2016**, *182*, 151–154. [CrossRef]

27. Rietveld, H.M. A profile refinement method for nuclear and magnetic structures. *J. Appl. Crystallogr.* **1969**, *2*, 65–71. [CrossRef]

28. Xu, J.; Ke, H.; Jia, D.; Wang, W.; Zhou, Y. Low-temperature synthesis of $BiFeO_3$ nanopowders via a sol—gel method. *J. Alloy. Compd.* **2009**, *472*, 473–477. [CrossRef]

29. Troyanchuk, I.O.; Bushinsky, M.V.; Karpinsky, D.V.; Mantytskaya, O.S.; Fedotova, V.V.; Prochnenko, O.I. Structural transformations and magnetic properties of $Bi_{1-x}Ln_xFeO_3$ (Ln = La, Nd, Eu) multiferroics. *Phys. Status Solidi Basic Res.* **2009**, *246*, 1901–1907. [CrossRef]

30. Ravindran, P.; Vidya, R.; Kjekshus, A.; Fjellvåg, H.; Eriksson, O. Theoretical investigation of magnetoelectric behavior in BiFeO₃. *Phys. Rev. B* **2006**, *74*, 224412. [CrossRef]

31. Iorgu, A.I.; Maxim, F.; Matei, C.; Ferreira, L.P.; Ferreira, P.; Cruz, M.M.; Berger, D. Fast synthesis of rare-earth (Pr^{3+}, Sm^{3+}, Eu^{3+} and Gd^{3+}) doped bismuth ferrite powders with enhanced magnetic properties. *J. Alloys Compd.* **2015**, *629*, 62–68. [CrossRef]

32. Khomchenko, V.A.; Troyanchuk, I.O.; Bushinsky, M.V.; Mantytskaya, O.S.; Sikolenko, V.; Paixão, J.A. Structural phase evolution in $Bi_{7/8}Ln_{1/8}FeO_3$ (Ln = La–Dy) series. *Mater. Lett.* **2011**, *65*, 1970–1972. [CrossRef]

33. Shannon, R.D. Revised effective ionic radii and systematic studies of interatomic distances in halides and chalcogenides. *Acta Crystallogr. Sect. A* **1976**, *32*, 751–767. [CrossRef]

34. Rao, T.D.; Karthik, T.; Asthana, S. Investigation of structural, magnetic and optical properties of rare earth substituted bismuth ferrite. *J. Rare Earths* **2013**, *31*, 370–375. [CrossRef]

35. Fukumura, H.; Harima, H.; Kisoda, K.; Tamada, M.; Noguchi, Y.; Miyayama, M. Raman scattering study of multiferroic BiFeO₃ single crystal. *J. Magn. Magn. Mater.* **2007**, *310*, 2006–2008. [CrossRef]

36. Kothari, D.; Reddy, V.R.; Sathe, V.G.; Gupta, A.; Banerjee, A.; Awasthi, A.M. Raman scattering study of polycrystalline magnetoelectric BiFeO₃. *J. Magn. Magn. Mater.* **2008**, *320*, 548–552. [CrossRef]

37. Gupta, H.C.; Singh, M.K.; Tiwari, L.M. A lattice dynamical investigation of Raman and infrared wavenumbers at the zone center of the orthorhombic NdNiO₃ perovskite. *J. Phys. Chem. Solids* **2003**, *64*, 531–533. [CrossRef]

38. Sati, P.C.; Kumar, M.; Chhoker, S.; Jewariya, M. Influence of Eu substitution on structural, magnetic, optical and dielectric properties of BiFeO₃ multiferroic ceramics. *Ceram. Int.* **2015**, *41*, 2389–2398. [CrossRef]

39. Qian, F.Z.; Jiang, J.S.; Guo, S.Z.; Jiang, D.M.; Zhang, W.G. Multiferroic properties of $Bi_{1-x}Dy_xFeO_3$ nanoparticles. *J. Appl. Phys.* **2009**, *106*, 084312. [CrossRef]

40. Prado-Gonjal, J.; Ávila, D.; Villafuerte-Castrejón, M.E.; González-García, F.; Fuentes, L.; Gómez, R.W.; Pérez-Mazariego, J.L.; Marquina, V.; Morán, E. Structural, microstructural and Mössbauer study of BiFeO₃ synthesized at low temperature by a microwave-hydrothermal method. *Solid State Sci.* **2011**, *13*, 2030–2036. [CrossRef]

41. Zhu, Y.; Quan, C.; Ma, Y.; Wang, Q.; Mao, W.; Wang, X.; Zhang, J.; Min, Y.; Yang, J.; Li, X.; et al. Effect of Eu, Mn co-doping on structural, optical and magnetic properties of BiFeO₃ nanoparticles. *Mater. Sci. Semicond. Process.* **2017**, *57*, 178–184. [CrossRef]

42. Dai, H.; Xue, R.; Chen, Z.; Li, T.; Chen, J.; Xiang, H. Effect of Eu, Ti co-doping on the structural and multiferroic properties of BiFeO₃ ceramics. *Ceram. Int.* **2014**, *40*, 15617–15622. [CrossRef]

43. Pandey, D.K.; Modi, A.; Pandey, P.; Gaur, N.K. Variable excitation wavelength photoluminescence response and optical absorption in BiFeO₃ nanostructures. *J. Mater. Sci. Mater. Electron.* **2017**, *28*, 17245–17253. [CrossRef]

44. Yu, X.; An, X. Enhanced magnetic and optical properties of pure and (Mn, Sr) doped BiFeO₃ nanocrystals. *Solid State Commun.* **2009**, *149*, 711–714. [CrossRef]

45. Moubah, R.; Schmerber, G.; Rousseau, O.; Colson, D.; Viret, M. Photoluminescence investigation of defects and optical band gap in multiferroic BiFeO₃ single crystals. *Appl. Phys. Express* **2012**, *5*, 035802. [CrossRef]

46. Chen, X.; Zhang, H.; Wang, T.; Wang, F.; Shi, W. Optical and photoluminescence properties of BiFeO₃ thin films grown on ITO-coated glass substrates by chemical solution deposition. *Phys. Status Solidi Appl. Mater. Sci.* **2012**, *209*, 1456–1460. [CrossRef]

47. Prashanthi, K.; Gupta, M.; Tsui, Y.Y.; Thundat, T. Effect of annealing atmosphere on microstructural and photoluminescence characteristics of multiferroic BiFeO₃ thin films prepared by pulsed laser deposition technique. *Appl. Phys. A* **2013**, *110*, 903–907. [CrossRef]

48. Hauser, A.J.; Zhang, J.; Mier, L.; Ricciardo, R.A.; Woodward, P.M.; Gustafson, T.L.; Brillson, L.J.; Yang, F.Y. Characterization of electronic structure and defect states of thin epitaxial BiFeO₃ films by UV-visible absorption and cathodoluminescence spectroscopies. *Appl. Phys. Lett.* **2008**, *92*, 222901. [CrossRef]

49. Irfan, S.; Rizwan, S.; Shen, Y.; Li, L.; Asfandiyar; Butt, S.; Nan, C.W. The Gadolinium (Gd^{3+}) and Tin (Sn4+) Co-doped BiFeO₃ Nanoparticles as New Solar Light Active Photocatalyst. *Sci. Rep.* **2017**, *7*, 42493. [CrossRef]

50. Chauhan, S.; Kumar, M.; Chhoker, S.; Katyal, S.C.; Singh, H.; Jewariya, M.; Yadav, K.L. Multiferroic, magnetoelectric and optical properties of Mn doped BiFeO₃ nanoparticles. *Solid State Commun.* **2012**, *152*, 525–529. [CrossRef]

51. Das, R.; Khan, G.G.; Varma, S.; Mukherjee, G.D.; Mandal, K. Effect of quantum confinement on optical and magnetic properties of Pr—Cr-codoped bismuth ferrite nanowires. *J. Phys. Chem. C* **2013**, *39*, 20209–20216. [CrossRef]

52. Kumar, K.S.; Ramanadha, M.; Sudharani, A.; Ramu, S.; Vijayalakshmi, R.P. Structural, magnetic, and photoluminescence properties of BiFeO$_3$: Er-doped nanoparticles prepared by sol-gel technique. *J. Supercond. Nov. Magn.* **2019**, *32*, 1035–1042. [CrossRef]

nanomaterials

MDPI

Article

Cultivar-Dependent Anticancer and Antibacterial Properties of Silver Nanoparticles Synthesized Using Leaves of Different *Olea Europaea* Trees

Valeria De Matteis [1,*], Loris Rizzello [2,3], Chiara Ingrosso [4], Eva Liatsi-Douvitsa [2], Maria Luisa De Giorgi [1], Giovanni De Matteis [1] and Rosaria Rinaldi [1]

[1] Department of Mathematics and Physics "Ennio De Giorgi", University of Salento, Via Arnesano, 73100 Lecce, Italy; marialuisa.degiorgi@unisalento.it (M.L.D.G.); dematteis.giovanni@virgilio.it (G.D.M.); ross.rinaldi@unisalento.it (R.R.)
[2] Department of Chemistry, University College London, 20 Gordon Street, London WC1H 0AJ, UK; lrizzello@ibecbarcelona.eu (L.R.); eva.liatsi-douvitsa.14@ucl.ac.uk (E.L.-D.)
[3] Institute for Bioengineering of Catalonia (IBEC), The Barcelona Institute of Science and Technology, Baldiri Reixac 10-12, 08028 Barcelona, Spain
[4] CNR-IPCF S.S. Bari, c/o Department of Chemistry, Università degli Studi di Bari, via Orabona 4, I-70126 Bari, Italy; c.ingrosso@ba.ipcf.cnr.it
[*] Correspondence: valeria.dematteis@unisalento.it

Received: 3 October 2019; Accepted: 25 October 2019; Published: 30 October 2019

Abstract: The green synthesis of nanoparticles (NPs) is currently under worldwide investigation as an eco-friendly alternative to traditional routes (NPs): the absence of toxic solvents and catalysts make it suitable in the design of promising nanomaterials for nanomedicine applications. In this work, we used the extracts collected from leaves of two cultivars (*Leccino* and *Carolea*) belonging to the species *Olea Europaea*, to synthesize silver NPs (AgNPs) in different pH conditions and low temperature. NPs underwent full morphological characterization with the aim to define a suitable protocol to obtain a monodispersed population of AgNPs. Afterwards, to validate the reproducibility of the mentioned synthetic procedure, we moved on to another Mediterranean plant, the *Laurus Nobilis*. Interestingly, the NPs obtained using the two olive cultivars produced NPs with different shape and size, strictly depending on the cultivar selected and pH. Furthermore, the potential ability to inhibit the growth of two woman cancer cells (breast adenocarcinoma cells, MCF-7 and human cervical epithelioid carcinoma, HeLa) were assessed for these AgNPs, as well as their capability to mitigate the bacteria concentration in samples of contaminated well water. Our results showed that toxicity was stronger when MCF-7 and Hela cells were exposed to AgNPs derived from *Carolea* obtained at pH 7 presenting irregular shape; on the other hand, greater antibacterial effect was revealed using AgNPs obtained at pH 8 (smaller and monodispersed) on well water, enriched with bacteria and coliforms.

Keywords: Green synthesis; Silver nanoparticles; *Olea Europaea*; *Leccino*; *Carolea*; cytotoxicity; genotoxicity; antibacterial activity

---

1. Introduction

The growing use of nanotechnology-based materials is providing new solutions to previously unsolved issues [1]. AgNPs represent the 24 % [2] of the whole materials present in the current textiles, plastics, food, and other countless commercial products [3–6]. Their use is even higher in electronics [7], medicine [8], and materials sciences [9]. These widespread applications are mainly due to their unique physico-chemical properties that ranged from plasmonic [10] to antibacterial [11] and anticancer activity [12]. Inevitably, their potential impact on human health as well as on the environment, especially in terms of dangerous chemicals used for producing any nanomaterial were investigated. AgNPs can be synthesized by means of physical or chemical methods [13]. The latter chemical approach is based on the use of either organic or water solutions containing metal precursors and reducing agents. The typical drawback here is the toxicity induced by the chemicals, that thus required the implementation of long and time-consuming post-synthesis purification procedures [14]. Despite chemical and physical routes represent the most common methods used to synthetize high quality AgNPs, finding new "green" solutions for their development will represent a serious step towards a decrease of environmental impacts [15]. Green chemistry is an eco-friendly and valid alternative due to the use of natural agents for the production of NPs [16]. This eco-friendly alternative employs natural products and non-toxic agents, which are usually in the form of plant extracts or even microorganisms and water [17]. Green synthesis routes show several pros, such as: (*i*) reducing the waste products, (*ii*) decreasing the energy-associated cost of productions and (*iii*) eliminating the environmental toxicity [18]. In many green syntheses, the silver ions (Ag^+) reduction, that is the crucial step for the assembling of AgNPs, is mediated by the biomolecules presents in the plants extract (i.e., proteins, alkaloids, saponins, amino polysaccharides proteins, tannins, enzymes, vitamins etc)[19,20]. Another advantage consists in the Ag^+ reduction in plant extracts-mediated syntheses, with a faster rate compared to the process performed with microorganisms [21]. Many works are available on the bio-synthesis of AgNPs from plant leaves extracts [22], such as *Rosa rugosa* [23], *Psidium guajava* [24], *Magnolia Kobus, Ginko biloba, Pinus desiflora, Diopyros kaki, Pllatanus orientalis* [25], *Stevia rebaudiana* [26], *Cocos nucifera coir* [27], *Chenopodium album* [28], *Gliricidia sepium* [29], *Ocimum sanctum* [30], *Cycas* [31]. *Olea Europaea* is one of the most important tree in the Mediterranean countries, due to their capability to produce oil, a relevant source of nutrition and medicine [32]; the leaves have always been used as antimalarial, antibacterial and anti-mycoplasma agent, thanks to their inherently high concentration of antioxidants and anti-inflammatory molecules [33]. Biological molecules such as luteolin-7-glucoside, luteolin, cafeic acid, p-coumaric acid, vanillin, diosmetin, rutin, apigenin-7-glucoside, diosmetin-7-glucoside and vanillic acid [34,35] are present in high concentrations [36]. Secoiridoids like oleuropein only exist in plants belonging to the family of *Oleaceae*. These compounds are both terpenic and hydroxy aromatic secondary metabolites. Olive leaves are also rich in mannitol, which is a sugar alcohol, synthesised only by the *Oleaceae* plants [37]. Therefore, in this study, we wondered if two *Olea Europaea* cultivar leaves, *Leccino* and *Carolea*, might be used as natural sources to synthesize AgNPs with controlled properties. We used the leaves of these trees resulting from pruning and other agricultural activities, which were destined to become waste. We investigated the physico-chemical properties of the AgNPs achieved, from the two selected cultivar leaves, in low temperature conditions and different pH conditions (pH 7 and pH 8). Such a green synthetic route, optimized for achieving AgNPs from olive oil extracts, was found to be fast, low-cost, reproducible and environmentally friendly. In addition, its reproducibility was assessed by synthesizing AgNPs from another Mediterranean plant, the *Laurus Nobilis*. Finally, the antibacterial and anticancer properties of the obtained AgNPs, were tested against bacteria colonized well water and two cancer cell lines, respectively.

2. Materials and Methods

2.1. Preparation of Leaves Extract

Leaves of two *Olea Europaea* cultivar (*Leccino* and *Carolea*), were collected in winter from olive trees. Several washes with MilliQ water were carried out in order to remove pollution or other contaminants from leaves. Leaves were then dried at room temperature overnight. There was 50 g of leaves finely cut and added in 500 mL of MilliQ water. The mix was boiled for 15 min. The extract was cooled down to room temperature, and afterwards filtered with Whatman No. 1 filter paper. The filtrate was further centrifuged at 4000 rpm for 10 min and the supernatant was used for the synthesis. The same procedure was implemented for the preparation of *Laurus Nobilis* extract.

2.2. Synthesis of Green AgNPs

There was 5 mL of extract added to 100 mL of AgNO$_3$ (1 mM) and the reaction was heated to 60 °C for 45 min. In this time, the reaction colour changed from clear yellow to dark brown, indicating reduction of Ag$^+$ ions to Ag^0 NPs at pH 7. We also used NaOH to increase the pH of the mixture from 7 to 8. The solution was centrifuged at 12.000 rpm for 30 min in order to obtain concentrated AgNPs for the next characterizations steps.

2.3. Characterization of Green AgNPs

2.3.1. Transmission Electron Microscopy (TEM), Dynamic Light Scattering (DLS) and ζ-Potential

TEM characterizations were carried out with a JEOL Jem 1011 microscope, operating at an accelerating voltage of 100 kV (JEOL USA, Inc, Peabody, MA, USA). TEM samples were prepared by dropping a dilute solution of NPs in water on carbon-coated copper grids (Formvar/Carbon 300 Mesh Cu). DLS and ζ-potential measurements were performed on a Zetasizer Nano-ZS equipped with a 4.0 mW HeNe laser operating at 633 nm and an avalanche photodiode detector (Model ZEN3600, Malvern Instruments Ltd., Malvern, UK). Measurements were performed at 25 °C in aqueous solution (pH 7).

ImageJ open source software (NIH image) version 1.47 was used with a suite of analysis routines used for particle analysis to test the circularity values of NPs measured on TEM acquisition. Sorting based on circularity and including only those with circularity values > 0.8 will ensure any aggregates are not included in the measurement [38].

2.3.2. UV-Vis Spectroscopy

The UV–Vis absorption spectra of AgNPs samples were collected at room temperature by means of a Cary 5000 UV/Vis/NIR spectrophotometer (Varian, Palo Alto, CA, USA).

2.3.3. Energy-Dispersive X-ray Spectroscopy (EDS)

EDS analyses were recorded with a Phenom ProX microscope (Phenom-World B. V., Eindhoven, Germany), at an accelerating voltage of 10 kV. The samples were prepared by dropping a solution of NPs in water onto monocrystalline silicon wafer.

2.3.4. Attenuated Total Reflection (ATR) Fourier Transform Infrared Spectroscopy (FTIR)

Mid-infrared spectra were acquired with a Varian 670-IR spectrometer equipped with a DTGS (deuterated tryglicine sulfate) detector. The spectral resolution used for all experiments was 4 cm^{-1}. For attenuated total reflection (ATR) measurements, a one-bounce 2 mm diameter diamond microprism was used as the internal reflection element (IRE). Films were directly cast onto the internal reflection element by depositing the solution or suspension of interest onto the upper face of the diamond crystal and allowing the solvent to evaporate.

2.3.5. Cell Culture

MCF-7 and Hela were maintained in high glucose DMEM with 50 μM of glutamine, supplemented with 10% FBS, 100 U/mL of penicillin and 100 mg/mL of streptomycin. Cells were incubated in a humidified controlled atmosphere with a 95 % to 5 % ratio of air/CO$_2$, at 37 °C.

2.3.6. WST-8 Assay

MCF-7 and Hela cells were seeded in 96 well microplates at the concentration of 5 × 10^3 cells/well after 24 h of stabilization. NPs stock solutions (AgNPs from the *Leccino*, *Carolea* at pH 7 and AgNPs from the *Leccino*, *Carolea* and *Laurus Nobilis* at pH 8) were added to the cell media at 20 μg/mL and 50 μg/mL. Cells were incubated for 48 and 96 h. At the endpoint, cell viability was determined using a standard WST-8 assay (Sigma Aldrich). Assays were performed following the procedure previously described in De Matteis et al [39]. Data were expressed as mean ± SD.

2.3.7. Lactate Dehydrogenase (LDH) Assay

MCF-7 and Hela cells were seeded in 96 well microplates (Constar) and treated with NPs stock solutions (AgNPs from the *Leccino*, *Carolea* at pH 7 and AgNPs from the *Leccino*, *Carolea* and *Laurus Nobilis* at pH 8) at 20 µg/mL and 50 µg/mL of concentration. After 48 and 96 h of cell–AgNP interaction, the LDH leakage assay was performed onto microplates by applying the CytoTox-ONE Homogeneous Membrane Integrity Assay reagent (Promega) following the manufacturer's instructions. The culture medium was collected, and the level of LDH was measured by reading absorbance at 490 nm using a Bio-Rad microplate spectrophotometer (Biorad, Hercules, CA, USA). Data were expressed as mean ± SD.

2.3.8. Determination of the Intracellular Uptake of Green AgNP$_S$ by Inductively Coupled Plasma Atomic Emission Spectroscopy (ICP-AES)

1×10^5 of MCF-7 and HeLa cells were seeded in 1 mL of medium in a 6-well plate. After 24 h of incubation at 37 °C, the medium was replaced with fresh medium containing the green AgNPs obtained at pH 7 and pH 8, at the concentrations of 20 µg/mL and 50 µg/mL. After 48 h and 96 h of incubation at 37 °C, the culture medium was removed, and the MCF-7 and Hela washed with PBS buffer to eliminate non-internalized NPs. Cells were detached with trypsin and counted by sing automatic cell counting chamber. 360.000 cells were suspended in 200 µL of milliQ, and treated with HNO$_3$ and diluted to 5 mL: the solution was analysed to evaluate Ag content. Elemental analysis was carried out by ICP-AES, Varian Vista AX spectrometer (Varian Inc., Palo Alto, CA, USA).

2.3.9. Comet Assay (Single Gel Electrophoresis)

HeLa cells were exposed to 50 µg/mL of AgNPs obtained from *Leccino* and *Carolea* for 96 h, at density of 5×10^4 in each well of 12-well plates in a volume of 1.5 mL. After treatments, cells were centrifuged and suspended in 10 µL of PBS at concentration of 1000 cells/µL. The cell pellets were mixed with 75 µL of 0.75 % low-melting-point agarose (LMA) and then layered onto microscope slides pre-coated with 1% normal melting agarose (NMA) and dried at room temperature. Subsequently, the slides were immersed in an alkaline solution (300 mM of NaOH, 1 mM of Na$_2$EDTA, pH 13) for 20 min to allow for unwinding of the DNA. The electrophoresis was carried out in the same buffer for 25 min at 25 V and 300 mA (0.73 V/cm). After electrophoresis, cellular DNA was neutralized by successive incubations in a neutralized solution (0.4 MTris–HCl, pH 7.5) for 5 min at room temperature. The slides were stained with 80 µL SYBR Green I (Invitrogen). Comets derived from single cells were photographed under a Nikon Eclipse Ti fluorescence microscope, and head intensity/tail length of each comet were quantified using Comet IV program (Perceptive Instruments).

2.3.10. Determination of Ag$^+$ Release

Ag$^+$ release was quantified using 50 µg/mL of AgNPs from *Leccino* and *Carolea* cultivar obtained at pH 7 and AgNPs from *Leccino*, *Carolea* cultivar and *Laurus Nobilis* obtained at pH 8. The release was studied upon 24 h, 48 h and 96 h of incubation time in water (pH 7) and acidic buffer (pH 4.5). After the time points, the NPs were collected by centrifugation at 13.000 rpm for 1 h and digested by the addition of HNO$_3$ solution (10% v/v). The number of free ions was measured by ICP-AES (Varian Inc., Palo Alto, CA, USA).

2.3.11. Confocal Measurements

HeLa cells were seeded at concentration of 8×10^4 cells/mL in glass Petri dishes (Sarstedt, Germany). After 24 h of stabilization, the culture media was supplemented with AgNPs derived from *Leccino* and *Carolea* (50 µg/mL) for 96 h. After exposure, the medium was removed; then three washes with Phosphate Buffered Saline (PBS, D1408, Sigma Aldrich) were performed. Samples were fixed by using glutaraldehyde (G5882, Sigma Aldrich) at 0.25 % in PBS for 10 min. After two washes with PBS, Triton X-100 (Sigma Aldrich) at 0.1 % for 5 min was used to permeabilize the cell membrane of fixed cells before staining the nuclei by 1 µg/mL of DAPI (D9542, Sigma Aldrich) for 5 min. Acquisitions were performed by Leica TCS SPE-II confocal microscope using a 100× objective (water immersion, HCX PL APO, 1.10NA). The fluorescent images were obtained exciting fluorescent dyes by means laser radiation having wavelength at 405 nm. The nuclear morphology was quantified in terms of shape descriptor parameter: circularity. Circularity parameter compares an object to a circle; it is ranges from 0 to 1 (for a perfect circle). All results were obtained as means calculated on 15 cells and data were statistically analysed by means of a paired two-tailed t-test. The statistical difference of results was considered significant for p-value < 0.05*.

2.4. Antibacterial Activity of Green AgNPs

Collection of Water Samples

Water samples were collected from coliforms contaminated artesian well in south of Italy with bottles previously sterilized. The collection was done in the early morning, because it was reported that the coliforms could increase in warm pulled water [40].

2.5. Total Bacteria Detection by Plate Count Techniques

The count of viable bacteria was performed by plate count techniques [41].Well water (1 mL) was dropped on petri dishes previously filled with 9 mL of Plate Count Agar (Liofilchem) using spread plate technique for control. For treated samples, 50 µg/mL of green AgNPs derived both from *Leccino* and *Carolea* at pH 7 and from *Leccino*, *Carolea* and *Laurus Nobilis* at pH 8 were added. The inoculated plates were incubated at 22 °C for 72 h and 37 °C for 48 h after which the plates were observed. Bacteria growth and numbers of colonies were counted using a colony counter. Colony counts were expressed as Colony Forming Units (CFU/ml) of the sample:

No. of CFU/ml = No. of colonies counted × Dilution factor × Volume of sample taken.

2.6. Most Probable Number (MPN) to detect Coliforms and Faecal Coliforms

Most Probable Number (MPN) method [42] permitted to evaluate the number of coliforms bacteria in well water by means of replicate liquid broth growth in ten-fold dilutions. Contaminated well water was diluted serially and inoculated in Lactose Broth (Merck) at 37 °C for 24 h. The treated well water samples were represented by water with 50 µg/mL of green AgNPs derived both from *Leccino* and *Carolea* at pH 7 and from *Leccino*, *Carolea* and *Laurus Nobilis* at pH 8 before the inoculation at 37 °C for 24 h. The presence of total coliforms in water was detected by the ability of these bacteria to produce acid and gas using lactose. The acid production was detected by color change and the gas with the gas bubbles in the inverted Durham tube. To evaluate the AgNPs-induced bacteria reduction, the total number of coliforms, in terms of MPN index (estimated number of coliforms in 100 mL of water), was obtained by counting the tubes within two reactions taking place and comparing them with standard statistical tubes. This involved the presumptive, confirmed and completed test for coliform bacteria [43]. Incubation at high temperature was used to distinguish organisms of the total coliforms from faecal coliform group. In order to detect faecal coliforms, 1 mL of liquid medium from the tubes that underwent a color change, indicating the presence of coliforms, was added to EC broth (10 mL) at 44.5 °C for 24 h. Gas production with growth within 24 ± 2 h of incubation at 44.5 ± 0.2 °C is considered positive for the presence of faecal coliforms in water. Absence of gas production is considered a negative test for the presence of faecal coliforms [44,45].

3. Results and Discussion

The use of plants or microorganisms to synthetize NPs is a method particularly suitable for achieving metal NPs, such as AgNPs [46]. In our work, we used plants extracts obtained from leaves of typical Mediterranean tree, which can be easily found in Italy, namely *Olea Europaea*. The olive is an evergreen tree and its leaves are by products of olive farming that are stored during the pruning process [47]. Among different cultivar, differences in leaves length can be observed ranging from 30 to 80 mm [48]. The leaves from *Olea Europaea*, *Leccino* and *Carolea* cultivar, growing in the same pedoclimatic conditions (Figure 1a), were collected in winter when the bioactive compounds, such as amino acids, tannins and carbohydrates content was abundant [49]. In particular, the production of NPs was favoured by the high concentration of phenols in cold season that helped the Ag^+ clustering, which was the seeding event in the growth of NPs [50]. Leaves were used to prepare plants extracts (Figure 1a) useful to obtain AgNPs with easy and not-toxic reproducible synthetic route with the addition of 1 mM of $AgNO_3$ at two different pH (7 and 8) and at low temperature. The solution turned dark brown in 45 min confirming the AgNPs formation promoted by reduction of the Ag^+ (Figure 1b). The dark brown colour indicated the free conduction electrons oscillation induced by the surface plasmon resonance excitation phenomenon [51,52].

The difference between the newly suggested green synthetic route and the previous approach by some of the authors [39] consists in the non-use of in sodium citrate and tannic acid, added at high temperature, to boost the Ag^+ reduction to obtain stable and monodispersed spherical AgNPs with a size of (20 ± 3) nm. NPs obtained from leaves extracts were deeply characterized by means of TEM, DLS, ζ-Potential, UV-vis, EDS and FTIR-ATR in water. TEM analyses confirmed that AgNPs were different in shape and size when using the two cultivar extracts at pH 7 (Figure 1c,d,f,g). In detail, AgNPs derived from *Leccino* at pH 7 showed a quasi-spherical morphology and a mean size of (35 ± 8) nm (Figure 1e), whereas AgNPs obtained from *Carolea* were mainly triangular and hexagonal in shape, with a mean size of (60 ± 11) nm (Figure 1h). DLS measurements carried out in water were perfectly consistent with TEM analyses: in fact, the AgNPs showed a hydrodynamic radius compatible with the mean size values noticed in TEM acquisitions. In particular, at pH 7, the hydrodynamic radius recorded for *Leccino* was (31 ± 9) nm (Figure 1i) whereas it was (58 ± 14) nm for *Carolea* (Figure 1l). Interestingly, the increase of the pH reaction solution up to 8 triggered the formation of monodispersed AgNPs with comparable size and shape, though using different *Olea Europaea* cultivars; in fact, spherical shape and smaller mean size (10 - 22 nm) were observed in NPs synthesized using extracts from both cultivars. Such figures of merit were indeed different compared to those observed in NPs synthesized at pH 7 (Figure 2a,b,d,e). These results were correlated to pH 8 that influenced the stabilization NPs [22]. AgNPs from *Leccino* had a mean size of (15 ± 2) nm (Figure 2c). On the contrary, the same NPs derived from *Carolea* showed a mean size of (23 ± 7) nm (Figure 2f). DLS measurements allowed an estimation of the NP hydrodynamic radius of (12 ± 3) nm for NPs from *Leccino* (Figure 2g), and 20 ± 8 nm from *Carolea* (Figure 2h). Such values were in agreement with the TEM observations.

After demonstrating the potential to use leaves from *Olea Europaea* for the synthesis of AgNPs, we moved on to synthesize AgNPs from the extract of *Laurus nobilis*, in order to validate the results obtained by using the same procedure at pH 8 using extracts from different trees deriving from the same Mediterranean area. In this case, the achieved AgNPs showed a mean size of (20 ± 8) nm (Figure 3c), as also confirmed by the DLS peak at (22 ± 6) nm (Figure 3d). These data were consistent with those found for the above reported syntheses at the same pH, using olive leaves extracts.

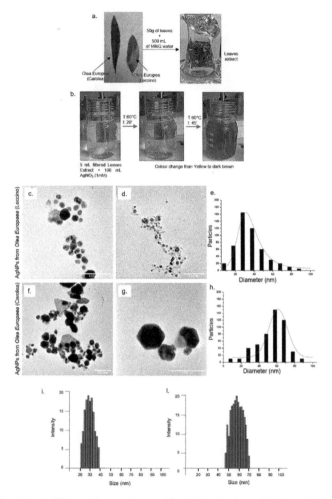

Figure 1. Morphologic differences between leaves collected from *Leccino* and *Carolea* cultivars and leaves extract preparation (**a**). The reaction started with AgNO₃ addition and the change of colour from yellow to dark brown indicated silver NPs (AgNPs) formation (**b**). Representative TEM images of AgNPs obtained from *Olea Europaea Leccino* (**c–d**) and from *Carolea* (**f–g**) at pH 7. Size distribution was measured on 500 AgNPs from *Leccino* (**e**) and *Carolea* (**h**) and fitted with a normal function (solid line). Dynamic Light Scattering (DLS) measurements in water of AgNPs from *Leccino* (**i**) and *Carolea* (**l**) at pH 7.

Afterwards, ζ-potential analyses revealed that negative surface charge was observed for all the obtained AgNPs in water (Table 1). This can be ascribed to proteins and other biological molecules present in leaves extracts, adsorbed on the NPs surface [53]. In fact, during green synthetic process, biomolecules such as proteins and peptides behave as capping agents and they are typically adsorbed during the NPs formation step, affecting the reaction dynamic and the NPs growth in different directions. These negatively charged natural capping agents are responsible of NPs stabilization due to their ability to control particles size, shape/morphology and to protect the surface from agglomeration phenomena that will influence their consequent uptake in cells [54].

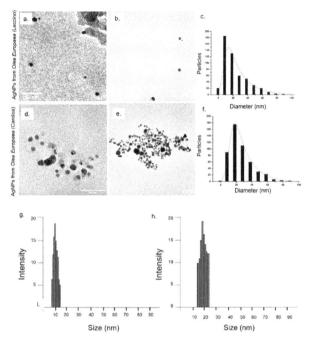

Figure 2. Representative TEM images of AgNPs obtained from *Olea Europaea* (**a–b**) *Leccino* and (**d–e**) *Carolea* at pH 8. Size distribution was measured on 500 AgNPs from *Leccino* (**c**) and *Carolea* (**f**) and fitted with a normal function (solid line). DLS measurements in water of AgNPs from *Leccino* (**g**) and *Carolea* (**h**) at pH 8.

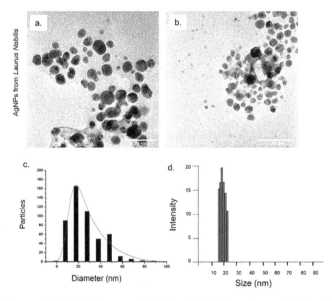

Figure 3. Representative TEM images of AgNPs obtained from *Laurus Nobilis* at pH 8 (**a–b**). Size distribution was measured on 500 AgNPs and fitted with a normal function (solid line) (**c**). DLS measurements in water of AgNPs from *Laurus Nobilis* at pH 8 (**d**).

Table 1. ζ-potential values of green silver NPs (AgNPs) achieved at different pH.

Green AgNPs	pH	Zeta Potential Value
AgNPs from *Olea europaea* (*Leccino*)	7	-15 ± 5
AgNPs from *Olea europaea* (*Carolea*)	7	-20 ± 3
AgNPs from *Olea europaea* (*Leccino*)	8	-30 ± 8
AgNPs from *Olea europaea* (*Carolea*)	8	-18 ± 2
AgNPs from *Laurus Nobilis*	8	-25 ± 2

Image J software was used to investigate the NPs sharpness by means the circularity parameter. We used high resolution TEM images to measure the geometrical parameters of the NPs; 50 random NPs from each type were investigated to obtain the average circularity distribution (Table 2). Circularity values near 1 indicate a perfect cycle, whereas near 0 an high sharpness degree [55]. The AgNPs obtained from *Carolea* cultivar at pH 7 had an average circularity of (0.28 ± 8) and showed a much higher degree of sharpness when compared to AgNPs from *Leccino* at the same pH 7, which presented an average circularity of (0.55 ± 4). NPs tended to assume a more spherical morphology at pH 8 even if different cultivars were used in basic synthesis, with an average circularity of (0.88 ± 3) for *Leccino*, and (0.63 ± 6) for *Carolea*. *Laurus nobilis* extract, on the other hand, induced the formation of AgNPs with a circularity value of (0.65 ± 4).

Table 2. Circularity values obtained using ImageJ software on TEM acquisitions.

Green AgNPs	Circularity Value (pH 7)	Circularity Value (pH 8)
AgNPs from *Olea Europaea*–(*Leccino*)	0.55 ± 4	0.88 ± 3
AgNPs from *Olea Europaea*–(*Carolea*)	0.28 ± 8	0.63 ± 6
AgNPs from *Laurus Nobilis*	-	0.65 ± 4

UV-vis absorption spectra of the AgNPs were recorded in the 300–800 nm range and compared with those of the corresponding leaves extracts (Figure 4a).

1 mg/mL of leaves extracts, prepared as reported in the experimental section was used, and two peaks in the UV region, namely at 280 nm and 350 nm, probably due to aromatic compounds, have been detected for all the extracts solutions, while no absorption signal was detected in the 400-800 nm range. The absorption spectra of the AgNPs obtained from *Leccino* and *Carolea* extracts at pH 7 showed a surface plasmon resonance peak at 463 nm and 458 nm, respectively, which was red-shifted compared to AgNPs synthesized by colloidal chemical routes [39], showing a peak at 400 nm. It is worthwhile to notice that in both samples; the plasmon peaks were rather broad, suggesting the formation of AgNPs with a broad size distribution (Figure 4b). In particular, when the sharpness degree increased like in the case of triangulary-shaped NPs, the spectrum underwent a pronounced red shifted with respect to the AgNPs produced by colloidal chemical reduction processes. When the synthesis was performed at pH 8, the absorption spectra were narrow and closer to the wavelength absorption peak of the AgNPs synthesized by colloidal chemical routes. Namely, a peak at 420 nm was observed for the AgNPs from *Leccino*, and at 417 nm for the AgNPs from *Carolea*. The absorption peak of the AgNPs from *Laurus nobilis* was at 415 nm (Figure 4c).

Elemental analyses were also performed to investigate the chemical composition of the NPs samples. The EDS analyses of the AgNPs deposited onto silicon substrate in the range of 0–5 keV (Figure 5) clearly showed a strong spectral signal in the silver region (3–3.5 keV), both for NPs derived from colloidal chemical route and from green NPs, processed both at pH 7 and pH 8. The signals related to Na, Mg, Cl, C, O suggested the presence of biomolecules (carbohydrates and proteins). The Na element can be originated from sodium citrate, which was used for the colloidal chemical synthesis (green spectrum in Figure 5a).

FTIR-ATR spectroscopy investigation was performed in order to study the chemical composition of the solution in which the AgNPs were synthesized by the here proposed "green" approach, and hence, to elucidate the chemical groups which could be involved in the stabilization of the NPs in aqueous solution upon their formation. For this purpose, the FTIR-ATR spectra of the AgNPs solutions were compared with those of the plant extracts achieved in the same experimental conditions used to synthesize AgNPs. The FTIR-ATR spectrum of AgNPs prepared by the chemical colloidal route, in the presence of sodium citrate and tannic acid surfactant was reported, as a suitable reference for the assignment of the FTIR-ATR peaks of the AgNPs, synthesized by the green route [39].

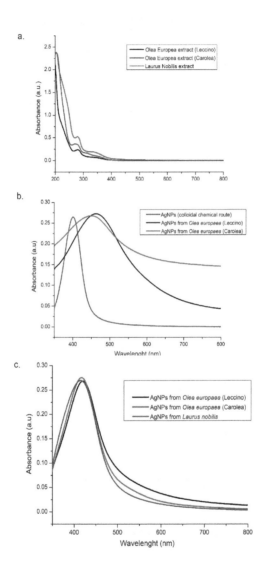

Figure 4. UV-vis spectra of *Olea Europaea* extracts (*Leccino, Carolea*) and *Laurus Nobilis* (**a**), UV-vis spectra of AgNPs derived from colloidal chemical routes and from green synthesis using *Olea Europaea* extracts (*Leccino* and *Carolea*) at pH 7 (**b**), AgNPs derived from green synthesis using *Olea Europaea* extracts (*Leccino* and *Carolea*) and *Laurus Nobilis* at pH 8 (**c**).

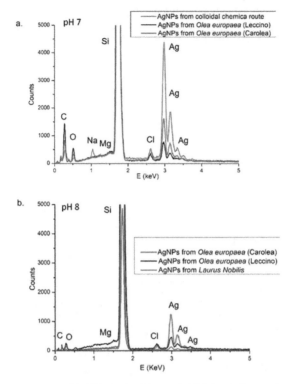

Figure 5. EDS measurements of AgNPs obtained from colloidal chemical route, from *Leccino* cultivar and *Carolea* at pH 7 (**a**) and from *Leccino* cultivar, *Carolea* cultivar and *Laurus Nobilis* at pH 8 (**b**).

The FTIR-ATR spectra of extracts (Figure 6a) showed, in the high wavenumber region, a broad band in the range of 3700–3014 cm^{-1} and the peaks at 2927 cm^{-1} and 2887 cm^{-1}, which were ascribed to the stretching vibrations of O-H and -C-H groups. Such vibrations attested for the presence of molecules containing alcohol, carboxylic acid and aliphatic groups in the plant extracts solutions. On the other hand, in the low wavenumber region, the peaks at ca. 1704 cm^{-1} and 1623 cm^{-1} were ascribed to carboxylic -C=O stretching and aromatic -C=C- ring stretching modes. The peak at 1383 cm^{-1}, with the shoulder at 1450 cm^{-1}, were aliphatic bending vibrations and the bands at 1261 cm^{-1}, 1198 cm^{-1} and 1050 cm^{-1} can be due to the -C-O bending of alcohols and carboxylic acid groups and –C-O-C- stretching vibrations of ethers. The FTIR-ATR spectra of the AgNPs synthetized from *Olea Europaea* (*Leccino* and *Carolea*) at pH 7 (Figure 6b) showed the broad band peak in the range of 3100–3014 cm^{-1} and the peaks at ca. 2927 and 2854 cm^{-1} due to stretching vibrations of the -O-H and aliphatic -C-H groups of the alcohols and carboxylic acids found in the plant extracts solutions (Figure 6a). In the low wavenumber region, they showed a shoulder at 1717 cm^{-1} which can be ascribed to the -COOH stretching of free carboxylic molecules, along with two strong bands, at 1640 cm^{-1} and 1387 cm^{-1} in the NPs from *Leccino*, and 1600 cm^{-1} and 1356 cm^{-1} in the NPs from *Carolea*. Such a double band can be accounted for by the signals of the antisymmetric and symmetric -COO- stretching of carboxylic molecules, respectively, coordinated to the surface Ag atoms [56,57] and thus responsible for the stabilization of the NPs in aqueous solution. Indeed, the same bands were observed also in the FTIR-ATR spectra of AgNPs, synthesized by the colloidal chemical route, by reduction of silver precursor in the presence of sodium citrate and tannic acid surfactants, which were coordinated to the AgNPs surface by their carboxyl groups. Finally, the characteristic peaks of the -C-O bending of alcohols and carboxylic acid groups and –C-O-C stretching vibrations of ethers present in the extracts solutions are still evident in the spectra of the green AgNPs. The AgNPs solutions synthesized at pH 8 (Figure 6c) presented the same vibrations of the NPs solutions achieved at pH 7 (Figure 6b), thus assessing the involvement of molecules containing carboxylic acid groups in the stabilization of the NPs achieved also in these synthesis conditions. The same evidence was obtained for the AgNPs achieved from *Laurus Nobilis* (Figure 6c) in the same experimental conditions, thus indicating that the same chemical moieties were responsible for the stabilization of the AgNPs, irrespectively of the plant from which the synthesis was performed.

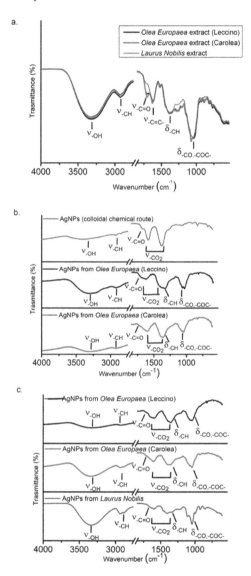

Figure 6. FTIR spectra of *Olea Europaea* (*Leccino* and *Carolea*) and *Laurus Nobilis* extracts (**a**), AgNPs derived from colloidal chemical routes and from green synthesis using *Olea Europaea* extracts (*Leccino* and *Carolea*) at pH 7 (**b**), AgNPs derived from green synthesis using *Olea Europaea* extracts (*Leccino* and *Carolea*) and *Laurus Nobilis* at pH 8 (**c**).

The potential toxicity of NPs against cells was then investigated, opting for the MCF-7 and HeLa cell lines. The interaction of AgNPs with MCF-7 was studied because these cells are a well-established model for the identification of adverse effects of NPs and have an epithelial and non-invasive phenotype, as reported elsewhere [58]. In addition, the human cervical carcinoma HeLa cells were selected because they are the most often used models for cytotoxicity studies [59]; in addition this cell line shows good growth and did not require growth factors for its proliferation [60,61].

We evaluated cell viability after exposing the MCF-7 and HeLa to AgNPs from *Leccino* and *Carolea* at pH 7 (Figure 7a) and to AgNPs from *Leccino*, *Carolea* and *Laurus Nobilis* obtained at pH 8, at concentrations of 20 µg/mL and 50 µg/mL, for 48 h and 96 h (Figure 7b). All the tested NPs induced toxicity in MCF-7 cells, with some differences observed between the AgNPs synthetized at pH 7 and those at pH 8. The cytotoxic effects were more evident when cells were treated with NPs produced at pH 7, especially for *Carolea*-derived NPs, in comparison with the same NPs, obtained at pH 8. In particular, the cells treated with 50 µg/mL of AgNPs prepared from *Carolea* at pH 7 showed a viability reduction of more than 30 % after 48 h, and only 48 % of cells were viable after 96 h. At pH 8, the cytotoxic effects were similar for the AgNPs derived from *Leccino*, *Carolea* and *Laurus Nobilis*.

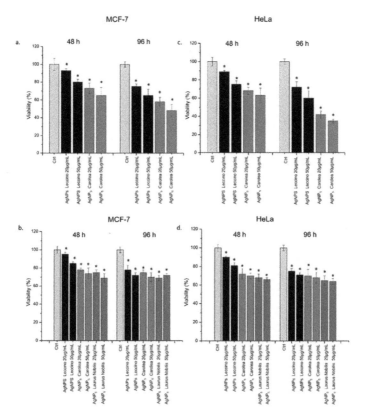

Figure 7. WST-8 of MCF-7 and HeLa cells after 48 h and 96 h exposure to two doses (20 µg/mL and 50 µg/mL) of AgNPs derived from *Oleaea Europeae* (*Leccino* and *Carolea*) at pH 7 (**a** and **c**) and from *Oleaea Europeae* (*Leccino* and *Carolea*) and *Laurus Nobilis* at pH 8 (**b** and **d**.) Viability of NPs-treated cells was normalized to non-treated control cells. As positive control (P), cells were incubated with 5 % Dimethyl Sulfoxide (DMSO) (data not shown). Data reported as mean ± SD from three independent experiments are considered statistically significant compared with control ($n = 8$) for *p* value < 0.05 (< 0.05 *).

The treatment against HeLa cells showed the same trend observed for MCF-7 (Figure 7c). Also, in this case, the AgNPs derived from *Carolea* at pH 7 were more toxic than the same obtained from *Leccino*, but the cytotoxic effect was stronger respect to MCF-7. Indeed, at 96 h, AgNPs obtained from *Carolea* induced a viability reduction of 62 % at 50 µg/mL of concentration. Using AgNPs obtained at pH 8, the effects followed the same trend obtained in MCF-7 but the effect resulted more visible in HeLa cells (Figure 7d).

These results suggested a selective toxicity that was dependent on the shape of the NPs, and hence, on the pH used for the synthesis reaction [62,63]. However, it was important to remark that the toxicity induced by these green NPs was lower compared to the same AgNPs obtained from colloidal chemical routes [39].

AgNPs induced cell membrane poration with a consequent LDH release in close agreement with the viability results (Figure 8). The effect was more evident in Hela cells with respect to MCF-7 especially upon AgNPs obtained from *Carolea* obtained at pH 7 and after 96 h at the higher concentration (Figure 8a–c). The LDH release percentage reached an increase of about 143 % with respect to the untreated (control) cells after 96 hours of exposure (Figure 8c). Using AgNPs obtained from the three plant extracts at pH 8 (Figure 8b,d), the effects on cell membrane of HeLa and MCF-7 were similar.

To understand whether the enhanced cytotoxicity of *Carolea*-derived AgNPs may be due to differences in uptake dynamics, we quantified the internalization of green AgNPs (20 µg/mL and 50 µg/mL) in MCF-7 and HeLa by elemental analyses (Figure 9). The uptake was slightly higher for AgNPs derived from *Carolea* compared to those derived from *Leccino* at pH 7 in the two cell lines. AgNPs obtained at pH 8 presented similar uptake levels, which were anyway lower compared to NPs produced at pH 7. In MCF-7, the detected intracellular amount of Ag was (5.93 ± 0.56) µg after incubation with 50 ug/mL of *Carolea*-derived AgNPs for 96 h. In the same conditions of exposure, the Ag content found for the *Leccino*-derived NPs, was of (4.5 ± 0.41) µg (Figure 9a). In HeLa cells, the uptake was more efficient than MCF-7: the effect was more evident for AgNPs from *Carolea* at pH 7: the intracellular Ag measured was (7.4 ± 0.67) µg for cells exposed to 50 ug/mL of NPs for 96 h. The Ag amount observed in HeLa cells exposed to AgNPs obtained from *Leccino* was (4.94 ± 0.78) µg (Figure 9c). The NPs derived from the three extracts at pH 8 shared similar trends of internalization, which were overall lower with respect to the same NPs produced at pH 7 in the two cell lines [64] (Figure 9b,d). The differences in uptake dynamics can be ascribed to the presence of differently shaped NPs, as each shape follows peculiar ways to interact with the cell plasma membrane [47,48]. For example, the high local curvature and irregular shape of *Carolea*-derived AgNPs, having circularity value of (0.55 ± 4), may explain their enhanced cellular internalization rate, compared to spherical NPs having a circularity value near 1. These results could support the idea that AgNPs from *Carolea* cultivar could be a preferable nano-tool for anticancer activity purpose respect to the AgNPs from *Leccino*.

Once observed that the stronger effect was induced by NPs obtained at pH 7 on HeLa cells, we investigated the effect on DNA using the Comet Assay. As showed in Figure 10a–c the AgNPs from *Leccino* and *Carolea* induced different genotoxicity on HeLa cells using the higher concentration (50 µg/mL) at 96 h. Figure 10b,c clearly showed the differences respect to control (Figure 10a): high level of DNA damage was found after AgNP achieved from *Leccino* and *Carolea* exposure, both in terms of tail length and DNA percentage in the head (Figure 10d,e), showing the typical comet morphology. AgNPs derived from *Carolea* induced a substantial DNA breaks that was evident in the tail length: (59 ± 5) µm after AgNPs-*Carolea* exposure and (35 ± 2) µm for NPs obtained from *Leccino* compared to control (15 ± 3) µm. The greater tail length corresponded to several DNA damage. Contrary, HeLa cells incubated with AgNPs from *Leccino* showed a more evident head percentage (50 ± 2 %) intensity respect to *Carolea* (25 ± 5%) because the DNA was mainly confined in the cellular nuclei. The chromatin remodelling induced an alteration of nuclear morphology that could be quantified in terms of nuclei circularity: in our case, the exposure to AgNPs provoked a circularity value reduction, indicating a pre-apoptotic condition [65]. Confocal acquisitions on Hela cells after the addition of AgNPs (50 µg/mL) up to 96 h showed an alteration of morphology: nuclei become less round and irregular (Figure 10g,h) in comparison with control cells (Figure 10f). Control HeLa cells presented nuclei circularity value of (0.83 ± 0.03). After green AgNPs treatment, a loss of circularity value was observed: the value changed from (0.69 ± 0.06) (AgNPs from *Leccino*) to 0.58 ± 0.05 (AgNPs from *Carolea*).

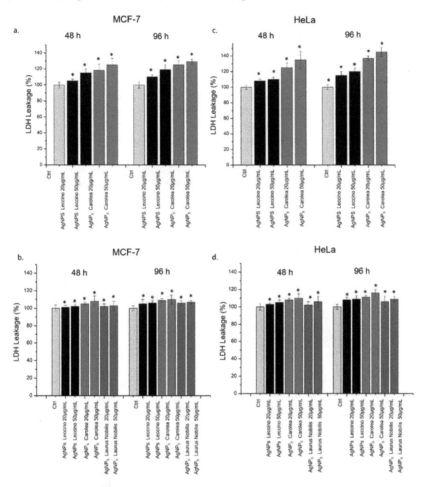

Figure 8. LDH assays on MCF-7 and HeLa cells after 48 h and 96 h exposure to two doses (20 µg/mL and 50 µg/mL) of AgNPs derived from *Oleaea Europeae* (*Leccino* and *Carolea*) at pH 7 (**a** and **c**) and from *Oleaea Europeae* (*Leccino* and *Carolea*) and *Laurus Nobilis* at pH 8 (**b** and **d**). Percent of LDH leakage of NP-treated cells are expressed relative to non-treated control cells. Positive controls (P) consisted in the treatment of cells with 0.9% Triton X-100 showing ca. 500 % LDH increase (data not shown). Data reported as mean ± SD from three independent experiments are considered statistically significant compared with control (*n* = 8) for *p* value < 0.05 (< 0.05 *).

We finally tested the antibacterial activities of these NPs. We thus incubated 50 µg/mL of AgNPs in water extracted from a well, containing typical bacteria of well water, e.g., coliforms and faecal coliforms (see the experimental section for details). We tested the total bacterial charge at 22 °C and 37 °C, to assess if the same trend obtained on human cells was maintained. In addition, the anti-coliforms activity was tested. The flora developed at 22 °C is autochthonous in water, while the one that developed at 37 °C can be considered an expression of the presence of bacteria hosted by warm-blooded animals [66]. As reported in EU directives [67], the water designated for human consumption requires absence of microorganisms and the quantitative determination of total colonies and coliforms to exclude the contamination, according to specific guidelines. The total bacteria growth at 37 °C was shown in Figure 11. The count of total bacteria charge at 22 °C and 37 °C, together with total coliforms and faecal coliforms quantification, was reported in Table 3. We observed a pronounced antibacterial activity, which was particularly remarked upon use of the spherical AgNPs, produced at pH 8. The incubation of AgNPs from *Carolea* and *Leccino*, obtained at pH 8, indeed inhibited the growth of bacteria present in the contaminated well water, having high bacterial titer (especially coliforms). These results were in line with those reported in literature [68–70], suggesting how the small spherical AgNPs (10–20 nm) showed enhanced antibacterial activity against several bacterial species, compared to the bigger and differently shaped AgNPs. The non-spherical AgNPs produced at pH 7 had lower antibacterial activity with respect to smaller spherical particles. This difference can be explained by the molecular mechanisms guiding the antibacterial effects of AgNPs. This was in fact ascribed to the ability of NPs to release Ag$^+$ ions from their surface, a process inducing damages at different level. In particular, they can induce membrane damage and cellular content leakage; in addition, AgNPs or Ag$^+$ can bind to the constitutive proteins of cell membrane, involved in transmembrane ATP generation [68]. Hence, the different antibacterial characteristics can be due to potential different rate of Ag$^+$ ions release from the particles surface.

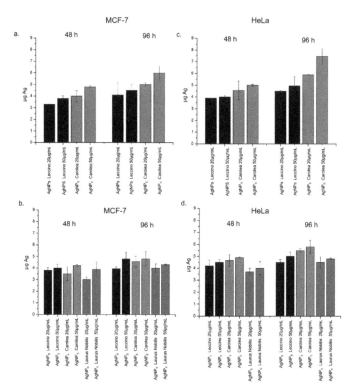

Figure 9. Green AgNPs accumulation in MCF-7 (**a,b**) and HeLa (**c,d**) cell lines exposed to 20 µg/mL and 50 µg/mL of AgNPs derived from *Leccino* and *Carolea* at pH 7 (**a,c**) and AgNPs from *Leccino, Carolea, Laurus Nobilis* at pH 8 (**b,d**) for 48 h and 96 h. Cells were then harvested, live cells were counted, and Ag content was measured in 360.000 cells (µg Ag). Data reported as mean ± SD from three independent experiments; statistical significance of exposed cells vs. control cells for *p* value < 0.05 (< 0.05 *).

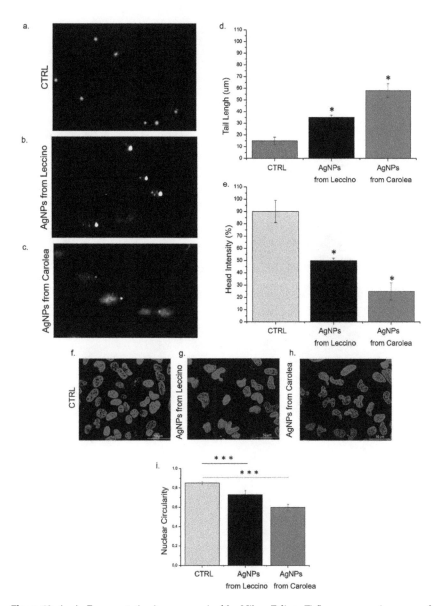

Figure 10. (**a–c**). Representative images acquired by Nikon Eclipse Ti fluorescence microscope of AgNPs (derived from *Leccino* and *Carolea*) effect on DNA damage in HeLa cells line. HeLa were treated with NPs (50 μg/mL) for 96 h. DNA damage was evaluated by (**d**) tail length (μm) and (**e**) head intensity (%). Values shown are means from 100 randomly selected comet images of each sample. As a positive control (P) cells were incubated with 500 μM H_2O_2 (data not shown). Data are reported as mean ±SD from three independent experiments; *$p < 0.05$ compared with control ($n = 3$). (**f–h**). Representative confocal images of HeLa nuclei: control (**f**) and after the exposure to 50 μg/mL of AgNPs from *Leccino* (**g**) and *Carolea* (**h**) after 96 h. (**i**) Histogram reported the mean values and their respective standard deviation of nuclear circularity. The statistical significance of results respect to control cells was evaluated by *t* test, and reported in histograms (*** $p < 0.005$.).

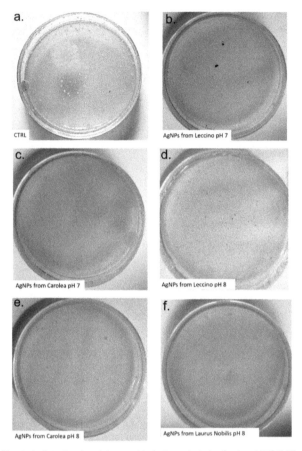

Figure 11. Total bacteria detection by plate count techniques to test colonies at 37 °C. 50 µg/mL of green AgNPs derived from *Leccino*, *Carolea* at pH 7 (**b**,**c**) and from *Leccino*, *Carolea* and *Laurus Nobilis* and pH 8 (**d**,**e**,**f**) were added in water before inoculation to test the antibacterial activity.

In order to understand if the toxicological profile of the green synthetized NPs were associated with their degradation, we have analysed the release of Ag^+ from AgNPs (50 µg/mL) in water and in acidic buffer that mimic the acidic lysosome environment (pH 4.5). Our results clearly showed a great ions release from *Carolea* obtained at pH 7 in acidic buffer (12.5 ± 1.9) µM after 96 h, whereas in water the release was few both from *Leccino* and *Carolea* AgNPs (Figure 12a): (1 ± 0.3) µM and (1.3± 0.4) µM respectively. The AgNPs obtained at pH 8 showed lower ionization trend even in acidic environment (Figure 12b) with similar results in the three AgNPs species. These results could explain the higher toxicity in cancer cells of AgNPs from *Carolea* obtained at pH 7 after cell internalization, whereas in well water the toxicity against bacteria could be verify due to the interaction of small amount of Ag^+ released in water. However, this kind of NPs obtained from plants extracts were more resistant to degradation with respect to the NPs obtained with standard chemical route as previously reported [12,39]. Probably the biomolecules adsorbed on NPs surface acted as protection and stabilization agents.

Table 3. Total colonies at 22 °C and 37 °C as UCF/ml and total coliforms and faecal coliforms as MPN/100ml after exposure to 50 µg/mL of green AgNPs.

Colonies and Coliforms	Contaminated Well Water	Well Water + 50µg/ml AgNPs from Leccino (pH 7)	Well Water + 50µg/ml AgNPs from Carolea (pH 7)	Well Water + 50 µg/ml AgNPs from Leccino (pH 8)	Well Water + 50 µg/ml AgNPs from Carolea (pH 8)	Well Water +50 µg/ml AgNPs from Laurus Nobilis (pH 8)
Total colonies (22 °C)(UCF/ml) 100 allowed	120	108	105	79	82	84
Total colonies (37 °C)(UCF/ml) 20 allowed	28	12	6	absent	1	3
Total Coliforms (MPN/100 mL) Absent allowed	19	11	9	absent	absent	absent
Total Faecal Coliforms (MPN/100 mL) Absent allowed	7	6	6	absent	absent	absent

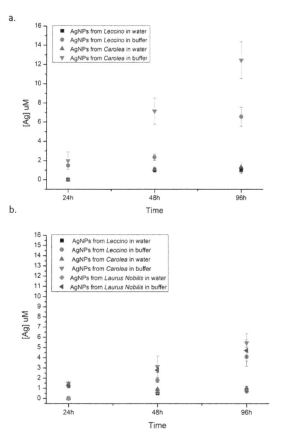

Figure 12. Effects of time and pH on silver ions release from AgNPs (50 µg/mL) derived from *Leccino* and *Carolea* obtained at pH 7 (**a**) and AgNPs from *Leccino*, *Carolea* and *Laurus Nobilis* at pH 8 (**b**). NPs degradation was evaluated both in buffer (pH 4.5) and in water (pH 7) up to 96 h. NPs degradation in neutral conditions was analysed also in culture medium, finding the same behaviour observed in water (data not shown).

4. Conclusions

In this work, a fully green method was reported for the production of AgNPs, completely free from both solvents and hazardous reagents. To do this, leaves extracts from two cultivars of *Olea Europaea* (*Leccino* and *Carolea*) were used, in order to produce AgNPs with a full control of physico-chemical characteristics. In particular, we demonstrated, for the first time, that using either *Leccino* or *Carolea* induces a clear difference in size and shape of AgNPs, in neutral environment; whereas at pH 8, using the same cultivar extracts, the NPs resulted smaller, with more regular morphology and monodispersed. Furthermore, the uptake dynamics and cytotoxicity of these AgNPs were studied in breast cancer cell lines, allowing to prove them as good antibacterial agents, with a further evidence of AgNPs' different behaviour to induce toxicity in cells and bacteria when obtained at pH 7 or pH 8. Moreover, another strength of this method consists in the theoretically unlimited source of reducing agent (i.e., the leaves extract obtained from agricultural processing waste), as well as its negligible environmental impact.

Author Contributions: V.D.M. conceived and designed the experimental activity. V.D.M. synthetized the nanomaterials. V.D.M. and L.R. performed the experiments in cell lines, E.L.-D. performed TEM images, C.I.

performed FTIR and UV-vis analysis, V.D.M. and M.L.D.G. performed the EDS analysis, V.D.M. and G.D.M. performed the microbiological analysis; V.D.M. wrote paper., V.D.M., L.R., R.R. edited and drafted the work. V.D.M.; L.R.; C.I., R.R. analyzed and explained the experimental data. V.D.M. and R.R. supervised work.

Funding: This research received no external funding.

Acknowledgments: VDM kindly acknowledges Programma Operativo Nazionale (PON) Ricerca e Innovazione 2014-2020 Asse I "Capitale Umano", Azione I.2, Avviso "A.I.M: Attraction and International Mobility" CUP F88D18000070001.

Conflicts of Interest: The authors declare no conflict of interest.

References

1. Marzo, J.L.; Jornet, J.M.; Pierobon, M. Nanotechnology Derived Nanotools in Biomedical Perspectives: An Update. *Curr. Drug Targets* **2019**, *20*, 800–807. [CrossRef]
2. Calderón-Jiménez, B.; Johnson, M.E.; Bustos, A.R.M.; Murphy, K.E.; Winchester, M.R.; Baudrit, J.R.V. Silver Nanoparticles: Technological Advances, Societal Impacts, and Metrological Challenges. *Front. Chem.* **2017**, *5*, 6. [CrossRef]
3. Bhattacharya, S.; Zhang, Q.; Carmichael, P.L.; Boekelheide, K.; Andersen, M.E. Toxicity Testing in the 21st Century: Defining New Risk Assessment Approaches Based on Perturbation of Intracellular Toxicity Pathways. *PLoS ONE* **2011**, *6*, e20887. [CrossRef] [PubMed]
4. Osório, I.; Igreja, R.; Franco, R.; Cortez, J. Incorporation of silver nanoparticles on textile materials by an aqueous procedure. *Mater. Lett.* **2012**, *75*, 200–203. [CrossRef]
5. Von Goetz, N.; Fabricius, L.; Glaus, R.; Weitbrecht, V.; Hungerbühler, K. Migration of silver from commercial plastic food containers and implications for consumer exposure assessment. *Food Addit. Contam. Part A* **2013**, *303*, 612–620. [CrossRef] [PubMed]
6. Chaudhry, Q.; Scotter, M.; Blackburn, J.; Ross, B.; Boxall, A.; Castle, L.; Aitken, R.; Watkins, R. Applications and implications of nanotechnologies for the food sector. *Food Addit. Contam. Part A* **2008**, *25*, 241–258. [CrossRef] [PubMed]
7. Karim, N.; Afroj, S.; Tan, S.; Novoselov, K.S.; Yeates, S.G. All Inkjet-Printed Graphene-Silver Composite Ink on Textiles for Highly Conductive Wearable Electronics Applications. *Sci. Rep.* **2019**, 8035. [CrossRef] [PubMed]
8. Burduşel, A.-C.; Gherasim, O.; Grumezescu, A.M.; Mogoantă, L.; Ficai, A.; Andronescu, E. Biomedical Applications of Silver Nanoparticles: An Up-to-Date Overview. *Nanomaterials* **2018**, *8*, 681. [CrossRef] [PubMed]
9. Campos, A.; Troc, N.; Cottancin, E.; Pellarin, M.; Weissker, H.; Lermé, J.; Kociak, M.; Hillenkamp, M. Plasmonic quantum size effects in silver nanoparticles are dominated by interfaces and local environments. *Nat. Phys.* **2019**, *15*, 275–280. [CrossRef]
10. Barnard, A.S. Size, Shape, Stability, and Color of Plasmonic Silver Nanoparticles. *J. Phys. Chem. C* **2014**, *118*, 9128–9136.
11. Tang, S.; Zheng, J. Antibacterial Activity of Silver Nanoparticles: Structural Effects. *Adv. Health Mater.* **2018**, *7*, 1–10. [CrossRef] [PubMed]
12. De Matteis, V.; Rizzello, L.; Di Bello, M.P.; Rinaldi, R. One-step synthesis, toxicity assessment and degradation in tumoral pH environment of $SiO_2@Ag$ core/shell nanoparticles. *J. Nanoparticle Res.* **2017**, *19*, 14. [CrossRef]
13. Iravani, S.; Korbekandi, H.; Mirmohammadi, S.V.; Zolfaghari, B. Mirmohammadi and BZ. Synthesis of silver nanoparticles: Chemical, physical and biological methods. *Res. Pharm. Sci.* **2014**, *9*, 385–406. [PubMed]
14. Farooqi, Z.H.; Khalid, R.; Begum, R.; Farooq, U.; Wu, Q.; Wu, W.; Ajmal, M.; Irfan, A.; Naseem, K. Facile synthesis of silver nanoparticles in a crosslinked polymeric system by in situ reduction method for catalytic reduction of 4-nitroaniline. *Environ. Technol.* **2019**, *40*, 2027–2036. [CrossRef] [PubMed]
15. Duan, H.; Wang, D.; Li, Y. Green chemistry for nanoparticle synthesis. *Chem. Soc. Rev.* **2015**, *44*, 5778–5792. [CrossRef] [PubMed]
16. Jime, V.M. The greener synthesis of nanoparticles. *Trends Biotechnol.* **2013**, *31*.
17. Das, R.K.; Pachapur, V.L.; Lonappan, L.; Naghdi, M.; Pulicharla, R.; Maiti, S.; Cledon, M.; Dalila, L.M.A.; Sarma, S.J.; Brar, S.K. Biological synthesis of metallic nanoparticles: Plants, animals and microbial aspects. *Nanotechnol. Environ. Eng.* **2017**, *2*, 1–21. [CrossRef]

18. Beach, E.S.; Cui, Z.; Anastas, P.T. Green Chemistry: A design framework for sustainability. *Energy Environ. Sci.* **2009**, *2*, 1038–1049. [CrossRef]

19. Ahmed, S.; Ahmad, M.; Swami, B.L.; Ikram, S. A review on plants extract mediated synthesis of silver nanoparticles for antimicrobial applications: A green expertise. *J. Adv. Res.* **2016**, *7*, 17–28. [CrossRef]

20. Khandel, P.; Yadaw, R.K.; Soni, D.K.; Kanwar, L.; Shahi, S.K. Biogenesis of metal nanoparticles and their pharmacological applications: Present status and application prospects. *J. Nanostructure Chem.* **2018**, *8*, 217–254. [CrossRef]

21. Gardea-Torresdey, J.L.; Gomez, E.; Peralta-Videa, J.R.; Parsons, J.G.; Troiani, H.; Jose-Yacaman, M. Alfalfa Sprouts: A Natural Source for the Synthesis of Silver Nanoparticles. *Langmuir* **2003**, *19*, 1357–1361. [CrossRef]

22. Khalil, M.M.; Ismail, E.H.; El-Baghdady, K.Z.; Mohamed, D. Green synthesis of silver nanoparticles using olive leaf extract and its antibacterial activity. *Arab. J. Chem.* **2014**, *7*, 1131–1139. [CrossRef]

23. Prabha, S.; Lahtinen, M.; Sillanpää, M. Green synthesis and characterizations of silver and gold nanoparticles using leaf extract of Rosa rugosa. *Colloids Surf. A Physicochem. Eng. Asp.* **2010**, *364*, 34–41.

24. Bose, D.; Chatterjee, S. Biogenic synthesis of silver nanoparticles using guava (*Psidium guajava*) leaf extract and its antibacterial activity against Pseudomonas aeruginosa. *Appl. Nanosci.* **2016**, *6*, 895–901. [CrossRef]

25. Song, J.Y.; Kim, B.S. Rapid biological synthesis of silver nanoparticles using plant leaf extracts. *Bioprocess Biosyst. Eng.* **2009**, 79–84.

26. Yılmaz, M.; Turkdemir, H.; Kiliç, M.A.; Bayram, E.; Cicek, A.; Mete, A.; Ulug, B.; Yılmaz, M. Biosynthesis of silver nanoparticles using leaves of Stevia rebaudiana. *Mater. Chem. Phys.* **2011**, *130*, 1195–1202. [CrossRef]

27. Roopan, S.M.; Rohit; Madhumitha, G.; Rahuman, A.; Kamaraj, C.; Bharathi, A.; Surendra, T.; Kamaraj, D. Low-cost and eco-friendly phyto-synthesis of silver nanoparticles using *Cocos nucifera* coir extract and its larvicidal activity. *Ind. Crop. Prod.* **2013**, *43*, 631–635. [CrossRef]

28. Dwivedi, A.D.; Gopal, K. Biosynthesis of silver and gold nanoparticles using Chenopodium album leaf extract. *Colloids Surf. A Physicochem. Eng. Asp.* **2010**, *369*, 27–33. [CrossRef]

29. Raut Rajesh, W.; Lakkakula Jaya, R.; Kolekar Niranjan, S.; Mendhulkar Vijay, D.; Kashid Sahebrao, B. Phytosynthesis of Silver Nanoparticle Using Gliricidia sepium (Jacq.). *Curr. Nanosci.* **2009**, *5*, 117–122.

30. Singhal, G.; Bhavesh, R.; Kasariya, K.; Sharma, A.R.; Singh, R.P. Biosynthesis of silver nanoparticles using Ocimum sanctum (Tulsi) leaf extract and screening its antimicrobial activity. *J. Nanoparticle Res.* **2011**, *13*, 2981–2988. [CrossRef]

31. Jha, A.K.; Prasad, K. Green Synthesis of Silver Nanoparticles Using Cycas Leaf Green Synthesis of Silver Nanoparticles Using Cycas Leaf. *Int. J. Green Nanotechnol. Phys. Chem.* **2010**, 111–117.

32. Parvaiz, M.; Hussain, K.; Shoaib, M.; William, G. A Review: Therapeutic Significance of Olive Olea europaea L. (Oleaceae Family). *Glob. J. Pharmacol.* **2013**, *7*, 333–336.

33. Khan, Y.; Panchal, S.; Vyas, N.; Butani, A.; Kumar, V. Olea europaea: A Phyto-Pharmacological Review. *Pharmacogn. Rev.* **2007**, *1*, 112–116.

34. Bianco, A.; Uccella, N. Biophenolic components of olives. *Food Res. Int.* **2000**, *33*, 475–485. [CrossRef]

35. Farag, R.S.; Basuny, A.M. Safety evaluation of olive phenolic compounds as natural antioxidants. *Int. J. Food Sci. Nutr.* **2009**, *54*, 159–174. [CrossRef] [PubMed]

36. Ghanbari, R.; Anwar, F.; Alkharfy, K.M.; Gilani, A. Valuable Nutrients and Functional Bioactives in Different Parts of Olive (*Olea europaea* L.)—A Review. *Int. J. Mol. Sci.* **2012**, 3291–3340. [CrossRef] [PubMed]

37. Şahin, S.; Bilgin, M. Olive tree (*Olea europaea* L.) leaf as a waste by-product of table olive and olive oil industry: A review. *Sci. Food Agric.* **2018**, *98*, 1271–1279. [CrossRef]

38. Rice, S.B.; Chan, C.; Brown, S.C.; Eschbach, P.; Han, L.; Ensor, D.S.; Stefaniak, A.B.; Bonevich, J.; Vladár, A.E.; Walker, A.R.H.; et al. Particle size distributions by transmission electron microscopy: An interlaboratory comparison case study. *Metrologia* **2013**, *50*, 663–678. [CrossRef]

39. De Matteis, V.; Malvindi, M.A.; Galeone, A.; Brunetti, V.; De Luca, E.; Kote, S.; Kshirsagar, P.; Sabella, S.; Bardi, G.; Pompa, P.P. Negligible particle-specific toxicity mechanism of silver nanoparticles: The role of Ag + ion release in the cytosol. Nanomedicine Nanotechnology. *Biol. Med.* **2015**, *11*, 731–739. [CrossRef]

40. Richard, F.; Micheal, G.M.; Mera, D. *Water Waste and Health Management in Hot Climates*; Ohn Wiley and Sons Ltd.: London, UK, 1979.

41. Talaro, K.P.; Chess, B. *Foundations in Microbiology*, 2nd ed.; WCB/McGraw Hill: Boston, MA, USA, 1996.

42. Bartram, J.; Ballance, R. (Eds.) *Water Quality Monitoring—A Practical Guide to the Design and Implementation of Freshwater Quality Studies and Monitoring Programmes*; United Nations Environment Programme and the World Health Organization: Geneva, Switzerland, 1996.

43. Hajna, A.A.; Perry, C.A.; Sc, D. Comparative Study of Presumptive and the Coliform Group and for Fecal Streptococci. *Am. J. Public Health Nations Health* **1943**, *33*, 550–556. [CrossRef]

44. Rand, M.C.; Greenberg, A.E.; Taras, M.J. *Standard Methods for the Examination of Water and Wastewater*, 14th ed.; American Public Health Association: Washington, DC, USA, 1976; ISBN 0875530788.

45. Fishbein, M.; Surkiewicz, B.F. Comparison of the Recovery of Escherichia coli from Frozen Foods and Nutmeats by Confirmatory Incubation in EC Medium at 44.5 and 45.5 C. *Appl. Microbiol.* **1964**, *12*, 127–131. [PubMed]

46. Singh, P.; Kim, Y.-J.; Zhang, D.; Yang, D.-C. Biological Synthesis of Nanoparticles from Plants and Microorganisms. *Trends Biotechnol.* **2016**, *34*, 588–599. [CrossRef] [PubMed]

47. Altemimi, A.B. A Study of the Protective Properties of Iraqi Olive Leaves against Oxidation and Pathogenic Bacteria in Food Applications. *Antioxidants* **2017**, *6*, 34. [CrossRef] [PubMed]

48. Baldoni, L.; Tosti, N.; Ricciolini, C.; Belaj, A.; Arcioni, S.; Pannelli, G.; Germana, M.A.; Mulas, M.; Porceddu, A. Genetic Structure of Wild and Cultivated Olives in the Central Mediterranean Basin. *Ann. Bot.* **2006**, *98*, 935–942. [CrossRef]

49. Fabbri, A.; Galaverna, G.; Ganino, T. Polyphenol composition of olive leaves with regard to cultivar, time of collection and shoot type. *Acta Hortic.* **2008**. [CrossRef]

50. Talhaoui, N.; Taamalli, A.; Gómez-Caravaca, A.M.; Fernández-Gutiérrez, A.; Segura-Carretero, A. Phenolic compounds in olive leaves: Analytical determination, biotic and abiotic influence, and health benefits. *Food Res. Int.* **2015**, *77*, 92–108. [CrossRef]

51. Pirtarighat, S.; Ghannadnia, M.; Baghshahi, S. Green synthesis of silver nanoparticles using the plant extract of Salvia spinosa grown in vitro and their antibacterial activity assessment. *J. Nanostructure Chem.* **2019**, *9*, 1–9. [CrossRef]

52. Kumar, P.; Selvi, S.S.; Govindaraju, M. Seaweed-mediated biosynthesis of silver nanoparticles using Gracilaria corticata for its antifungal activity against *Candida* spp. *Appl. Nanosci.* **2013**, *3*, 495–500. [CrossRef]

53. Casals, E.; Pfaller, T.; Duschl, A.; Oostingh, G.J.; Puntes, V. Time Evolution of the Nanoparticle Protein Corona. *ACS Nano* **2010**, *4*, 3623–3632. [CrossRef]

54. Qi, C.; Musetti, S.; Fu, L.-H.; Zhu, Y.-J.; Huang, L. Biomolecule-assisted green synthesis of nanostructured calcium phosphates and their biomedical applications. *Chem. Soc. Rev.* **2019**, *48*, 2698–2737. [CrossRef]

55. Cruz-Matías, I.; Ayala, D.; Hiller, D.; Gutsch, S.; Zacharias, M.; Estradé, S.; Peiró, F. Sphericity and roundness computation for particles using the extreme vertices model. *J. Comput. Sci.* **2019**, *30*, 28–40. [CrossRef]

56. Singh, C.; Kumar, J.; Kumar, P.; Chauhan, B.S.; Tiwari, K.N.; Mishra, S.K.; Srikrishna, S.; Saini, R.; Nath, G.; Singh, J. Green synthesis of silver nanoparticles using aqueous leaf extract of Premna integrifolia (L.) rich in polyphenols and evaluation of their antioxidant, antibacterial and cytotoxic activity. *Biotechnol. Biotechnol. Equip.* **2019**, *33*, 359–371. [CrossRef]

57. Prakash, P.; Gnanaprakasam, P.; Emmanuel, R.; Arokiyaraj, S.; Saravanan, M. Green synthesis of silver nanoparticles from leaf extract of Mimusops elengi, Linn. for enhanced antibacterial activity against multi drug resistant clinical isolates. *Colloids Surf. B Biointerfaces* **2013**, *108*, 255–259. [CrossRef]

58. Van der Zande, M.; Undas, A.K.; Kramer, E.; Monopoli, M.P.; Peters, R.J.; Garry, D.; Antunes Fernandes, E.C.; Hendriksen, P.J.; Marvin, H.J.; Peijnenburg, A.A.; et al. Different responses of Caco-2 and MCF-7 cells to silver nanoparticles are based on highly similar mechanisms of action. *Nanotoxicology* **2016**, *10*, 1431–1441. [CrossRef] [PubMed]

59. Butler, M. *Animal Cell Culture and Technology*, 2nd ed.; Taylor & Francis: London, UK, 2003; ISBN 9780203427835.

60. Lu, B.; Hu, M.; Liu, K.; Peng, J. Cytotoxicity of berberine on human cervical carcinoma HeLa cells through mitochondria, death receptor and MAPK pathways, and in-silico drug-target prediction. *Toxicol. In Vitro* **2010**, *24*, 1482–1490. [CrossRef]

61. Moradhaseli, S.; Mirakabadi, A.Z.; Sarzaeem, A.; Kamalzadeh, M.; Hosseini, R.H. Cytotoxicity of ICD-85 NPs on Human Cervical Carcinoma HeLa Cells through Caspase-8 Mediated Pathway. *Iran J. Pharm. Res.* **2013**, *12*, 155–163. [PubMed]

62. Bannunah, A.M.; Vllasaliu, D.; Lord, J.; Stolnik, S.S. Mechanisms of Nanoparticle Internalization and Transport across an Intestinal Epithelial Cell Model: Effect of Size and Surface Charge. *Mol. Pharm.* **2014**, *11*, 4363–4373. [CrossRef]

63. Sukhanova, A.; Bozrova, S.; Sokolov, P.; Berestovoy, M.; Karaulov, A.; Nabiev, I. Dependence of Nanoparticle Toxicity on Their Physical and Chemical Properties. *Nanoscale Res. Lett.* **2018**, *13*, 44. [CrossRef]

64. Nambara, K.; Niikura, K.; Mitomo, H.; Ninomiya, T.; Takeuchi, C.; Wei, J.; Matsuo, Y.; Ijiro, K. Reverse Size Dependences of the Cellular Uptake of Triangular and Spherical Gold Nanoparticles. *Langmuir* **2016**, *32*, 12559–12567. [CrossRef]

65. Helmy, I.M.; Azim, A.M.A. Efficacy of ImageJ in the assessment of apoptosis. *Diagn. Pathol.* **2012**, *7*, 15. [CrossRef]

66. João, P.S. Cabral. Water Microbiology. Bacterial Pathogens and Water. *Int. J. Environ. Res. Public Health* **2010**, *7*, 3657–3703.

67. Council of the European Union. Council Directive 80/778/EEC of 15 July 1980 relating to the quality of water intended for human consumption. *Off. J.Eur. Communities* **1980**.

68. Rizzello, L.; Pompa, P.P. Nanosilver-based antibacterial drugs and devices: Mechanisms, methodological drawbacks, and guidelines. *Chem. Soc. Rev.* **2014**, *43*, 1501–1518. [CrossRef] [PubMed]

69. Raza, M.A.; Kanwal, Z.; Rauf, A.; Sabri, A.N.; Riaz, S.; Naseem, S. Size- and Shape-Dependent Antibacterial Studies of Silver Nanoparticles Synthesized by Wet Chemical Routes. *Nanomaterials* **2016**, *6*, 74. [CrossRef] [PubMed]

70. Zhang, J.; Zhou, P.; Liu, J.; Yu, J. New understanding of the difference of photocatalytic activity among anatase, rutile and brookite TiO$_2$. *Phys. Chem. Chem. Phys.* **2014**, *16*, 20382–20386. [CrossRef]

nanomaterials

MDPI

Article

Strong Biomimetic Immobilization of Pt-Particle Catalyst on ABS Substrate Using Polydopamine and Its Application for Contact-Lens Cleaning with H_2O_2

Yuji Ohkubo [1,*], Tomonori Aoki [1], Daisuke Kaibara [1], Satoshi Seino [1], Osamu Mori [2], Rie Sasaki [2], Katsuyoshi Endo [1] and Kazuya Yamamura [1]

[1] Graduate School of Engineering, Osaka University, Suita, Osaka 565-0871, Japan; t-aoki@div1.upst.eng.osaka-u.ac.jp (T.A.); d-kaibara@div1.upst.eng.osaka-u.ac.jp (D.K.); seino@mit.eng.osaka-u.ac.jp (S.S.); endo@upst.eng.osaka-u.ac.jp (K.E.); yamamura@prec.eng.osaka-u.ac.jp (K.Y.)

[2] Menicon Co., Ltd., Kasugai, Aichi 487-0032, Japan; o-mori@menicon.co.jp (O.M.); r-sasaki@menicon.co.jp (R.S.)

* Correspondence: okubo@upst.eng.osaka-u.ac.jp; Tel.: +81-6-6879-7294

Received: 13 November 2019; Accepted: 2 January 2020; Published: 7 January 2020

Abstract: Polydopamine (PDA)—a known adhesive coating material—was used herein to strongly immobilize a Pt-particle catalyst on an acrylonitrile–butadiene–styrene copolymer (ABS) substrate. Previous studies have shown that the poor adhesion between Pt particles and ABS surfaces is a considerable problem, leading to low catalytic durability for H_2O_2 decomposition during contact-lens cleaning. First, the ABS substrate was coated with PDA, and the PDA film was evaluated by X-ray photoelectron spectroscopy. Second, Pt particles were immobilized on the PDA-coated ABS substrate (ABS-PDA) using the electron-beam irradiation reduction method. The Pt particles immobilized on ABS-PDA (Pt/ABS-PDA) were observed using a scanning electron microscope. The Pt-loading weight was measured by inductively coupled plasma atomic emission spectroscopy. Third, the catalytic activity of the Pt/ABS-PDA was evaluated as the residual H_2O_2 concentration after immersing it in a 35,000-ppm H_2O_2 solution (the target value was less than 100 ppm). The catalytic durability was evaluated as the residual H_2O_2 concentration after repeated use. The PDA coating drastically improved both the catalytic activity and durability because of the high Pt-loading weight and strong adhesion among Pt particles, PDA, and the ABS substrate. Plasma treatment prior to PDA coating further improved the catalytic durability.

Keywords: catalytic durability; polydopamine (PDA); strong adhesion; supported catalyst; H_2O_2 decomposition

1. Introduction

Mussels can strongly adhere to several surfaces using their body fluid, regardless of whether the surfaces are dry or wet [1–3]. This phenomenon of adhesion to wet surfaces is unusual in the adhesives industry. The body fluid of mussels was examined, and it was found that 3,4-dihydroxy-L-phenylalanine (DOPA) and lysine-enriched proteins contributed to its strong adhesion [4–6]. As a result, polydopamine (PDA) has attracted the attention of many scientists because its structure is similar to that of DOPA. Surface chemical composition affects adhesion properties. Both DOPA and PDA have hydroxyl and amino groups and a benzene ring, so they can interact with various materials such as metal oxides, metals, and polymers, not only through van der Waals forces, but also via hydrogen or coordinate bonding or π–π stack interaction. Since PDA has been reported as a novel adhesive coating for several materials such as Pt, Cu, TiO_2, SiO_2, and Al_2O_3 [7], it has received even more attention. For example, there are reports of PDA being utilized at sites for growing hydroxyapatite (HAp) [8]; PDA-coated polystyrene (PS) particles have been used to prepare a structural-color-controlled ink [9]; a PDA-grafted hydrogel has been demonstrated to adhere to a wet mucous membrane [10]; and a PDA coating has been used as a seed layer for a TiO_2–polytetrafluoroethylene (PTFE) nanocomposite coating [11]. Recently, PDA was utilized to improve the adhesion between a polydimethylsiloxane (PDMS) nanosheet and a living body [12]. This research demonstrated that a PDA coating combines both high adhesion and biocompatibility.

The number of contact-lens wearers has increased in recent years and is currently estimated to be approximately 140 million [13]. There are two types of contact-lens wearers, namely, those who use one-day disposable lenses and those who prefer repeatable-use (e.g., monthly) contact lenses. Although one-day disposable contact lenses do not need cleaning and disinfecting, their high cost is a serious disadvantage. In contrast, repeatable-use contact lenses are less expensive in the long term, but it is essential to clean and disinfect them properly once per day to prevent eye infections. There are three types of methods for cleaning and disinfecting contact lenses: the first method is boil cleaning, the second is H_2O_2 cleaning, and the third is cleaning with a multipurpose solution (MPS). MPS cleaning has the advantage of being simple because only one solution is required for cleaning, disinfection, and storage. However, contact-lens wearers applying the MPS cleaning method are likely to have eye problems if they do not clean their contact lenses carefully enough. Thus, the number of contact-lens wearers using MPS cleaning has gradually decreased since 2009, while that of users applying the H_2O_2 cleaning method has increased [14]. The reason for this is that eye problems are unlikely to occur in the case of H_2O_2 cleaning because a 35,000-ppm H_2O_2 solution exhibits a high disinfecting performance. However, when H_2O_2 cleaning is applied, there is a risk of the eyes becoming bloodshot or painful—even of blindness—if the 35,000-ppm H_2O_2 solution enters the eyes without being decomposed to a concentration of 100 ppm [15]. Electroless platinum (Pt) plating has been performed to give an acrylonitrile–butadiene–styrene copolymer (ABS) substrate catalytic performance for accelerating the H_2O_2 decomposition process (Figure 1a). Pt is very expensive, so there is a strong need to decrease the amount of Pt used in contact-lens cleaners. We have suggested replacing the Pt film with Pt particles, which results in a drastic decrease in the amount of Pt required (Figure 1b) [16]. However, some problems remain regarding the catalytic durability, although we pretreated the surface by etching, electric charge control, or both [17]. As mentioned above, PDA has the potential to adhere metal particles to resin substrates. In this study, we used a PDA coating as a pre-treatment to strongly and safely immobilize Pt particles on an ABS substrate to improve the catalytic durability of the material (Figure 1c). The effects of the PDA coating on the properties of the ABS surface and the catalytic activity, Pt-loading weight, and catalytic durability of the system were investigated.

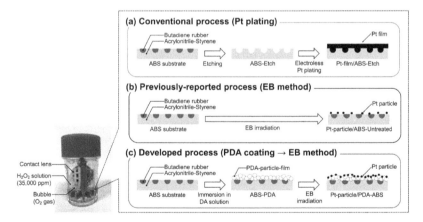

Figure 1. Schematic of the processes for preparing (**a**) a Pt-film/ABS sample by the electroless plating method, (**b**) a Pt-particle/ABS sample by EBIRM without pre-treatment, and (**c**) a Pt-particle/ABS-PDA sample by EBIRM adding a PDA coating as a pre-treatment.

2. Results and Discussion

2.1. External Appearance

The changes in the external appearances of the ABS samples were monitored to confirm that PDA coating had occurred. Figure 2 presents photographs of ABS samples with a masking using polyimide (PI) tape before and after PDA coating. The color of the PDA-coated area changed from cream to gray, and this gray color remained after ultrasonic cleaning, which confirmed that the PDA film was strongly attached to the ABS surface.

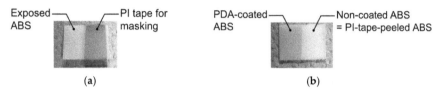

(a)   (b)

Figure 2. Photographs of ABS samples with a masking using polyimide (PI) tape (**a**) before PDA coating and (**b**) after PDA coating for 24 h.

2.2. Confirmation of the Formation of a PDA Film by X-Ray Photoelectron Spectroscopy (XPS)

To examine the effects of the PDA coating on the chemical composition of ABS substrates, we analyzed pretreated ABS surfaces that did not contain Pt particles by XPS. Figure 3 presents the XPS spectra of the surface of an ABS substrate before and after PDA coating at different immersion times in a dopamine (DA) solution. When the ABS substrates were immersed in a DA solution, the intensities of the peaks assigned to C–H and C–C (285 eV) decreased, whereas those of the peaks assigned to C–N and C–O (286.5 eV) increased, as illustrated in Figure 3a. The intensities of the signals in the N1s-XPS spectra did not increase because the ABS substrate originally contained a C≡N bond, as illustrated in Figure 3b. In addition, when the ABS substrates were immersed in a DA solution, the intensities of the signals in the O1s-XPS spectra also increased, as illustrated in Figure 3c. The calculated N/C atomic ratio is presented in Figure 3d, where it can be seen that it increased with increasing immersion time in the DA solution. When the ABS substrates were immersed in a DA solution for 3 and 24 h, the N/C ratios were 0.129 and 0.120, respectively. These ratios were roughly consistent with the theoretical value of N/C = 0.125. These results indicate that the ABS surfaces were coated with a PDA film.

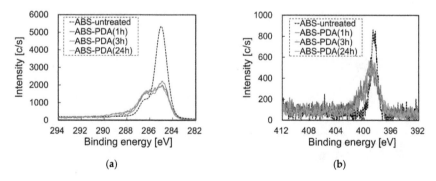

(a)   (b)

Figure 3. *Cont.*

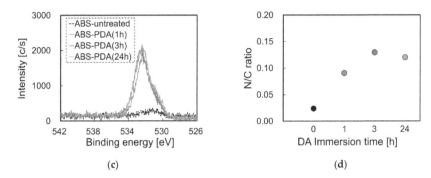

(c) (d)

Figure 3. X-ray photoelectron spectroscopy (XPS) spectra of an ABS surface before and after PDA coating at different immersion times in a DA solution: (**a**) C1s-XPS, (**b**) N1s-XPS, (**c**) O1s-XPS, and (**d**) the N/C ratio.

2.3. Observation of Pt Particles by SEM

The Pt particles immobilized on the ABS surfaces were analyzed by SEM to confirm the deposition of Pt particles. Figure 4 presents SEM micrographs of the surfaces of Pt/ABS samples with or without PDA coating at a low magnification. The small white spots in the images are Pt particles. The high dispersibility of Pt particles was confirmed for all samples. The number of Pt particles clearly increased upon PDA coating. Moreover, the number of Pt particles increased with increasing DA immersion time. Some holes (with diameters of 50–200 nm) were observed in the Pt/ABS-untreated sample (Figure 4a), but they were absent in the Pt/ABS-PDA samples (Figure 4b–d). This indicates that the holes originally present on the ABS surface were successfully coated with the PDA film.

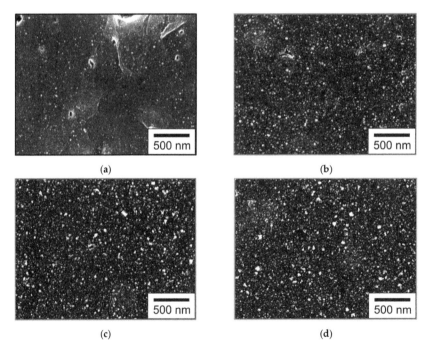

Figure 4. Scanning electron microscope (SEM) images of the surface of an ABS substrate before and after PDA coating and Pt immobilization: (**a**) Pt/ABS-untreated, (**b**) Pt/ABS-PDA(1h), (**c**) Pt/ABS-PDA(3h), and (**d**) Pt/ABS-PDA(24h).

2.4. Pt-Loading Weight Determined by Inductively Coupled Plasma Atomic Emission Spectrometry (ICP-AES)

The effects of the PDA coating on the Pt-loading weights of the Pt/ABS samples were also examined. Figure 5 illustrates the Pt-loading weights of substrates with or without PDA coating. It can be seen that the values were higher for the Pt/ABS-PDA samples than for the Pt/ABS-untreated ones, thus indicating that PDA coating increased the Pt-loading weight. In addition, the Pt-loading weights for the Pt/ABS-PDA samples increased with increasing DA immersion time. These results of Pt-loading weight are consistent with the SEM images illustrated in Figure 4. When an ABS substrate is coated with a PDA film, the number of sites for immobilizing Pt particles also increases, resulting in an increased Pt-loading weight. The Pt-loading weight of the Pt/ABS-PDA(24h) material (11.2 µg/substrate) was approximately twice that of the Pt/ABS-untreated sample, but at least 130 times lower than that of an ABS substrate coated with an electroless-plated Pt film (Pt-film/ABS) (1500 µg/substrate), which had been studied earlier [16].

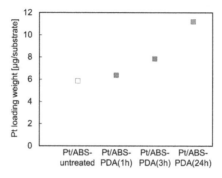

Figure 5. Pt-loading weights of Pt/ABS samples with or without the PDA coating.

2.5. Catalytic Activity for H_2O_2 Decomposition

To evaluate the catalytic activity of the materials for H_2O_2 decomposition, the residual H_2O_2 concentration was measured in the system after immersing Pt/ABS samples with or without the PDA coating in a 35,000-ppm H_2O_2 solution for 360 min. Briefly, the lower the residual H_2O_2 concentration, the higher the catalytic activity. Figure 6 illustrates the catalytic activity of Pt/ABS samples with or without the PDA coating. The untreated ABS sample without Pt particles did not decompose H_2O_2 within 360 min, whereas all the samples with Pt immobilized on the ABS substrate significantly decreased the residual H_2O_2 concentration from 35,000 to less than 400 ppm. Moreover, the residual H_2O_2 concentrations for the Pt/ABS-PDA samples became lower than that for the Pt/ABS-untreated sample. It is clear that the PDA coating improved the catalytic activity for H_2O_2 decomposition, as well as that the residual H_2O_2 concentration decreased with increasing DA immersion time. This result is consistent with those obtained in the SEM (Figure 4) and ICP-AES (Figure 5) studies. In summary, an increase in the Pt-loading weight contributed to the improvement of the catalytic activity of the resulting material. The target value for the residual H_2O_2 concentration (i.e., <100 ppm) was successfully reached after 3 h of immersion in DA.

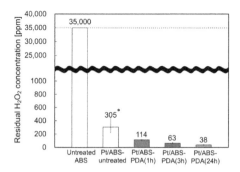

Figure 6. Catalytic activity of Pt/ABS samples with or without PDA coating: residual H_2O_2 concentration after immersion for 360 min. * The data for the Pt/ABS-untreated are the same as in our previous report [17].

2.6. Catalytic Durability during H_2O_2 Decomposition

To examine the effect of the PDA coating on the catalytic durability of the system, the relation between the number of repeated uses and the residual H_2O_2 concentration was examined. Figure 7 illustrates the catalytic durability of Pt/ABS samples with or without the PDA coating. The residual H_2O_2 concentration for the Pt/ABS-untreated sample increased significantly with increasing usage, thus resulting in low catalytic durability. In the case of the Pt/ABS-PDA material, the residual H_2O_2 concentrations measured after using the samples 10 times were 935, 194, and 54 ppm for Pt/ABS-PDA(1h), Pt/ABS-PDA(3h), and Pt/ABS-PDA(24h), respectively; this indicates that the residual H_2O_2 concentrations decreased with increasing DA immersion time. This result demonstrates that the PDA coating effectively improved the catalytic durability of the material for H_2O_2 decomposition. Moreover, the residual H_2O_2 concentration for Pt/ABS-PDA(24h) was still below 100 ppm after the catalyst had been used 10 times.

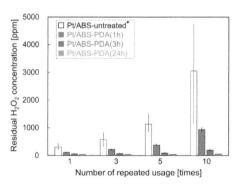

Figure 7. Catalytic durability of Pt/ABS samples with or without PDA coating: the relationship between the number of usage cycles and the residual H_2O_2 concentration. * The data for the Pt/ABS-untreated are the same as in our previous report [17].

2.7. Effect of Plasma Treatment on Catalytic Durability

Although Pt/ABS-PDA(24h) exhibited high catalytic durability, the residual H_2O_2 concentration mildly increased from 38 to 51 ppm with the number of usage cycles. Thus, to further improve the catalytic durability, plasma treatment was applied before the PDA coating. Figure 8 illustrates the catalytic durability of Pt/ABS-PDA(24h) samples with or without plasma treatment. Please note that the range of the vertical axis is from 0 to 100 ppm. The residual H_2O_2 concentration of the

Pt/ABS-plasma-PDA(24h) sample after it had been used 10 times was 35 ppm, which indicates that the residual H_2O_2 concentration barely changed and that the plasma treatment before the PDA coating further improved the catalytic durability of the material. The catalytic durability of the Pt/ABS-plasma-PDA(24h) sample was sufficient for use in practical applications.

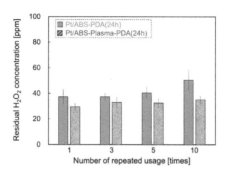

Figure 8. Catalytic durability of Pt/ABS-PDA(24h) samples with or without plasma treatment before PDA coating: the relationship between the number of usage cycles and the residual H_2O_2 concentration.

To investigate the effect of the plasma treatment on the morphology of the PDA coating, the surfaces of the ABS-PDA(24h) and ABS-plasma-PDA(24h) samples were observed and compared by SEM. Figure 9 presents SEM images of the surface of an ABS substrate before and after PDA coating, with or without plasma treatment. Although many cracks were observed on the as-received ABS surface, no cracks were observed on the PDA-coated one, regardless of the plasma treatment. This result indicates that the ABS surfaces were coated with a PDA film. A comparison between the ABS-plasma-PDA(24h) and ABS-PDA(24h) samples indicates that the surface roughness of the plasma-treated material was larger than that of the untreated one. The Pt-loading weights of Pt/ABS-plasma-PDA(24h) and Pt/ABS-PDA(24h) were 24.0 and 11.2 μg, respectively. As can be seen, the plasma treatment induced an increase in the Pt-loading weight, which is one of the reasons for the improved catalytic activity and durability. In addition, XPS measurements were also carried out for the ABS samples after Pt immobilization to investigate the effect of plasma treatment on the Pt state, such as Pt(0), Pt(II), and Pt(IV). Figure 10 presents the Pt4f-XPS spectra of the Pt/ABS samples with or without plasma treatment and with or without PDA coating. Two broad peaks indexed to Pt4f$_{7/2}$ and Pt4f$_{5/2}$ were observed for all the Pt4f-XPS spectra. The Pt4f$_{7/2}$ peak was resolved into three peaks indexed to Pt(0), Pt(II), and Pt(IV) at ca. 71.6, ca. 72.6, and 74.4 eV, respectively [18,19]. The ratios of Pt(0), Pt(II), and Pt(IV) are listed in Table 1 according to the references. Although the main state was Pt(0), minor states of Pt(II) and Pt(IV) also were detected. These results indicate two likelihoods: the first is a coordinate bond of Pt ions, and the second is an oxidation of Pt. In summary, it is possible that, firstly, Pt in an ionic state interacted with C–O groups on the electron-beam-irradiated ABS substrate in water or NH$_2$ and N–H groups in the PDA film; secondly, a coordinate bond was formed; thirdly, Pt clusters grew to form Pt particles; and finally, the surface of the Pt particles were oxidized to change the surface to Pt oxide (PtO or PtO$_2$). The PDA coating increased the ratio of the Pt metallic state to >80%, while plasma treatment mostly did not affect the ratio of the Pt metallic state.

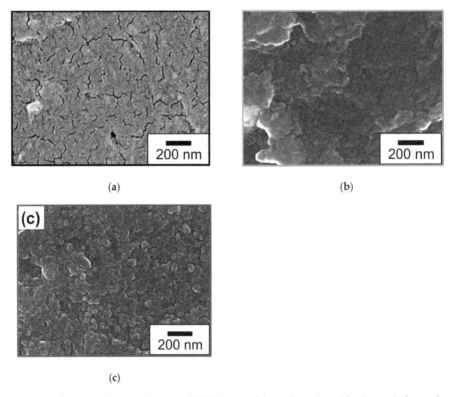

Figure 9. Scanning electron microscope (SEM) images of the surface of an ABS substrate before and after PDA coating with or without plasma treatment: (**a**) ABS-untreated, (**b**) ABS-PDA(24h), and (**c**) Pt/ABS-plasma-PDA(24h).

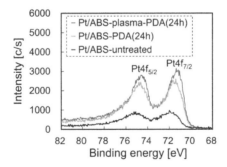

Figure 10. Pt4f-XPS spectra of the Pt/ABS samples with or without plasma treatment and with or without PDA coating.

Table 1. Ratios of Pt(0), Pt(II), and Pt(IV) calculated from peak resolution of the Pt4f$_{7/2}$ in Figure 10.

Sample ID	Pt(0)	Pt(II)	Pt(IV)
Pt/ABS-untreated	69.1	16.4	14.5
Pt/ABS-PDA(24h)	81.1	11.0	7.9
Pt/ABS-plasma-PDA(24h)	86.7	7.1	6.2

3. Materials and Methods

3.1. PDA Coating and Immobilization of Pt Particles on ABS

Pt particles were immobilized on an ABS substrate by the electron-beam-irradiation reduction method (EBIRM) because of the advantages of this procedure, which include a low processing temperature, highly uniform deposition, and a high throughput [20]. Although there are many methods for preparing and immobilizing particles (e.g., sonolytic [21–24], polyol [25–27], and impregnation [28–30] methods), they all have disadvantages such as high processing temperatures, nonuniform deposition of the metal particles, and low throughput. Therefore, EBIRM was selected in this study. The mechanism of metal ion reduction by a radiochemical approach has been described in previous reports [31,32], and the methods applied for washing the ABS substrate and depositing the Pt particles on its surface were the same here as those reported previously [16,17]. The differences between this report and previous ones are the plasma pre-treatment and the use of a PDA coating to improve the catalytic durability during H_2O_2 decomposition.

A commercially available ABS sheet (thickness t = 1 mm, 2-9229-01, AS-ONE, Nishi-ku, Osaka, Japan) was cut into substrates with a size of 20 × 15 mm^2. The ABS substrates were first washed with ethanol (99.5%, Kishida Chemical, Chuo-ku, Osaka, Japan) and pure water for 10 min each in an ultrasonic cleaner (USK-1R, AS-ONE) and then dried through blowing N_2 gas (99.99%, Iwatani Fine Gas, Amagasaki, Hyogo, Japan). Prior to immobilizing the Pt particles, the washed substrates were pretreated either by only PDA coating or both plasma treatment and PDA coating. Table 2 presents the sample conditions and IDs.

Low-pressure plasma treatment was applied at 100 Pa using a plasma chamber (PR-501A, Yamato Scientific, Chuo-ku, Tokyo, Japan) with a radio frequency power source of 13.56 MHz. Before the plasma treatment, the washed ABS substrates were placed in the plasma chamber; subsequently, the pressure in the chamber was decreased to 5 Pa using a rotary vacuum pump (2012AC, Alcatel Vacuum Technology, Annecy, France). Then, helium gas (99.99%, Iwatani Fine Gas) flowed into the chamber until its pressure reached 100 Pa. The applied power for plasma generation was 100 W, and the standing wave ratio was controlled at less than 1.1 through impedance matching. The plasma treatment time was 60 s. XPS measurements were performed to confirm that the surface of the ABS substrate was modified by the plasma treatment. The results confirmed that oxygen-containing functional groups (–O–C=O and –C–O) were generated by the plasma treatment, as illustrated in Figure S1.

Aqueous solutions (20 mL) containing 2 mg/mL of DA were prepared using 2-(3,4-dihydroxyphenyl) ethylamine hydrochloride ($C_8H_{12}NO_2$·HCl; 98%, Fujifilm Wako Pure Chemical, Chuo-ku, Osaka, Japan) as the dopamine precursor and Tris hydrochloride acid buffer, controlled at pH = 8.5 (1 mol/L, Fujifilm Wako Pure Chemical), as the solvent. The ABS substrates were immersed in the DA solution for different times (i.e., 1, 3, or 24 h), and the DA solutions were neither stirred nor bubbled with oxygen gas during the immersion. After PDA coating, the substrates were removed from the DA solution and washed with pure water for 10 min using an ultrasonic cleaner to remove the unreacted DA and PDA. Finally, they were dried with N_2 gas. Precleaned slide glasses (S7213, Matsunami Glass, Kishiwada, Osaka, Japan) were also used and coated with a PDA film to confirm that the PDA coating was completed because ABS substrates originally contain carbon and nitrogen atoms. The results demonstrated that the intensity of the peaks in the Si2p-XPS spectra decreased with increasing DA immersion time and eventually disappeared when the glass slide was immersed for 24 h, as illustrated in Figure S2. This result indicates that immersion in a DA solution for 24 h is enough to uniformly cover the substrate's surface.

Aqueous solutions (5 mL) containing 4 mM of Pt ions were prepared using hexachloroplatinic acid hexahydrate (H_2PtCl_6·$6H_2O$; 98.5%, Fujifilm Wako Pure Chemical) in cylindrical PS containers (diameter $\varnothing$ = 33 mm and height h = 16 mm). Then, 2-propanol (IPA; 99.7%, Kishida Chemical) was added to the Pt ion solution (to be controlled at 1 vol %), and the pretreated ABS substrate was immersed in the Pt precursor solutions. A high-energy electron beam (of 4.8 MeV) was irradiated

on these Pt precursor solutions containing the pretreated ABS substrate for 7 s using a Dynamitron®
accelerator from SHI-ATEX Co. Ltd. (Izumiotsu, Osaka, Japan). During electron-beam irradiation,
H radicals and hydrated electrons were generated by the radiolysis of water. The reductive species
reduced Pt^{4+} ions to Pt^{3+}, Pt^{2+}, Pt^{1+}, and Pt^0, as described in a previous report [31]. Subsequently,
clusters of Pt atoms were formed, which grew to produce Pt particles. The Pt particles were formed
not only on the substrate, but also in the solution. When the Pt particles grew from Pt clusters on the
substrate, they were immobilized on it. When the Pt particles grew from Pt clusters in the solution,
they fell down and were deposited on the substrate, but were not immobilized on it. To remove the
unimmobilized Pt particles, the ABS substrates were taken out of the solution, washed with pure water
for 10 min using an ultrasonic cleaner, and finally dried with N_2 gas. In a previous report [16], XPS and
water contact angle (WCA) measurements were performed for untreated ABS substrates before and
after electron beam irradiation to investigate the immobilization mechanism; however, this mechanism
is not yet clear. Those XPS and WCA results suggest two models: the first model assumes the chemical
adhesion of Pt nanoparticles through a chemical reaction of functional groups (C–O) and/or carbon
radicals with Pt ions and/or Pt^0, and the second model assumes the unreactive immobilization of Pt
nanoparticles through a C–C crosslinking network under EB irradiation. To comprehensively clarify
the immobilization mechanism, further experiments using a simplex polymer such as polyethylene
should be conducted.

Table 2. Sample conditions and IDs.

Sample ID	Plasma pre-treatment	PDA coating	Pt-particle deposition
Glass-untreated	—	—	—
Glass-PDA(1h)	—	○	—
Glass-PDA(3h)	—	○	—
Glass-PDA(24h)	—	○	—
ABS-untreated	—	—	—
ABS-PDA(1h)	—	○	—
ABS-PDA(3h)	—	○	—
ABS-PDA(24h)	—	○	—
ABS-plasma	○	—	—
ABS-plasma-PDA(24h)	○	○	—
Pt/ABS-untreated	—	—	○
Pt/ABS-PDA(1h)	—	○	○
Pt/ABS-PDA(3h)	—	○	○
Pt/ABS-PDA(24h)	—	○	○
Pt/ABS-plasma-PDA(24h)	○	○	○

"—" indicates no operation and "○" indicates operation.

3.2. Characterization

To confirm that the ABS surface was coated with a PDA film, its chemical composition was
determined by XPS using Quantum 2000 equipment (Ulvac-Phi, Chigasaki, Kanagawa, Japan) attached
to an Al-$K\alpha$ source at 15 kV. The diameter of X-ray irradiation was ø = 100 μm; the pass energy and
step size were 23.50 and 0.05 eV, respectively; and the take-off angle was 45°. To neutralize the electric
charges on the surfaces, the measured samples were irradiated with a low-speed electron beam and an
Ar ion beam during the XPS measurements.

Secondary electron images using a field-emission scanning electron microscope (FE-SEM; S-4800,
Hitachi High-Technologies Corporation, Minato-ku, Tokyo, Japan) at 5 kV of accelerated voltage were
obtained to monitor the deposition behavior of the Pt particles on the ABS and/or ABS-PDA surfaces.
Prior to the observations, osmium (Os) was coated on the Pt/ABS surfaces by plasma chemical vapor
deposition using an osmium plasma coater (OPC60AL, Filgen, Nagoya, Aichi, Japan) to prevent the
generation of electrostatic charges during the measurements. The same FE-SEM instrument was also
used to investigate the effect of the plasma treatment on the film-forming state of PDA.

To measure the Pt-loading weight, the Pt particles on the ABS and/or PDA surfaces were dissolved in aqua regia, which was prepared by mixing hydrochloric acid (HCl; 35%, Sigma-Aldrich Japan, Meguro-ku, Tokyo, Japan) and nitric acid (HNO$_3$; 69%, Sigma-Aldrich Japan) at a ratio of 3:1. Then, the Pt concentrations were measured by ICP-AES (ICPE-9000, Shimadzu, Chukyo-ku, Kyoto, Japan) using the diluted aqua regia solutions containing Pt ions. A calibration curve prepared with a standard Pt solution (1000 ppm, Fujifilm Wako Pure Chemical) was used to calculate the amount of Pt in the Pt/ABS samples, as illustrated in Figure S3.

Figure 11 is a schematic of the process for evaluating the catalytic activity and durability of the system using representative H$_2$O$_2$ decomposition curves. First, the Pt/ABS samples were immersed in 5 mL of a 35,000-ppm solution of H$_2$O$_2$ (30 wt%, Kishida Chemical) at 25 °C for 360 min in an incubator (i-CUBE FCI-280, AS-ONE, Nishi-ku, Osaka, Japan). H$_2$O$_2$ decomposition occurred with increasing immersion time. The residual H$_2$O$_2$ concentrations were measured after immersion times of 1, 2, 5, 10, 30, 60, 120, 240, and 360 min to obtain the decomposition curves. The method for measuring the H$_2$O$_2$ concentration was the same as that reported in a previous article [16]. The optical absorbance of a H$_2$O$_2$ solution, colored using diluted (5 wt%) titanium sulfate (Ti(SO$_4$)$_2$; 30 wt%, Fujifilm Wako Pure Chemical), was measured using a deuterium–halogen and tungsten lamp (DH-2000, Ocean Optics, Largo, FL, USA), a fiber multichannel spectrometer (HR-4000, Ocean Optics), and optical fiber (P600-1-UV/VIS, Ocean Optics). The absorbance at 407 nm was used to calculate the residual H$_2$O$_2$ concentration from the calibration curve, as presented in Figure S4. The catalytic activity for H$_2$O$_2$ decomposition was evaluated from the value of the residual H$_2$O$_2$ concentration after immersing the Pt/ABS samples in the H$_2$O$_2$ solution for 360 min. This process of immersing the catalyst in the H$_2$O$_2$ solution for 360 min and drying it was repeated 10 times. The residual H$_2$O$_2$ concentrations were measured after one, three, five, and 10 immersions. The catalytic durability for H$_2$O$_2$ decomposition was evaluated from the value of the residual H$_2$O$_2$ concentration after immersing the Pt/ABS samples in the H$_2$O$_2$ solution 10 times (for 360 min each). The target value was less than 100 ppm, which means that if the residual H$_2$O$_2$ concentration was below 100 ppm after repeated use (i.e., after 10 uses), the Pt/ABS sample had long catalytic durability.

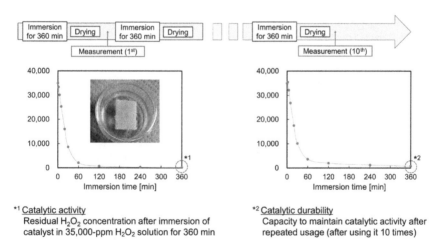

*1 Catalytic activity
  Residual H$_2$O$_2$ concentration after immersion of catalyst in 35,000-ppm H$_2$O$_2$ solution for 360 min

*2 Catalytic durability
  Capacity to maintain catalytic activity after repeated usage (after using it 10 times)

Figure 11. Schematic of the process for evaluating the catalytic activity and durability of the system using typical H$_2$O$_2$ decomposition curves. *1 Value of the residual H$_2$O$_2$ concentration for evaluating the catalytic activity. *2 Value of the residual H$_2$O$_2$ concentration for evaluating the catalytic durability.

4. Conclusions

We introduced PDA coating as a pre-treatment for the strong immobilization of Pt-particle catalysts on ABS substrates and investigated the effect of the PDA coating on the deposition behavior of the Pt particles, the Pt-loading weight, and the catalytic activity and durability of the material. We found that the PDA coating improved both the catalytic activity and durability of the Pt-based material. Moreover, introducing a plasma treatment before the PDA coating was effective for further improving the catalytic durability. Finally, in the case of the Pt/ABS-plasma-PDA(24h) catalyst, the residual H_2O_2 concentrations were 30, 33, 33, and 35 ppm after using the material 1, 3, 5, and 10 times, respectively. Although the PDA coating also increased the Pt-loading weight (from 5.9 to 24.0 µg/substrate), the value measured for the Pt-particle/ABS-plasma-PDA(24h) catalyst (i.e., 24.0 µg/substrate) was significantly below that determined for a Pt-film/ABS catalyst (1500 µg/substrate) prepared by electroless plating. In summary, we successfully achieved a decrease in Pt usage while maintaining the high catalytic activity and durability. In addition, the developed process, which includes a combination of plasma treatment, PDA coating, and EBIRM, does not require etching of the ABS surface using dangerous chemical solutions, as is the case for electroless plating, where previous etching is necessary to obtain high adhesion between the ABS substrate and the Pt film. Therefore, the developed process is more ecofriendly. Although the PDA coating was used to improve the catalytic durability of a Pt-based catalyst in this study, this type of coating is useful as a pre-treatment for the strong immobilization of metal particles on several substrates or microparticles.

Supplementary Materials: The following are available online at http://www.mdpi.com/2079-4991/10/1/114/s1. Figure S1: XPS spectra of the ABS surface before and after plasma treatment, Figure S2: Results of XPS analysis of the glass surface before and after PDA coating, Figure S3: Calibration curve for calculating the amount of Pt on Pt/ABS samples, and Figure S4: Calibration curve for calculating the H_2O_2 concentration.

Author Contributions: Y.O., K.E., and K.Y. supervised the work. T.A. and S.S. prepared the Pt particle/ABS samples. Y.O., T.A., and D.K. performed the XPS analysis and SEM observation. Y.O., T.A., and S.S. measured and calculated the Pt-loading weights of the samples using ICP-AES. T.A. evaluated the catalytic activity and durability. O.M. and R.S. helped with the evaluations. All authors contributed to the scientific discussion and manuscript preparation. Y.O. wrote the manuscript. All authors have read and agreed to the published version of the manuscript.

Funding: This research was funded by the Japan Science and Technology Agency, grant numbers JST No. MP27215667957 and JST No. VP29117941540.

Acknowledgments: We thank the staff of the SHI-ATEX Co., Ltd. for their assistance with the electron-beam irradiation experiments. We also thank Tohru Sekino, Sunghun Cho, and project researcher Hideki Hashimoto for their assistance with the Os coating prior to SEM observations.

Conflicts of Interest: The authors declare no conflict of interest.

References

1. Waite, J.H. Nature's underwater adhesive specialist. *Int. J. Adhes. Adhes.* **1987**, *7*, 9–14. [CrossRef]
2. Vreeland, V.; Waite, J.H.; Epstein, L. Polyphenols and oxidases in substratum adhesion by marine algae and mussels. *J. Phycol.* **1998**, *34*, 1–8. [CrossRef]
3. Silverman, H.G.; Roberto, F.F. Understanding marine mussel adhesion. *Mar. Biotechnol.* **2007**, *9*, 661–681. [CrossRef]
4. Waite, J.H.; Tanzer, M.L. Polyphenolic substance of Mytilus edulis: Novel adhesive containing L-dopa and hydroxyproline. *Science* **1981**, *212*, 1038–1040. [CrossRef]
5. Waite, J.H.; Qin, X. Polyphosphoprotein from the adhesive pads of Mytilus edulis. *Biochemistry* **2001**, *40*, 2887–2893. [CrossRef]
6. Waite, J.H. Adhesion a la Moule. *Integr. Comp. Biol.* **2002**, *42*, 1172–1180. [CrossRef]
7. Korbel, J.O.; Urban, A.E.; Affourtit, J.P.; Godwin, B.; Grubert, F.; Simons, J.F.; Kim, P.M.; Palejev, D.; Carriero, N.J.; Du, L.; et al. Paired-end mapping reveals extensive structural variation in the human genome. *Science* **2007**, *318*, 420–426. [CrossRef] [PubMed]

8. Ryu, J.; Ku, S.H.; Lee, H.; Park, C.B. Mussel-inspired polydopamine coating as a universal route to hydroxyapatite crystallization. *Adv. Funct. Mater.* **2010**, *20*, 2132–2139. [CrossRef]

9. Kawamura, A.; Kohri, M.; Morimoto, G.; Nannichi, Y.; Taniguchi, T.; Kishikawa, K. Color Biomimetic Photonic Materials with Iridescent and Non-Iridescent Structural Colors. *Sci. Rep.* **2016**, *6*, 33984. [CrossRef] [PubMed]

10. Ryu, J.H.; Lee, Y.; Kong, W.H.; Kim, T.G.; Park, T.G.; Lee, H. Catechol-functionalized chitosan/pluronic hydrogels for tissue adhesives and hemostatic materials. *Biomacromolecules* **2011**, *12*, 2653–2659. [CrossRef] [PubMed]

11. Zhang, S.; Liang, X.; Gadd, G.M.; Zhao, Q. Advanced titanium dioxide-polytetrafluoroethylene (TiO$_2$-PTFE) nanocomposite coatings on stainless steel surfaces with antibacterial and anti-corrosion properties. *Appl. Surf. Sci.* **2019**, *490*, 231–241. [CrossRef]

12. Yamagishi, K.; Kirino, I.; Takahashi, I.; Amano, H.; Takeoka, S.; Morimoto, Y.; Fujie, T. Tissue-adhesive wirelessly powered optoelectronic device for metronomic photodynamic cancer therapy. *Nat. Biomed. Eng.* **2019**, *3*, 27–36. [CrossRef] [PubMed]

13. Muntz, A.; Subbaraman, L.N.; Sorbara, L.; Jones, L. Tear exchange and contact lenses: A review. *J. Optom.* **2015**, *8*, 2–11. [CrossRef] [PubMed]

14. Nichols, J.J.; Fisher, D. Contact Lenses 2018. *Contact Lens Spectr.* **2019**, *34*, 18–23.

15. Paugh, J.R.; Brennan, N.A.; Efron, N. Ocular Response to Hydrogen Peroxide. *Am. J. Optom. Physiol. Opt.* **1988**, *65*, 91–98. [CrossRef]

16. Ohkubo, Y.; Aoki, T.; Seino, S.; Mori, O.; Ito, I.; Endo, K.; Yamamura, K. Radiolytic Synthesis of Pt-Particle/ABS Catalysts for H$_2$O$_2$ Decomposition in Contact Lens Cleaning. *Nanomaterials* **2017**, *7*, 235. [CrossRef]

17. Ohkubo, Y.; Aoki, T.; Seino, S.; Mori, O.; Ito, I.; Endo, K.; Yamamura, K. Improved Catalytic Durability of Pt-Particle/ABS for H$_2$O$_2$ Decomposition in Contact Lens Cleaning. *Nanomaterials* **2019**, *9*, 342. [CrossRef]

18. Cahen, D.; Lester, J.E. Mixed and partial oxidation states. Photoelectron spectroscopic evidence. *Chem. Phys. Lett.* **1973**, *18*, 108–111. [CrossRef]

19. Isaifan, R.J.; Ntais, S.; Baranova, E.A. Particle size effect on catalytic activity of carbon-supported Pt nanoparticles for complete ethylene oxidation. *Appl. Catal. A Gen.* **2013**, *464*, 87–94. [CrossRef]

20. Ohkubo, Y.; Kageyama, S.; Seino, S.; Nakagawa, T.; Kugai, J.; Ueno, K.; Yamamoto, T.A. Mass production of highly loaded and highly dispersed PtRu/C catalysts for methanol oxidation using an electron-beam irradiation reduction method. *J. Exp. Nanosci.* **2016**, *11*, 123–137. [CrossRef]

21. Mizukoshi, Y.; Oshima, R.; Maeda, Y.; Nagata, Y. Preparation of platinum nanoparticles by sonochemical reduction of the Pt(II) ion. *Langmuir* **1999**, *15*, 2733–2737. [CrossRef]

22. Ziylan-Yavas, A.; Yuya, M.; Ono, T.; Nakagawa, T.; Seino, S.; Okitsu, K.; Mizukoshi, Y.; Emura, S.; Yamamoto, T.A. Sonochemically synthesized core-shell structured Au-Pd nanoparticles supported on γ-Fe$_2$O$_3$ particles. *J. Nanoparticle Res.* **2006**, *15*, 875–880.

23. Mizukoshi, Y.; Tsuru, Y.; Tominaga, A.; Seino, S.; Masahashi, N.; Tanabe, S.; Yamamoto, T.A. Sonochemical immobilization of noble metal nanoparticles on the surface of maghemite: Mechanism and morphological control of the products. *Ultrason. Sonochem.* **2008**, *15*, 875–880. [CrossRef] [PubMed]

24. Ziylan-Yavas, A.; Mizukoshi, Y.; Maeda, Y.; Ince, N.H. Supporting of pristine TiO$_2$ with noble metals to enhance the oxidation and mineralization of paracetamol by sonolysis and sonophotolysis. *Appl. Catal. B Environ.* **2015**, *172*, 7–17. [CrossRef]

25. Toshima, N.; Wang, Y. Preparation and Catalysis of Novel Colloidal Dispersions of Copper/Noble Metal Bimetallic Clusters. *Langmuir* **1994**, *10*, 4574–4580. [CrossRef]

26. Toshima, B.N.; Wang, Y. Polymer-Protected Cu/Pd Bimetallic Clusters. *Adv. Mater.* **1994**, *100080*, 245–247. [CrossRef]

27. Daimon, H.; Kurobe, Y. Size reduction of PtRu catalyst particle deposited on carbon support by addition of non-metallic elements. *Catal. Today* **2006**, *111*, 182–187. [CrossRef]

28. Liao, P.C. Activity and XPS Studies Catalysts Bimetallic. *J. Catal.* **1982**, *316*, 307–316. [CrossRef]

29. Mohamed, R.M.; Mkhalid, I.A. Characterization and catalytic properties of nano-sized Ag metal catalyst on TiO$_2$-SiO$_2$ synthesized by photo-assisted deposition and impregnation methods. *J. Alloys Compd.* **2010**, *501*, 301–306. [CrossRef]

30. Rahsepar, M.; Pakshir, M.; Piao, Y.; Kim, H. Preparation of highly active 40 wt.%pt on multiwalled carbonnanotube by improved impregnation method for fuel cell applications. *Fuel Cells* **2012**, *12*, 827–834. [CrossRef]
31. Belloni, J. Nucleation, growth and properties of nanoclusters studied by radiation chemistry: Application to catalysis. *Catal. Today* **2006**, *113*, 141–156. [CrossRef]
32. Seino, S.; Kinoshita, T.; Nakagawa, T.; Kojima, T.; Taniguci, R.; Okuda, S.; Yamamoto, T.A. Radiation induced synthesis of gold/iron-oxide composite nanoparticles using high-energy electron beam. *J. Nanoparticle Res.* **2008**, *10*, 1071–1076. [CrossRef]

MDPI

Article

Photoluminescent Hydroxylapatite: Eu^{3+} Doping Effect on Biological Behaviour

Ecaterina Andronescu [1,2,3], Daniela Predoi [4], Ionela Andreea Neacsu [1,2], Andrei Viorel Paduraru [1], Adina Magdalena Musuc [1,5], Roxana Trusca [2,3], Ovidiu Oprea [1,2], Eugenia Tanasa [2,3], Otilia Ruxandra Vasile [2,3], Adrian Ionut Nicoara [1,2], Adrian Vasile Surdu [1,2], Florin Iordache [6], Alexandra Catalina Birca [1,2], Simona Liliana Iconaru [4] and Bogdan Stefan Vasile [1,2,3,*]

[1] Faculty of Applied Chemistry and Materials Science, Department of Science and Engineering of Oxide Materials and Nanomaterials, University Politehnica of Bucharest, 060042 Bucharest, Romania
[2] National Centre for Micro and Nanomaterials, University Politehnica of Bucharest, 060042 Bucharest, Romania
[3] National Research Center for Food Safety, University Politehnica of Bucharest, 060042 Bucharest, Romania
[4] Multifunctional Materials and Structures Laboratory, National Institute of Materials Physics, 077125 Magurele, Romania
[5] Ilie Murgulescu Institute of Physical Chemistry, 060021 Bucharest, Romania
[6] Faculty of Veterinary Medicine, Department of Biochemistry, University of Agronomic Science and Veterinary Medicine, 011464 Bucharest, Romania
* Correspondence: bogdan.vasile@upb.ro; Tel.: +40727589960

Received: 20 July 2019; Accepted: 18 August 2019; Published: 22 August 2019

Abstract: Luminescent europium-doped hydroxylapatite (Eu$_X$HAp) nanomaterials were successfully obtained by co-precipitation method at low temperature. The morphological, structural and optical properties were investigated by scanning electron microscopy (SEM), transmission electron microscopy (TEM), X-ray diffraction (XRD), Fourier Transform Infrared (FT-IR), UV-Vis and photoluminescence (PL) spectroscopy. The cytotoxicity and biocompatibility of Eu$_X$HAp were also evaluated using MTT (3-(4,5-dimethylthiazol-2-yl)-2,5-diphenyltetrazolium bromide)) assay, oxidative stress assessment and fluorescent microscopy. The results reveal that the Eu^{3+} has successfully doped the hexagonal lattice of hydroxylapatite. By enhancing the optical features, these Eu$_X$HAp materials demonstrated superior efficiency to become fluorescent labelling materials for bioimaging applications.

Keywords: europium doped hydroxylapatite; photoluminescence; MTT assay; oxidative stress assessment; fluorescent microscopy

1. Introduction

Bioceramics can be defined as the category of ceramics used in repairing and replacing processes of damaged and diseased parts of skeletal system [1,2]. Their biocompatibility varies from inert ceramic oxides to bioresorbable materials. One of the most used bioresorbable ceramics for biomedical applications are calcium orthophosphates (CaP's) [3,4]. The CaP's give hardness and stability to the tissues and can be found in teeth, bones, and tendons. Starting with Ca/P molar ratio equal to 0.5 and finishing with 2.0, there are 11 known non-ion substituted calcium orthophosphates, among which the most used one is hydroxylapatite (HAp) [5,6].

HAp synthesis, with its diverse morphologies, structures and textures, has attracted much interest in academic and industrial research for many heterogeneous catalysis applications [7–9]. Numerous synthetic routes for obtaining hydroxylapatite were developed over time, and can be divided in four main categories: (1) wet methods [10–12], (2) dry methods [13], (3) microwave-assisted methods [14–16], ball-milling [17–19] or ultrasound methods [20,21], and (4) miscellaneous methods [22]. Depending on the reagents and conditions, each category offers several variations [23,24].

Stoichiometric hydroxylapatite ($Ca_{10}(PO_4)_6(OH)_2$) is the most similar material to the mineral component of human hard tissues and therefore it is considered the ideal substance for bone defects restorations [25–27]. Keeping the same geometry while accepting a big variety of anions and cations is one of the most important structural characteristics of hydroxylapatite [28,29]. Synthetic HAp can also be doped with several metal ions in order to improve its properties, like bioactivity, degradation rate, antibacterial characteristics, luminescence and magnetic properties [30–34].

A photoluminescent material is the most promising candidate for clinical applications and implantation. The biocompatibility is not the only important feature, a longer lifetime of luminescence being also an significant benefit in practical applications [35,36].

Photoluminescence is a very important and useful mechanism in in situ investigations for tissue engineering, surgery, tissue restoration, etc. Using of organic fluorescent molecules for labelling it was a popular practice in clinical trials for many years. Recent studies use inorganic components, even in form of nanoparticles, to replace photoluminescent organic compounds. Because of the toxicity and the nano-size of this particles, the usage of these materials represents a challenge yet [37,38].

Due to high values of the Stokes shift and long lifetime of the excited state, lanthanide coordination compounds are the perfect materials for bioimaging. Europium complexes possess this unique luminescent properties and beside this Europium combine high photoluminescence quantum yields (PLQYs) with the emission in the long wavelength range, which can easily penetrate through the tissues [39–41].

The luminescence of the europium (III) and terbium (III) complexes is especially sensitive to changes in the structure and coordination environment of ions and depends substantially on the interaction with the analyte. The intensive luminescence and characteristic Stark structure of Eu^{3+} luminescence spectra allow registering fine changes in the structure of the coordination sphere of a rare-earth ion upon surrounding impact. Luminescent lanthanide-containing complex compounds can be applied as optical chemosensors in detection of anions, cations, gases, etc. [42].

The new generation of biomaterials with multifunctional europium (III)-doped HAp scaffolds has shown remarkable development. The luminescent multifunctional biomaterials show potential for use in various biomedical applications such as smart drug delivery, bioimaging, and photothermal therapy. In this generation, the Eu^{3+} ion has been widely used as traceable fluorescence probe due to well-known dopant narrow emission spectral lines by visible-light excitation caused by shielding by the 5 s and 5 p orbitals. The spectral shapes depend on the local ion symmetry and forbidden f–f transitions. Thus, Eu^{3+} is a sensitive optical probe for the dopant site environment because of its characteristic luminescence properties [43].

2. Experimental

2.1. Materials

All the reagents for synthesis, including calcium nitrate tetrahydrate ($Ca(NO_3)_2 \cdot 4H_2O$, 99.0%, Sigma-Aldrich, St. Louis, MI, USA), europium-(III) nitrate hexahydrate ($Eu(NO_3)_3 \cdot 6H_2O$, 99.9%, Alfa Aesar, Haverhill, MA, USA), ammonium phosphate dibasic ($(NH_4)_2HPO_4$, 99.0%, Alfa Aesar, Haverhill, MA, USA), sodium hydroxide (NaOH, 25% solution, Alfa Aesar, Haverhill, MA, USA) were used as received, without further purification. Deionised water was used for the experiment.

2.2. Synthesis

The biocompatible photoluminescent europium-doped hydroxylapatite (Eu_XHAp) nanomaterials have been synthesized by co-precipitation method. In order to obtain Eu_XHAp powders, appropriate amounts of calcium nitrate tetrahydrate and europium-(III) nitrate hexahydrate were dissolved in deionized water, under vigorous stirring at room temperature, thus obtaining solution A. Meanwhile, a solution B was prepared by dissolving an appropriate amount of ammonium phosphate dibasic in deionized water, under vigorous stirring at room temperature. The atomic ratio Ca/P and $[Ca^+Eu]/P$

was 1.67, while the atomic ratio Eu/(Eu$^+$Ca) was varied between 0 and 50%. Solution B was added dropwise to solution A, at 80 °C, under vigorous stirring, while adjusting and maintaining the pH of the resulting suspension at 10, by adding NH$_4$OH (25%) solution, for 2 h. Pure HAp was synthesized following the same methodology, except the Eu^{3+} precursor addition. The resulting suspensions were matured for 24 h. The precipitate was then filtered and washed several times with deionized water, until the pH values were close to 7. Finally, the resulting precipitates were dried at 80 °C for 50 h in an air oven.

2.3. Samples Notation

Eu$_X$HAp represents Eu–doped hydroxylapatite, where X is equal to the used europium content (X$_{Eu}$ = 0.05, 0.1, 0.15, 0.2, 1.0, 5.0). The correspondent Eu-free hydroxylapatite is noted HAp (X$_{Eu}$ = 0).

2.4. Morphological and Structural Characterization

X-ray diffraction (XRD) studies were carried out using a PANalytical Empyrean diffractometer at room temperature, with a characteristic Cu X-ray tube (λ Cu K$_{\alpha 1}$ = 1.541874 Å) with in-line focusing, programmable divergent slit on the incident side and a programmable anti-scatter slit mounted on the PIXcel3D detector on the diffracted side. The samples were scanned in a Bragg - Brentano geometry with a scan step increment of 0.02° and a counting time of 255 s/step. The XRD patterns were recorded in the 2θ angle range of 20°–80°. Lattice parameters were refined by the Rietveld method, using the HighScore Plus 3.0 e software. The morphology of the samples was analyzed using a Quanta Inspect F50 FEG (field emission gun) scanning electron microscope with 1.2 nm resolution, equipped with an energy-dispersive X-ray (EDX) analyzer (resolution of 133 eV at MnK$_\alpha$, Thermo Fisher, Waltham, MA, USA) on sample covered with a thin gold layer. The high-resolution TEM images of the samples were obtained on finely powdered samples using a Tecnai G^2 F30 S-Twin high-resolution transmission electron microscope from Thermo Fisher (former FEI) (Waltham, MA, USA).The microscope operated in transmission mode at 300 kV acceleration voltage with a TEM resolution of 1.0 Å. FTIR spectra were recorded with a Nicolet iS50R spectrometer (Thermo Fisher Waltham, MA, USA), at room temperature, in the measurement range 4000–400 cm^{-1}. Spectral collection was carried out in ATR mode at 4 cm^{-1} resolution. For each spectrum, 32 scans were co-added and converted to absorbance using OmincPicta software (Thermo Scientific, Waltham, MA, USA). Raman spectra were recorded at room temperature on a Horiba Jobin-Yvon LabRam HR spectrometer equipped with nitrogen cooled detector. The near–infrared (NIR) line of a 785 nm laser was employed for excitation and the spectral range went from 500 to 1200 cm^{-1}. UV-Vis diffused reflectance spectra were obtained using an Able Jasco V-560 spectrophotometer (PW de Meern, Netherlands,) with a scan speed of 200 nm/s, between 200 and 850 nm. The fluorescence spectra were measured by using a Perkin Elmer LS 55 fluorescence spectrophotometer (Arkon, OH, USA). Spectra were recorded with a scan speed of 200 nm/s between 350 and 800 nm, and with excitation and emission slits widths of 7 and 5 nm, respectively. An excitation wavelength of 320 nm was used.

2.5. Cellular Viability Assays

2.5.1. Quantitative In-Vitro Evaluation of Biocompatibility—MTT Assay

MTT assay is a quantitative colorimetric method, which allows evaluation of cell viability and proliferation, and cytotoxicity of different compounds. The method is based on reduction of MTT tetrazolium salt (3-(4,5-dimethylthiazolyl)-2,5-diphenyltetrazolium bromide) to dark blue formazan. Reduction by mitochondrial enzymes (especially succinate dehydrogenase) is an indication of cell/mitochondrial integrity. Formazan, insoluble in water, can be solubilized with isopropanol, dimethylsulfoxide or other organic solvent. The optical density (DO) of solubilized formazan is evaluated spectrophotometrically, resulting in a color-absorbent-color-counting function of the number of metabolic active cells in the culture.

The human mesenchymal amniotic fluid stem cells (AFSC) were used to evaluate the biocompatibility of Eu_XHAp nanoparticles. The cells were cultured in DMEM medium (Sigma-Aldrich, Saint Luis, MI, USA) supplemented with 10% fetal bovine serum, 1% penicillin and 1% streptomycin antibiotics (Sigma-Aldrich, Saint Luis, MI, USA). To maintain optimal culture conditions, medium was changed twice a week. The biocompatibility was assessed using MTT assay (Vybrant®MTT Cell Proliferation Assay Kit, Thermo Fischer Scientific, Waltham, MA, USA). Briefly, the AFSC were grown in 96-well plates, with a seeding density of 3000 cells/well in the presence of Eu_XHAp for 72 h. Then 15 mL Solution I (12 mM MTT) was added and incubated at 37 °C for 4 h. Solution II (1 mg Sodium Dodecyl Sulphate + 10 ml HCl, 0,01M) was added and pipettes vigorously to solubilize formazan crystals. After 1 h the absorbance was read using spectrophotometer at 570 nm (TECAN Infinite M200, Männedorf, Switzerland).

2.5.2. Oxidative Stress Assessment—GSH-Glo Glutathione Assay

The GSH-Glo Assay is a luminescent-based assay for the detection and quantification of glutathione (GSH) in cells or in various biological samples. A change in GSH levels is important in the assessment of toxicological responses and is an indicator of oxidative stress, potentially leading to apoptosis or cell death. The assay is based on the conversion of a luciferin derivative into luciferin in the presence of GSH. The reaction is catalyzed by a glutathione S-transferase (GST) enzyme supplied in the kit. The luciferin formed is detected in a coupled reaction using Ultra-Glo Recombinant Luciferase that generates a glow type luminescence that is proportional to the amount of glutathione present in cells. The assay provides a simple, fast and sensitive alternative to colorimetric and fluorescent methods and can be adapted easily to high-throughput applications.

AFSC were seeded at a density of 3000 cells in 300 µL of Dulbecco's Modified Eagle's medium (DMEM) supplemented with 10% fetal bovine serum and 1% antibiotics (penicillin, streptomycin/neomycin) in 96-well plates. Twenty-four hours after seeding, cells are treated with Eu_XHAp and incubated for 72 h.

The working protocol consisted of adding 100 µL 1X GSH-Glo Reagent and incubating at 37 °C for 30 min. Then, 100 µL Luciferin DeectionReagent was added and incubated at 37 °C for an additional 15 min. At the end of the time, the wells were well homogenized and then the plate was read on the luminometer (MicroplateLuminometerCentro LB 960, Berthold, Germany).

2.5.3. Qualitative In-Vitro Evaluation of Biocompatibility—Fluorescent Microscopy

The biocompatibility of the Eu_XHAp was also evaluated by fluorescent microscopy, using RED CMTPX fluorophore (Thermo Fischer Scientific, Waltham, MA, USA), a cell tracker for long-term tracing of living cells. The CMTPX tracker was added in cell culture treated with Eu_XHAp nanomaterials and the viability and morphology of the AFSC was evaluated after 5 days. The CMTPX fluorophore was added in the culture medium at a final concentration of 5 µM, incubated for 30 min in order to allow the dye penetration into the cells. Next, the AFSC were washed with PBS and visualized by fluorescent microscopy. The photomicrographs were taken with Olympus CKX 41 digital camera driven by CellSense Entry software (Olympus, Tokyo, Japan).

3. Results and Discussions

3.1. X-ray Diffraction

The X-ray diffraction patterns of Eu-doped hydroxylapatite, with different concentration of europium and with pure HAp are shown in Figure 1.

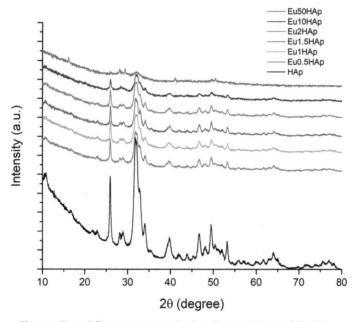

Figure 1. *X*-ray diffraction patterns of hydroxylapatite (HAp) and Eu$_X$HAp.

Pristine HAp (Figure 1) is shown as a single phase calcium phosphate material (ICDD PDF4+ no.04-021-1904 [44]), with a crystallinity of 33.9% and an average crystallite size of 10.84 nm (Table 1). The XRD patterns of all Eu$_X$HAp samples indicate only the pure crystalline hexagonal HAp phase (according to ICDD PDF4+ no.04-021-1904 [44] of the space group P6$_3$/m, in consistence with literature [45] up to a substitution degree of 10%. For 50% substitution, Eu(OH)$_3$ secondary phase in a mass ratio of 27.6% has occurred (according to ICDD PDF4+ no. 01-083-2305 [46]). This means that for 50% substitution, the limit of solubility of Eu in HAp lattice was exceeded. The intensities of *X*-ray peaks decrease when the Eu-doping level increases up to 2%, indicating an interference of Eu^{3+} with HAp crystal structure. Also, the peak position is influenced by Eu for Ca substitution as they shift to higher angle values which suggest decreasing of unit cell parameters. Table 2 and Figure 2 illustrate the values of unit cell parameters *a*, *c* and *V* and the agreement indices of the Rietveld analysis (R_{exp}, R_p, R_{wp} and χ^2), which indicate the quality of the fit.

Table 1. Calculated crystallite size (*D*) values and degree of crystallinity (X_c) of pure HAp and europium doped hydroxylapatite with various amount of Eu.

No.	Samples	D/nm	S/%	X_c/%
1	HAp	9.62 ± 1.46	0.95 ± 0.29	33.93
2	Eu0.5HAp	14.88 ± 2.96	0.74 ± 0.66	26.25
3	Eu1HAp	12.82 ± 2.72	0.84 ± 0.66	24.11
4	Eu1.5HAp	17.3 ± 4.92	0.65 ± 0.55	22.97
5	Eu2HAp	14.44 ± 3.56	0.77 ± 0.63	22.63
6	Eu10HAp	10.78 ± 3.04	1.04 ± 0.85	23.03
7	Eu50HAp	3.55 ± 0.34	2.66 ± 1.56	21.87

Table 2. Unit cell parameters *a*, *c*, *V* and agreement indices for hydroxylapatite HAp and Eu-doped HAp (with concentration of Eu^{3+} of 0.5, 1, 1.5, 2, 10 and 50%).

Sample	a [Å]	c [Å]	V [Å³]	R_{exp}	R_p	R_{wp}	χ^2
HAp	9.434979 ± 0.001401	6.8803397 ± 0.001042	530.6588	3.4323	3.99694	5.19277	2.28891
Eu0.5HAp	9.440372 ± 0.002067	6.878142 ± 0.001529	530.8601	3.96601	4.82349	6.554	2.7309
Eu1HAp	9.445147 ± 0.002444	6.879144 ± 0.001839	531.4745	3.92894	4.95602	6.61378	2.83367
Eu1.5HAp	9.439496 ± 0.001874	6.879717 ± 0.001413	530.883	3.7995	4.38638	5.76083	2.29889
Eu2HAp	9.440372 ± 0.002067	6.878142 ± 0.001529	530.8601	3.96601	4.82349	6.554	2.7309
Eu10HAp	9.443276 ± 0.002227	6.879882 ± 0.001665	531.321	3.88457	4.43662	5.82612	2.24943
Eu50HAp	9.544463 ± 0.021918	6.315845 ± 0.023992	498.2704	3.02385	3.11688	4.08783	1.82753

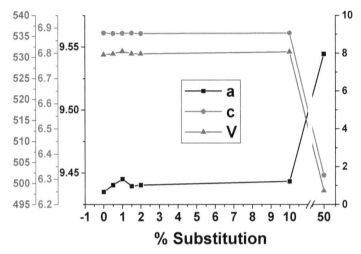

Figure 2. Unit cell parameters versus substitution degree for HAp and Eu_xHAp.

Hydroxylapatite crystallizes in a hexagonal symmetry with lattice parameters $a = b \neq c$ and space group $P6_3/m$. In this structure, PO_4 tetrahedral form basic structural units, while the coordination around the distinct Ca sites defines the $Ca1O13O23$ metaprism and the distorted $Ca2O1O2O3_4(OH)$ polyhedron. Thus, the formula per unit cell may be expressed as $Ca1_4Ca2_6(PO1O2O3_2)_6(O_HH)_2$ [47]. Taking into consideration that the ionic radius of Eu^{3+} (0.947 Å) is smaller than that of Ca^{2+} (1 Å) and that small ions are preferentially substituted at Ca1, in this case Eu^{3+} most probably substitutes Ca^{2+} from site 2 [48,49]. While a low concentration does not induce significant changes (Figure 2), when looking at higher substitution degree (10% and 50%) it may be concluded that there is an increase of *a*-axis and a decrease of *c*-axis parameters which is consistent with the results obtained for Al^{3+} substitution by Fahami et al. [49]. Moreover, the decrease of unit cell volume proves the incorporation of Eu^{3+} in hydroxylapatite lattice.

The estimation of crystallite size, lattice microstrain and degree of crystallinity are shown in Table 2 and Figure 3.

Up to 1.5%, the introduction of Eu^{3+} in HAp lattice induces an increase of crystallite size from 9.62 nm to 17.3 nm and a decrease of lattice microstrain from 0.95% to 0.65%. For a substitution degree of more than 1.5%, there may be observed a decrease of the crystallite size to 3.55 nm and an increase of lattice microstrain to 2.66% in the case of maximum substitution degree. In what concerns the crystallinity degree (Table 2), as it may be appreciated qualitatively from the profile of the peaks, the crystallinity decreases when the substitution degree increases.

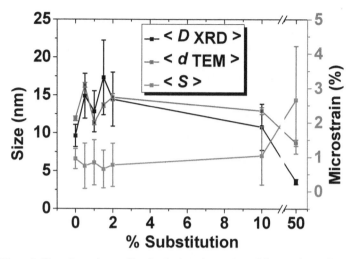

Figure 3. The estimated crystallite size, lattice microstrain and degree of crystallinity.

3.2. SEM Analysis

Figure 4 shows the SEM images (column A) and EDX spectra (column B) of pure HAp (Figure 4a) and Eu_xHAp samples (Figure 4b).

It can be seen that the doping with Eu^{3+} has little influence on the morphology of substituted HAp compared to the pure HAp. SEM images (Figure 4, column A) reveal quasi-spherical (at low dopant concentration) and acicular (mostly after 10% Eu) particles with width in the range of 6–16 nm. Due to high surface area, agglomerates are present in all samples. The EDX spectra of studied samples (Figure 4, column B) confirm the presence of all elements specific to Eu-doped HAp powders: calcium (Ca), phosphor (P), oxygen (O) and europium (Eu).

A   B

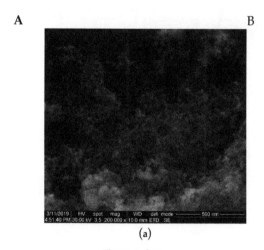

(a)

Figure 4. *Cont.*

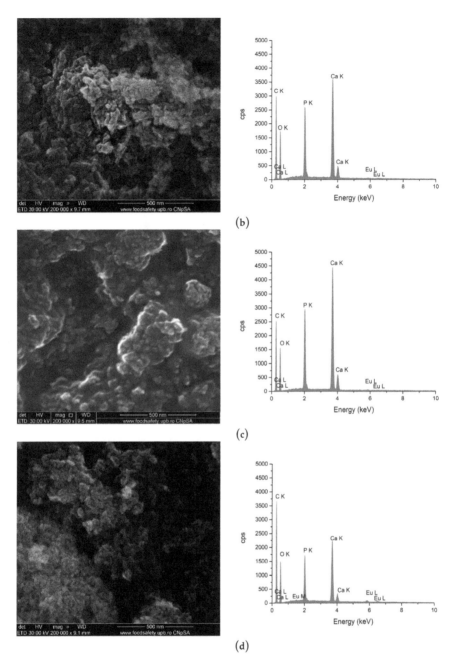

Figure 4. *Cont.*

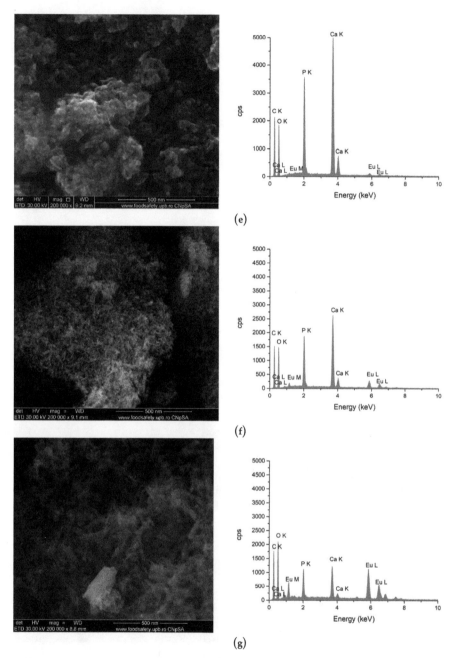

(e)

(f)

(g)

Figure 4. The SEM images (column **A**) and EDX spectra (column **B**) of pure HAp and Eu_XHAp samples (**a**) HAp, (**b**) Eu0.5HAp, (**c**) Eu1HAp, (**d**) Eu1.5HAp, (**e**) Eu2HAp, (**f**) Eu10HAp, (**g**) Eu50HAp.

3.3. TEM Analysis

Figure 5 shows the bright field TEM, HRTEM images, SAED patterns and particle size distribution of pure HAp and Eu_XHAp at different Eu^{3+} concentrations.

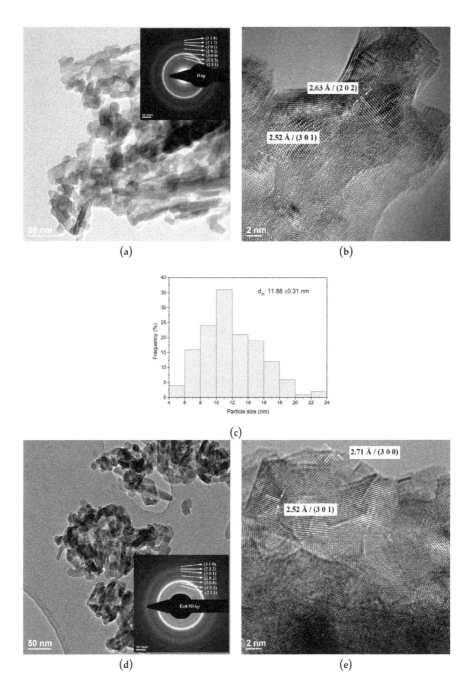

Figure 5. *Cont.*

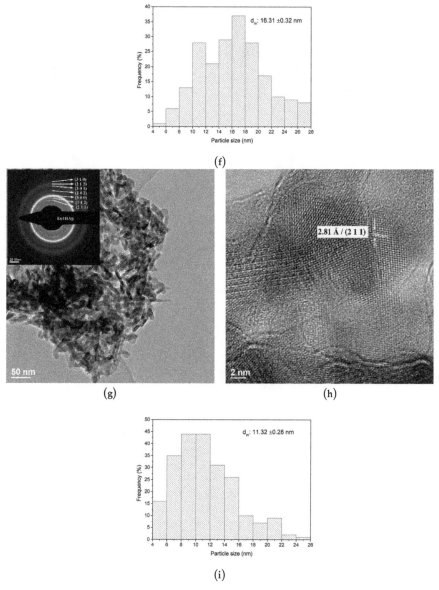

Figure 5. *Cont.*

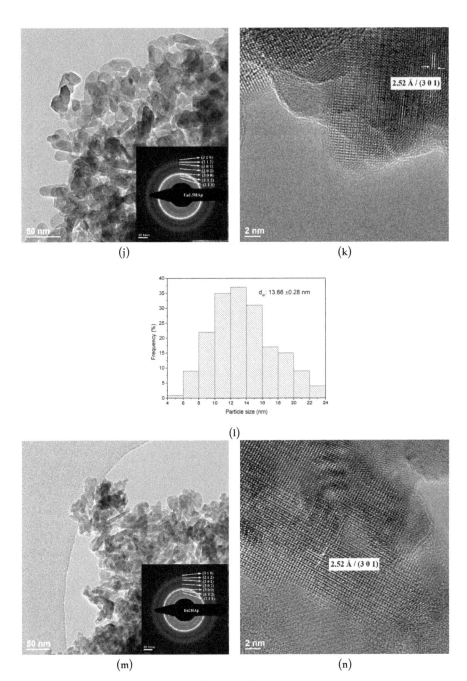

Figure 5. *Cont.*

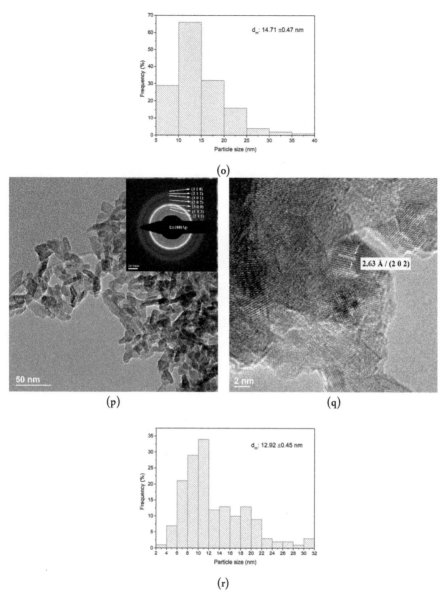

Figure 5. *Cont.*

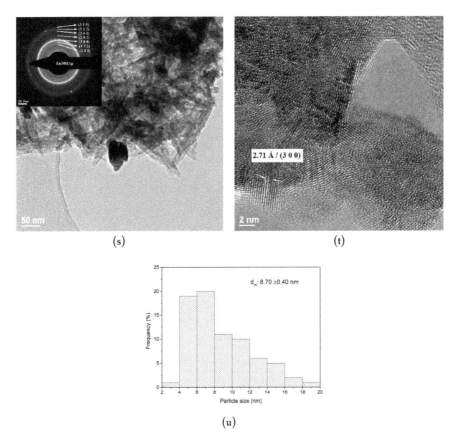

Figure 5. The TEM and HRTEM images, SAED patterns and particle size distribution of pure HAp and Eu$_X$HAp samples (**a–c**) HAp, (**d–f**) Eu0.5HAp, (**g–i**) Eu1HAp, (**j–l**) Eu1.5HAp, (**m–o**) Eu2HAp, (**p–r**) Eu10HAp, (**s–u**) Eu50HAp.

From the bright field TEM images presented in Figure 5a,d,g,j,m,p,s it can be observed that the doping level does not affect the the sample morphologies from 0 to 10% Eu^{3+} doping, the only visible modification if for the 50% Eu^{3+} doping and it is related to the width (smaller) and length (longer) of the acicular particles. The SEM micrographs together with TEM images confirmed the tendency to form agglomerate due to probably the nanometric dimensions of particles. The size distribution presented in Figure 5c,f,i,l,o,r,u is depended on the dopant concentration. The HRTEM images of the Eu-doped HAp powders are also presented in Figure 5 which allows us to see that the particles are well crystalize and the measured distance of the Miller indices correspond to the hexagonal hydroxylapatite.

3.4. FTIR Spectra

The FTIR spectra of Eu$_X$HAp samples with various europium concentrations are shown in Figure 6. The broad band in the region 3200–3400 cm^{-1} corresponds to [OH]$^-$ bands of adsorbed water. The strong band at 632 cm^{-1} (presented in all FTIR spectra) corresponds to [OH]$^-$ arising from stretching vibrational mode [50,51]. The typical bands attributed to [PO$_4$]$^{3-}$ can be also found. The bands at around 1090 cm^{-1} and about 1040 cm^{-1} may be assigned to the antisymetric stretching ν_3[PO$_4$]$^{3-}$ of P-O bond, while the band at 962 can be due to the symmetric stretching ν_1[PO$_4$]$^{3-}$. The 602 cm^{-1} and 564 cm^{-1} bands appear from ν_4[PO$_4$]$^{3-}$ of P-O bond. The band at 475 cm^{-1} can be attributed to the ν_2[PO$_4$]$^{3-}$ [52]. In all spectra of Eu$_X$HAp a band at 875 cm^{-1} was detected, and is due to [HPO$_4$]$^{2-}$ ions [53]. The intensity of the phosphate bands decreases with increase of europium concentration. The bands at 475 and 962 cm^{-1} progressively reduced their intensities with the increase of europium concentration, and disappear in Eu50HAp sample, as already suggested also by XRD results. The different possible mechanisms of Ca substitution with Eu are not fully comprehended and need further studies

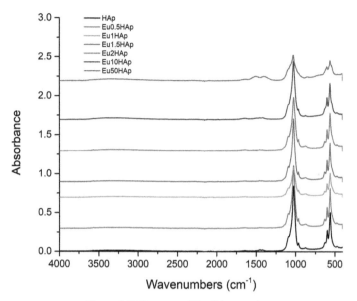

Figure 6. FTIR spectra of Eu$_X$HAp samples.

3.5. Raman Spectroscopy

The Raman spectra of Eu$_X$HAp samples are shown in Figure 7.

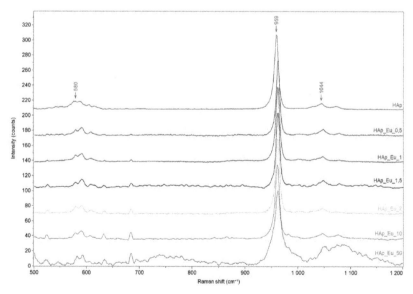

Figure 7. Raman spectra of Eu$_X$HAp samples.

As it can be seen from Figure 7 all analyzed powders present bands attributed to PO_4^{3-} group [54]. Raman spectrum of pure HAp shows a very intense band at 959 cm^{-1} attributed to symmetric stretching mode of PO_4^{3-}, which is the characteristic peak of HAp. The asymmetric stretching and symmetric bending modes of PO_4^{3-} are also observed at 580 cm^{-1} and 1044 cm^{-1} [53]. Raman spectra of Eu$_X$HAp powders show the internal modes of the frequency ν_1 PO_4^{3-} tetrahedral, which appears at 962 cm^{-1} and it corresponds to the symmetric stretching of P–O bonds [55]. There are observed several other bands at 580 cm^{-1} and 610 cm^{-1} attributed to ν_4 PO_4^{3-}, and 1048 cm^{-1} and 1080 cm^{-1}, respectively, attributed to ν_3 PO_4^{3-}. The intensity of the characteristic peak of HAp, compared to the other secondary peaks, decreases with increasing of the europium content.

3.6. UV-Vis and PL Spectra

The electronic spectra of Eu$_X$HAp samples (Figure 8) contain several bands, which increase in their intensity with increasing of europium content.

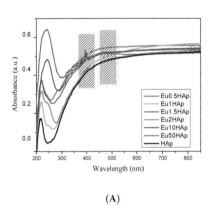
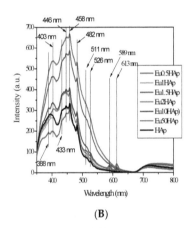

(A) (B)

Figure 8. (**A**) UV-Vis absorption spectra of Eu$_X$HAp at different Eu-concentrations (**B**) Room-temperature photoluminescence spectra of Eu$_X$HAp at different Eu-concentrations.

The pure HAp has an absorption peak in UV at 218 nm. The 242 nm (41322 cm^{-1}) absorption peak is absent in HAp, but is increasing in intensity from a weak shoulder in Eu0.5HAp to a strong, broad peak in Eu50HAp. This is a charge-transfer band characteristic to Eu^{3+} in oxides [56]. The band which appears at 395 nm, close to the visible light domain, corresponds to the $^7F_0 \rightarrow {}^5L_6$ transition, and is the most intense transition of europium (III) in UV-Vis absorption spectra [57]. Another three bands appear at 465, 526 and 535 nm, assignable to $^7F_0 \rightarrow {}^5D_2$; $^7F_0 \rightarrow {}^5D_1$; and $^7F_1 \rightarrow {}^5D_1$, transitions respectively [56]. The last peak, at 535 nm, is belonging to a group called "hot" bands because it can be observed only at room temperatures or higher since it requires the thermal population of $^7F_1$ level (at room temperature ~35% of ions are populating this level, rest being on $^7F_0$ ground state) [56].

Figure 8B presents the photoluminescence emission spectra of Eu$_X$HAp samples excited with 320 nm wavelength. Usually, HAp sample shows three strong emission peaks at 458 nm with two shoulders (433 nm and 446 nm), 403 nm and 482 nm, respectively, associated with various oxygen defects. The intensity of luminescence for HAp is enhanced by the presence of small quantities of Eu^{3+} ions. The effect is stronger for small numbers of dopant ions and is decreasing as the quantity of Eu^{3+} ions is increasing.

Over the fluorescence maxima of HAp, the emission spectra of europium are superposed. The peak at 389 nm is due to the $^7F_0 \rightarrow {}^5L_6$ electronic transition, the peak at 512 nm is due to the $^7F_0 \rightarrow {}^5D_1$ electronic transition and the peak at 526 nm is due to the $^7F_1 \rightarrow {}^5D_1$ electronic transition. The characteristic emission peaks at 589 nm and 613 nm are due to the $^5D_0 \rightarrow {}^7F_1$ and $^5D_0 \rightarrow {}^7F_2$

electronic transition, indicating a red fluorescence [58]. For higher Eu^{3+} concentration even the 648 nm emission peak of $^5D_0 \rightarrow {}^7F_3$ becomes visible. As expected, when the europium doping concentration increased, the characteristic emission intensity at 589 and 613 nm are also increased.

Beside the two dominant peaks, the weak $^5D_0 \rightarrow {}^7F_0$ transition from 579 nm (which can be observed in emission spectra of Eu50HAp) is related with Eu^{3+} ions distributed on Ca^{2+} sites of the apatitic structure [59]. The dominant emission peaks present no shift due to modification of Eu^{3+} concentration, the most intense being the hypersensitive $^5D_0 \rightarrow {}^7F_2$ transition from 613 nm. This feature shows the potential application of the Eu_xHAp compounds to be tracked or monitored by the characteristic luminescence.

4. Biocompatibility and Cytotoxicity of Eu_xHAp Photoluminescent Ceramic Materials

Cytotoxic effect of Eu_xHAp was evaluated by measuring the metabolic activity of AFSC using MTT assay. Eu_xHAp biomaterials did not have cytotoxic effect, the absorption values being near or higher compared to the control sample. Furthermore, the Eu_xHAp increases cellular metabolism, suggesting that it stimulates cell proliferation. The proliferation increases between 12%–58%, depending on the sample, highest increase being observed for Eu2HAp (58%) (Figure 9). The results are similar to other research groups that showed the viability of HGF-1 fibroblast was decreased in the Eu10Hap after 48 h. Naderi et al. 2012 [60] tested different concentrations of nanohydroxyapatite from 2 to 0.002 mg/mL on gingiva-derived fibroblast cell line (HGF-2) at 24, 48, and 72 h, and concluded that after 24 h high doses of nanohydroxyapatite have cytotoxic effect on gingival-derived fibroblasts suggesting that cytotoxicity is dependent on the cell line [60,61].

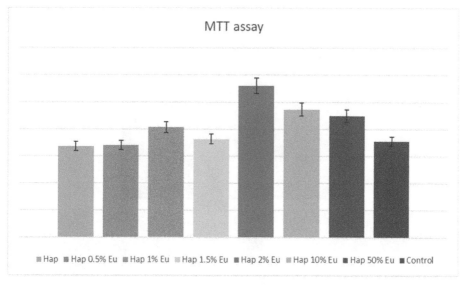

Figure 9. MTT assay showing the viability of AFSC in the presence of the Eu_xHAp ceramic materials: HAp, Eu05HAp, Eu1HAp, Eu1.5HAp, Eu2HAp, Eu10HAp, Eu50HAp, and control (cell only).

Glutathione, an oxidative stress marker, is capable of preventing cellular damage caused by reactive oxygen species, such as free radicals, peroxides, lipid peroxides and heavy metals. In the presence of Eu_xHAp biomaterials, AFSC responded similarly to control cells, indicating that the analysed materials did not induce cellular stress (Figure 10). Furthermore, the morphology of AFSC was investigated by fluorescence microscopy using CMTPX cell tracker for long-term tracing of living cells. Cellular metabolism is active, as shown in microscopy images, cells absorbing CMTPX fluorophore in the cytoplasm, suggesting that they are viable.

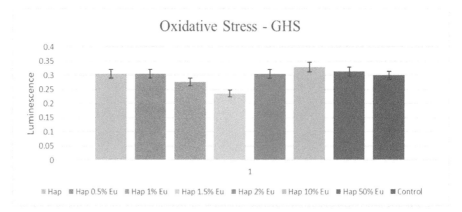

Figure 10. GSH assay showing the oxidative stress of AFSC in the presence of the $Eu_X HAp$ ceramic materials: HAp, Eu05HAp, Eu1HAp, Eu1.5HAp, Eu2HAp, Eu10HAp, Eu50HAp and control (cell only).

After 5 days in the presence of $Eu_X HAp$, AFSCs presents a normal morphology with fibroblastic-like characteristic appearance (Figure 11). Fluorescent images show that AFSC cells are viable, no dead cells or cell fragments are observed, more the cells spread filopodia to move and establish contacts with neighboring cells, suggesting that AFSC have an active phenotype.

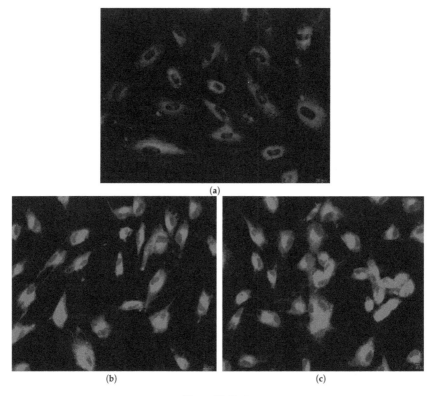

Figure 11. *Cont.*

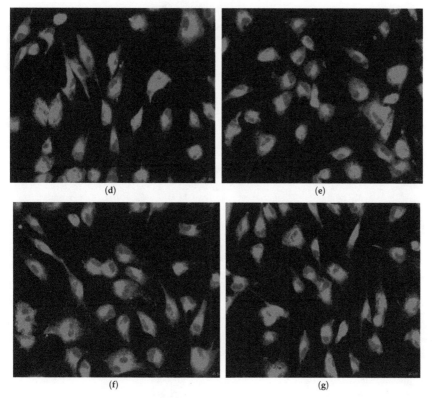

Figure 11. Fluorescence images of Eu$_X$HAp samples coloured with CMTPX fluorophore (**a**) Control sample, (**b**) Eu0.5HAp, (**c**) Eu1HAp, (**d**) Eu1.5HAp, (**e**) Eu2HAp, (**f**) Eu10HAp, (**g**) Eu50HAp.

5. Conclusions

In recent years, a special attention has been created regarding to multimodal imaging. This technique represents various imaging modalities in the manner of photoluminescence and magnetic resonance imaging that are generally connected in an individual diagnostic step. If we talk about multimodal contrast agents that can be used or marker for biomedical imaging with luminescence, literature data provides information on the use of doped hydroxyapatite with rare earth elements like europium of nanometric dimensions.

The aim of the present study was to obtain europium-doped nano hydroxylapatite by using a simple method of synthesis (co-precipitation method) and, thereafter characterize by physico-chemical techniques and also by biological point of view because of the fact that the obtained material has medical application.

The results after the XRD analysis shows that the obtained material is represented by pure hexagonal phase of hydroxylapatite, when using up to 10% dopant Eu^{3+} ions concentration, but there is a correlation between HAp and Eu^{3+} by the interference of Eu^{3+} ions with HAp crystal structure. About the morphology of HAp doped with Eu^{3+} it can be concluded that there is a limited modification after the addition of Eu^{3+} and also by the EDS spectra is confirmed the presence of Eu^{3+} in all obtained materials. Typically, the HAp sample shows three strong emission peaks, but the photoluminescence intensities is enhanced by the characteristic Eu^{3+} peaks with the presence of Eu^{3+} ions. This feature is required and necessary to have a good action in medical imaging.

From the biological point of view, it can be seen by the MTT assay results that the obtained material supports the proliferation process of the amniotic fluid stem cells. Also, the best results about the viability characteristic is offered by the Eu2HAp material which can be a standard for the doped HAp, because after the addition of more Eu^{3+} ions there is a small evidence about the cellular viability that can decrease. As a complete evaluation, the qualitative information by the fluorescence microscopy gives characteristic about the biocompatibility of the material in light of the fact that after 5 days of incubation with amniotic fluid stem cells, no dead cells are observed.

Therefore, there is a fine line about the using doped HAp with Eu^{3+} as fluorescent or multimodal contrast agent, but the results can be a promising start due to its characteristics but there is a need for the further investigation.

Author Contributions: The authors have participated to the paper as follows; conceptualization, B.S.V., D.P.; methodology, E.A., B.S.V., D.P., S.L.I.; validation, E.A., B.S.V.; formal analysis, I.A.N., A.P., R.T., O.O., E.T., O.R.V., A.I.N., A.V.S., F.I., A.C.B., S.L.I.; investigation, I.A.N., R.T., O.O., E.T., O.R.V., A.I.N., A.V.S., F.I., A.C.B.; data curation, A.V.P., A.M.M.; writing—original draft preparation, A.M.M, A.V.P.; writing—review and editing, I.A.N., B.S.V., A.M.M.; visualization, E.A.; supervision, E.A., B.S.V.; project administration, E.A.

Funding: This research was funded by project "Innovative biomaterials for treatment and diagnosis" BIONANOINOV—P3 grant number PN-IIIP1-1.2-PCCD-I2017-0629, and the support of the EU-funding project POSCCE-A2-O2.2.1-2013-1/Priority Axe 2, Project No. 638/12.03.2014, ID 1970, SMIS-CSNR code 48652 is gratefully acknowledged for the equipment's purchased from this project.

Conflicts of Interest: The authors declare no conflict of interest.

References

1. Zhou, Y.; Wu, C.; Chang, J. Bioceramics to regulate stem cells and their microenvironment for tissue regeneration. *Mater. Today* **2019**, *24*, 41–56. [CrossRef]

2. Ben-nissan, B.; Cazalbou, S.; Choi, A. Bioceramics. *Biomater. Sci. Eng.* **2019**, *1*, 16–33.

3. Dorozhkin, S.V. Calcium orthophosphates (CaPO4): Occurrence and properties (Les orthophosphates de calcium (CaPO4): Occurrence et Proprieties). *Morphologie* **2017**, *101*, 125–142. [CrossRef] [PubMed]

4. Dorozhkin, S.V. A history of calcium orthophosphates (CaPO4) and their biomedical applications. *Morphologie* **2017**, *101*, 143–153. [CrossRef] [PubMed]

5. Pelin, I.M.; Maier, V.; Suflet, D.M.; Popescu, I.; Darie-Nita, R.N.; Aflori, M.; Butnaru, M. Formation and characterization of calcium orthophosphates in the presence of two different acidic macromolecules. *J. Cryst. Growth* **2017**, *475*, 266–273. [CrossRef]

6. Sathiskumar, S.; Vanaraj, S.; Sabarinathan, D.; Bharath, S. Green synthesis of biocompatible nanostructured hydroxyapatite from Cirrhinus mrigala fish scale—A biowaste to biomaterial. *Ceram. Int.* **2019**, *45*, 7804–7810. [CrossRef]

7. Beau, S.; Rouillon, T.; Millet, P.; Le, J.; Weiss, P.; Chopart, J.; Daltin, A. Synthesis of calcium-deficient hydroxyapatite nanowires and nanotubes performed by template-assisted electrodeposition. *Mater. Sci. Eng. C Mater. Biol. Appl.* **2019**, *98*, 333–346.

8. Xia, X.; Chen, J.; Shen, J.; Huang, D.; Duan, P.; Zou, G. Synthesis of hollow structural hydroxyapatite with different morphologies using calcium carbonate as hard template. *Adv. Powder Technol.* **2018**, *29*, 1562–1570. [CrossRef]

9. Hashemi, S.; Javadpour, J.; Khavandi, A.; Erfan, M. Morphological evolution on the surface of hydrothermally synthesized hydroxyapatite microspheres in the presence of EDTMP. *Ceram. Int.* **2018**, *44*, 19743–19750.

10. Turan, İ.; Kaygili, O.; Tatar, C.; Bulut, N.; Koytepe, S.; Ates, T. The effects of Ni-addition on the crystal structure, thermal properties and morphology of Mg-based hydroxyapatites synthesized by a wet chemical method. *Ceram. Int.* **2018**, *44*, 14036–14043.

11. Wolff, J.; Hofmann, D.; Amelung, W.; Lewandowski, H.; Kaiser, K.; Bol, R. Applied Geochemistry Rapid wet chemical synthesis for 33 P-labelled hydroxyapatite—An approach for environmental research. *Appl. Geochem.* **2018**, *97*, 181–186. [CrossRef]

12. Yelten-yilmaz, A.; Yilmaz, S. Wet chemical precipitation synthesis of hydroxyapatite (HA) powders. *Ceram. Int.* **2018**, *44*, 9703–9710. [CrossRef]

13. Jiao, Y.; Lu, Y.; Xiao, G.; Xu, W.; Zhu, R. Preparation and characterization of hollow hydroxyapatite microspheres by the centrifugal spray drying method. *Powder Technol.* **2012**, *217*, 581–584. [CrossRef]

14. Sabu, U.; Logesh, G.; Rashad, M.; Joy, A.; Balasubramanian, M. Microwave assisted synthesis of biomorphic hydroxyapatite. *Ceram. Int.* **2019**, *45*, 6718–6722. [CrossRef]

15. Ghorbani, F.; Zamanian, A.; Behnamghader, A.; Daliri-joupari, M. Bone-like hydroxyapatite mineralization on the bio-inspired PDA nanoparticles using microwave irradiation. *Surf. Interfaces* **2019**, *15*, 38–42. [CrossRef]

16. Anwar, A.; Akbar, S. Novel continuous microwave assisted flow synthesis of nanosized manganese substituted hydroxyapatite. *Ceram. Int.* **2018**, *44*, 10878–10882. [CrossRef]

17. Wu, S.; Hsu, H.; Hsu, S.; Chang, Y.; Ho, W. Synthesis of hydroxyapatite from eggshell powders through ball milling and heat treatment. *J. Asian Ceram. Soc.* **2016**, *4*, 85–90. [CrossRef]

18. Hannora, A.E.; Ataya, S. Structure and compression strength of hydroxyapatite/titania nanocomposites formed by high energy ball milling. *J. Alloy. Compd.* **2016**, *658*, 222–233. [CrossRef]

19. Ferro, A.C.; Guedes, M. Mechanochemical synthesis of hydroxyapatite using cuttle fish bone and chicken eggshell as calcium precursors. *Mater. Sci. Eng. C Mater. Biol. Appl.* **2019**, *97*, 124–140. [CrossRef]

20. Bouyarmane, H.; Gouza, A.; Masse, S.; Saoiabi, S.; Saoiabi, A.; Coradin, T.; Laghzizil, A. Nanoscale conversion of chlorapatite into hydroxyapatite using ultrasound irradiation. *Colloids Surf. A Physicochem. Eng. Asp.* **2016**, *495*, 187–192. [CrossRef]

21. Nikolaev, A.L.; Gopin, A.V.; Severin, A.V.; Rudin, V.N.; Mironov, M.A.; Dezhkunov, N.V. Ultrasonic synthesis of hydroxyapatite in non-cavitation and cavitation modes. *Ultrason. Sonochem.* **2018**, *44*, 390–397. [CrossRef] [PubMed]

22. Valle, L.J.; Katsarava, R. *Other Miscellaneous Materials and Their Nanocomposites*; Elsevier Inc.: Amsterdam, The Netherlands, 2019; ISBN 9780128146156.

23. Fihri, A.; Len, C.; Varma, R.S.; Solhy, A. Hydroxyapatite: A review of syntheses, structure and applications in heterogeneous catalysis. *Coord. Chem. Rev.* **2017**, *347*, 48–76. [CrossRef]

24. Sadat-Shojai, M.; Khorasani, M.T.; Dinpanah-Khoshdargi, E.; Jamshidi, A. Synthesis methods for nanosized hydroxyapatite with diverse structures. *Acta Biomater.* **2013**, *9*, 7591–7621. [CrossRef]

25. Kubasiewicz-Ross, P.; Hadzik, J.; Seeliger, J.; Kozak, K.; Jurczyszyn, K.; Gerber, H.; Dominiak, M.; Kunert-Keil, C. New nano-hydroxyapatite in bone defect regeneration: A histological study in rats. *Ann. Anat.* **2017**, *213*, 83–90. [CrossRef] [PubMed]

26. Munhoz, M.A.S.; Hirata, H.H.; Plepis, A.M.G.; Martins, V.C.A.; Cunha, M.R. Use of collagen/chitosan sponges mineralized with hydroxyapatite for the repair of cranial defects in rats. *Injury* **2018**, *49*, 2154–2160. [CrossRef] [PubMed]

27. Solonenko, A.P.; Blesman, A.I.; Polonyankin, D.A. Preparation and in vitro apatite-forming ability of hydroxyapatite and β-wollastonite composite materials. *Ceram. Int.* **2018**, *44*, 17824–17834. [CrossRef]

28. Kaur, K.; Singh, K.J.; Anand, V.; Islam, N.; Bhatia, G.; Kalia, N.; Singh, J. Lanthanide (=Ce, Pr, Nd and Tb) ions substitution at calcium sites of hydroxyl apatite nanoparticles as fluorescent bio probes: Experimental and density functional theory study. *Ceram. Int.* **2017**, *43*, 10097–10108. [CrossRef]

29. Safarzadeh, M.; Ramesh, S.; Tan, C.Y.; Chandran, H.; Fauzi, A.; Noor, M.; Krishnasamy, S.; Alengaram, U.J.; Ramesh, S. Effect of multi-ions doping on the properties of carbonated hydroxyapatite bioceramic. *Ceram. Int.* **2019**, *45*, 3473–3477. [CrossRef]

30. Tesch, A.; Wenisch, C.; Herrmann, K.H.; Reichenbach, J.R.; Warncke, P.; Fischer, D.; Müller, F.A. Luminomagnetic Eu^{3+}- and Dy^{3+}-doped hydroxyapatite for multimodal imaging. *Mater. Sci. Eng. C Mater. Biol. Appl.* **2017**, *81*, 422–431. [CrossRef]

31. Zilm, M.E.; Yu, L.; Hines, W.A.; Wei, M. Magnetic properties and cytocompatibility of transition-metal-incorporated hydroxyapatite. *Mater. Sci. Eng. C Mater. Biol. Appl.* **2018**, *87*, 112–119. [CrossRef]

32. Li, H.; Sun, X.; Li, Y.; Li, B.; Liang, C.; Wang, H. Preparation and properties of carbon nanotube (Fe)/hydroxyapatite composite as magnetic targeted drug delivery carrier. *Mater. Sci. Eng. C Mater. Biol. Appl.* **2019**, *97*, 222–229. [CrossRef] [PubMed]

33. Riaz, M.; Zia, R.; Ijaz, A.; Hussain, T.; Mohsin, M.; Malik, A. Synthesis of monophasic Ag doped hydroxyapatite and evaluation of antibacterial activity. *Mater. Sci. Eng. C Mater. Biol. Appl.* **2018**, *90*, 308–313. [CrossRef] [PubMed]

34. Kim, H.; Mondal, S.; Jang, B.; Manivasagan, P. Biomimetic synthesis of metal—Hydroxyapatite (Au-HAp, Ag-HAp, Au-Ag- HAp): Structural analysis, spectroscopic characterization and biomedical application. *Ceram. Int.* **2018**, *44*, 20490–20500. [CrossRef]

35. Gonzalez, G.; Costa-vera, C.; Borrero, L.J.; Soto, D.; Lozada, L.; Chango, J.I.; Diaz, J.C.; Lascano, L. Effect of carbonates on hydroxyapatite self-activated photoluminescence response. *J. Lumin.* **2018**, *195*, 385–395. [CrossRef]

36. Neacsu, I.A.; Stoica, A.E.; Vasile, B.S. Luminescent Hydroxyapatite Doped with Rare Earth Elements for Biomedical Applications. *Nanomaterials* **2019**, *9*, 239. [CrossRef] [PubMed]

37. Liu, J.; Lécuyer, T.; Seguin, J.; Mignet, N.; Scherman, D.; Viana, B.; Richard, C. Imaging and therapeutic applications of persistent luminescence nanomaterials. *Adv. Drug Deliv. Rev.* **2019**, *138*, 193–210. [CrossRef] [PubMed]

38. Xing, Q.; Zhang, X.; Wu, D.; Han, Y.; Wickramaratne, M.N.; Dai, H.; Wang, X. Ultrasound-Assisted Synthesis and Characterization of Heparin-Coated Eu^{3+} Doped Hydroxyapatite Luminescent Nanoparticles. *Colloid Interface Sci. Commun.* **2019**, *29*, 17–25. [CrossRef]

39. Kudryashova, V.A. Europium and terbium pyrrole-2-carboxylates: Structures, luminescence, and energy transfer. *Inorg. Chim. Acta* **2019**, *492*, 1–7.

40. Corbin, B.A.; Hovey, J.L.; Thapa, B.; Schlegel, H.B.; Allen, M.J. Luminescence differences between two complexes of divalent europium. *J. Organomet. Chem.* **2018**, *857*, 88–93. [CrossRef]

41. Utochnikova, V.V.; Koshelev, D.S.; Medvedko, A.V.; Kalyakina, A.S.; Bushmarinov, I.S.; Grishko, A.Y.; Schepers, U.; Bräse, S.; Vatsadze, S.Z. Europium 2-benzofuranoate: Synthesis and use for bioimaging. *Opt. Mater.* **2017**, *74*, 191–196. [CrossRef]

42. Petrochenkova, N.V.; Mirochnik, A.G.; Emelina, T.B.; Sergeev, A.A.; Leonov, A.A.; Voznesenskii, S.S. Luminescent amine sensor based on europium(III) chelate. *Spectrochim. Acta Part A Mol. Biomol. Spectrosc.* **2018**, *200*, 70–75. [CrossRef] [PubMed]

43. Lima, T.A.R.M.; Valerio, M.E.G. X-ray absorption fine structure spectroscopy and photoluminescence study of multifunctional europium (III)-doped hydroxyapatite in the presence of cationic surfactant medium. *J. Lumin.* **2018**, *201*, 70–76. [CrossRef]

44. Mugnaioli, E.; Reyes-gasga, J.; Kolb, U.; Hemmerlø, J. Evidence of Noncentrosymmetry of Human Tooth Hydroxyapatite Crystals. *Chem. A Eur. J.* **2014**, *20*, 6849–6852. [CrossRef] [PubMed]

45. Chen, M.; Tan, J.; Lian, Y.; Liu, D. Preparation of Gelatin coated hydroxyapatite nanorods and the stability of its aqueous colloidal. *Appl. Surf. Sci.* **2008**, *254*, 2730–2735. [CrossRef]

46. Mullica, D.F.; Milligan, W.O.; Beall, G.W. Crystal structures of Pr(OH)$_3$, Eu(OH)$_3$ and Tm(OH)$_3$. *J. Inorg. Nucl. Chem.* **1979**, *41*, 525–532. [CrossRef]

47. Terra, J.; Dourado, E.R.; Eon, J.G.; Ellis, D.E.; Gonzalez, G.; Rossi, A.M. The structure of strontium-doped hydroxyapatite: An experimental and theoretical study. *Phys. Chem. Chem. Phys.* **2009**, *11*, 568–577. [CrossRef]

48. Shannon, R.D. Revised effective ionic radii and systematic studies of interatomic distances in halides and chalcogenides. *Acta Crystallogr. Sect. A* **1976**, *32*, 751–767. [CrossRef]

49. Fahami, A.; Nasiri-Tabrizi, B.; Beall, G.W.; Basirun, W.J. Structural insights of mechanically induced aluminum-doped hydroxyapatite nanoparticles by Rietveld refinement. *Chin. J. Chem. Eng.* **2017**, *25*, 238–247. [CrossRef]

50. Fowler, B. Infrared Studies of Apatites. I. Vibrational Assignments for Calcium, Strontium, and Barium Hydroxyapatites Utilizing Isotopic Substitution. *Inorg. Chem.* **1974**, *13*, 194–207. [CrossRef]

51. Iconaru, S.; Motelica-Heino, M.; Predoi, D. Study on Europium-Doped Hydroxyapatite Nanoparticles by Fourier Transform Infrared Spectroscopy and Their Antimicrobial Properties. *J. Spectrosc.* **2013**, *2013*, 284285. [CrossRef]

52. Sitarz, M.; Rokita, M.; Bułat, K. Infrared spectroscopy of different phosphates structures. *Spectrochim. Acta Part A Mol. Biomol. Spectrosc.* **2011**, *79*, 722–727.

53. Markovic, M.; Fowler, O.B. Preparation and Comprehensive Characterization of a Calcium Hydroxyapatite Reference Material Volume. *J. Res. Natl. Inst. Stand. Technol.* **2004**, *109*, 553–568. [CrossRef] [PubMed]

54. Cuscó, R.; Guitián, F.; de Aza, S.; Arttis, L. Differentiation between Hydroxyapatite and/I-Tricalcium Phosphate by Means of p-Raman Spectroscopy. *J. Eur. Ceram. Soc.* **1998**, *2219*, 1301–1305. [CrossRef]

55. Ciobanu, C.S.; Massuyeau, F.; Andronescu, E.; Stan, M.S.; Predoi, D. Biocompatibility study of europium doped crystalline hydroxyapatite bioceramics. *Dig. J. Nanomater. Biostructures* **2011**, *6*, 1639–1647.

56. Binnemans, K. Interpretation of europium (III) spectra. *Coord. Chem. Rev.* **2015**, *295*, 1–45. [CrossRef]

57. Al-kattan, A.; Dufour, P.; Dexpert-ghys, J.; Drouet, C. Preparation and Physicochemical Characteristics of Luminescent Apatite-Based Colloids. *J. Phys. Chem. C* **2010**, *114*, 2918–2924. [CrossRef]

58. Yang, C.; Yang, P.; Wang, W.; Gai, S.; Wang, J.; Zhang, M.; Lin, J. Synthesis and characterization of Eu-doped hydroxyapatite through a microwave assisted microemulsion process. *Solid State Sci.* **2009**, *11*, 1923–1928. [CrossRef]

59. Ciobanu, C.S.; Iconaru, S.L.; Massuyeau, F.; Constantin, L.V.; Costescu, A.; Predoi, D. Synthesis, Structure, and Luminescent Properties of Europium-Doped Hydroxyapatite Nanocrystalline Powders. *J. Nanomater.* **2012**, *2012*, 942801. [CrossRef]

60. Noushin Jalayer Naderi, R.Y.; Jamali, D.; Mohammad Rezvani, B. Cytotoxicity of nano-hydroxyapatite on gingiva-derived fibroblast cell line (HGF2): An in vitro study. *Daneshvar Med.* **2012**, *19*, 79–86.

61. Paulina-Guadalupe, M.-M.; Gabriel-Alejandro, M.-C.; Nereyda, N.-M.; Nuria, P.-M.; Miguel-Ángel, C.-S.; Brenda-Erendida, C.-S.; Facundo, R. Facile Synthesis, Characterization, and Cytotoxic Activity of Europium-Doped Nanohydroxyapatite. *Bioinorg. Chem. Appl.* **2016**, *2016*, 1057260.

Article

Novel Nanocomposites Based on Functionalized Magnetic Nanoparticles and Polyacrylamide: Preparation and Complex Characterization

Eugenia Tanasa [1,2], Catalin Zaharia [3,*], Ionut-Cristian Radu [3], Vasile-Adrian Surdu [1,2], Bogdan Stefan Vasile [1,4], Celina-Maria Damian [3] and Ecaterina Andronescu [1,4]

1   University Politehnica of Bucharest, Faculty of Applied Chemistry and Materials Science, 060042 Bucharest, Romania; eugenia.vasile27@gmail.com (E.T.); adrian.surdu@upb.ro (V.-A.S.); bogdan.vasile@upb.ro (B.S.V.); ecaterina.andronescu@upb.ro (E.A.)
2   National Centre for Micro and Nanomaterials, University Politehnica of Bucharest, 060042 Bucharest, Romania
3   Advanced Polymer Materials Group, University Politehnica of Bucharest, 060042 Bucharest, Romania; radu.ionucristian@gmail.com (I.-C.R.); celina.damian@yahoo.com (C.-M.D.)
4   National Research Center for Food Safety, University Politehnica of Bucharest, 060042 Bucharest, Romania
*   Correspondence: zaharia.catalin@gmail.com

Received: 1 September 2019; Accepted: 23 September 2019; Published: 27 September 2019

Abstract: This paper reports the synthesis and complex characterization of nanocomposite hydrogels based on polyacrylamide and functionalized magnetite nanoparticles. Magnetic nanoparticles were functionalized with double bonds by 3-trimethoxysilyl propyl methacrylate. Nanocomposite hydrogels were prepared by radical polymerization of acrylamide monomer and double bond modified magnetite nanoparticles. XPS spectra for magnetite and modified magnetite were recorded to evaluate the covalent bonding of silane modifying agent. Swelling measurements in saline solution were performed to evaluate the behavior of these hydrogels having various compositions. Mechanical properties were evaluated by dynamic rheological analysis for elastic modulus and vibrating sample magnetometry was used to investigate the magnetic properties. Morphology, geometrical evaluation (size and shape) of nanostructural characteristics and the crystalline structure of the samples were investigated by SEM, HR-TEM and selected area electron diffraction (SAED). The nanocomposite hydrogels will be further tested for the soft tissue engineering field as repairing scaffolds, due to their mechanical and magnetization behavior that can stimulate tissue regeneration.

Keywords: magnetic nanoparticles; polyacrylamide; functionalization; nanocomposite; hydrogel

---

1. Introduction

Polymeric hydrogel-like materials are a category of soft materials containing crosslinked hydrophilic networks with a high swelling ability. The hydrophilic nature of the macromolecular chains is based, in general, on side hydrophilic active groups [1–4]. The cross-linking reaction of hydrophilic chains is an absolute requirement for dissolution avoiding of polymeric material. The generation of a cross-linked network assumes formation of inter and intramolecular bridges, which do not allow the solvent molecules to solve and unfold the macromolecules. Thus, the solvent can only penetrate among polymeric molecules and swell the material [5,6]. In the swollen state, the polymeric hydrogel exhibits brittleness and obvious low mechanical properties. These disadvantages seriously limit their usage in special biomedical applications. The use of polymeric hydrogels is directly related to the intrinsic mechanical properties in the swollen state. A relatively new concept of polymeric nanocomposite hydrogels has started to overcome these problems by combining the advantage of polymeric hydrogels with the advantage of polymeric nanocomposites [7–15]. Nanocomposite hydrogels have been developed by various methods, such as in situ polymerization or pre-modified inorganic nanoparticles [16–21]. Modified inorganic nanomaterials have gained special attention, as they can be used as inorganic crosslinkers. These types of modified crosslinkers exhibit a unique flexible intrinsic structure with a serious contribution to improving mechanical properties [22]. The major limitation of the swollen hydrogels is related to the network generation process based on traditionally low molecular weight organic crosslinkers. The limitation of classic organic crosslinkers, due to their relative low number of available groups for reactions with polymeric chains, can be overcome by inorganic nanoparticles modified with multiple groups. A suitable modification involves designing molecular architectures with long and short intermolecular and intramolecular bridges at the same time [7,23–27]. The mechanical stress generates the fracture first of short chains to partially dissipate the elastic energy, while the long chains take the remaining loading. Meanwhile, the hydrogel is still intact [4,28–31]. Inorganic nanoparticles as crosslinkers possess high stretchability, elasticity and superior toughness for polymeric nanocomposite hydrogels, with potential use in soft tissue applications. Inorganic nanoparticles such as magnetite exhibit a high potential for modification with functional groups, due to the presence of hydroxyl groups. They show outstanding physico-chemical properties due to the presence of both species of iron [32–34]. Furthermore, magnetite has been used with great success for various biomedical applications [34–37], including cellular imaging [38] or cancer diagnosis, monitoring and treatment [39].

This research study is focused on the development of nanocomposite networks crosslinked by highly-functionality modified magnetite with enhanced stretchability and elasticity for biological tissue applications.

2. Materials and Methods

2.1. Materials

The reagents used for the synthesis of the magnetic iron oxide nanoparticles were iron chloride iron (III) chloride ($FeCl_3$, 97%), ferrous sulfate heptahydrate ($FeSO_4 \cdot 7H_2O$) and ammonium hydroxide solution (NH_4OH). The acrylamide monomer, 3-trimethoxysilyl propyl methacrylate modifier agent and potassium persulfate initiator were used for the preparation of hydrogels. All the reagents were supplied by Sigma-Aldrich, 3050 Spruce Street, St. Louis, MO, United States.

2.2. Synthesis of Magnetite (Fe_3O_4) Nanoparticles

The synthesis of the Fe_3O_4 nanoparticles (MNPs) was carried out at room temperature, by co-precipitation method, starting from iron (III) chloride, ferrous sulfate heptahydrate and ammonium hydroxide solution [40,41]. The iron chloride was dissolved in deionized water to give a clear solution. Under vigorous magnetic stirring, the $FeSO_4 \cdot 7H_2O$ was added to the solution (Fe^{2+}/Fe^{3+} = 1:2 molar ratio). Independently, an aqueous solution of ammonium hydroxide is prepared, and the mixture solution resulting from the iron chloride and ferrous sulfate heptahydrate was added to it. Magnetite nanoparticles formed and precipitated. The MNPs were separated from the reaction medium using a strong magnet. The powder was rinsed several times with distilled water until reaching a neutral pH (pH = 7) in the washing solution. After washing, the precipitate was dried for 12h in air oven, at 60 °C.

2.3. Synthesis of Double Bond Modified Magnetite Nanoparticles

The surface modification of the magnetic nanoparticles with double bonds was carried out in several steps, as follows (Figure 1). Briefly, 2 g of MNPs were reacted with 4 mL of 3-trimethoxysilyl propyl methacrylate (3-TPM) by dispersion in 40 mL of toluene for 24 hours at room temperature under magnetic stirring. The modified magnetic nanoparticles (denoted by MMNPs) were then washed several times with toluene to remove the unmodified MNPs and unreacted 3-TPM by centrifugation and then dried.

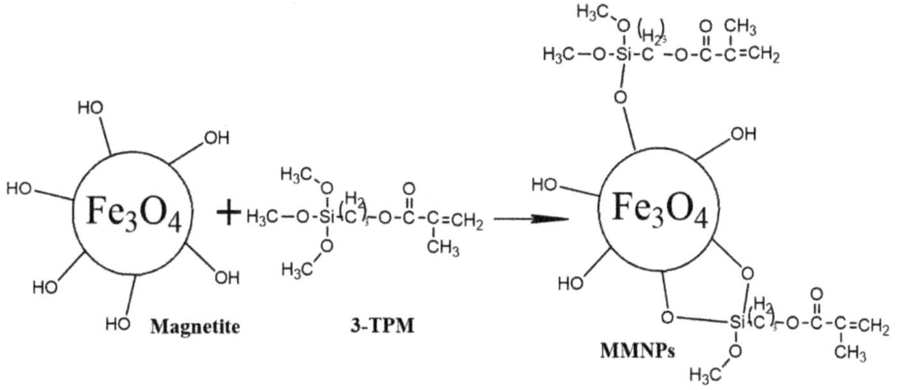

Figure 1. Modification of magnetite nanoparticles with double bonds.

2.4. Preparation of Polyacrylamide/MMNPs Nanocomposite Hydrogels (PAA/MMNPs)

Hydrogels were obtained by free-radical polymerization of acrylamide and MMNPs in aqueous solution (Figure 2). Briefly, various ratios between acrylamide monomer and MMNPs (90/10; 80/20; 70/30; 60/40 and 50/50 *w/w*) were prepared. The MMNPs were dispersed in water by sonication and added in a mixture of 15 wt. % aqueous acrylamide solution and initiator (potassium persulfate). The ratio between organic phase (acrylamide) and MMNPs was varied in order to enhance the mechanical properties of the hydrogels. The nanocomposite hydrogel samples were added in circular glass matrix and put at 60 °C for 24 h. Finally, the nanocomposite hydrogel samples were removed from the glass matrix and immersed in distilled water for 5 days to remove residual monomer and final purification. Hydrogels were cut as disks for further mechanical investigations (rheological measurements).

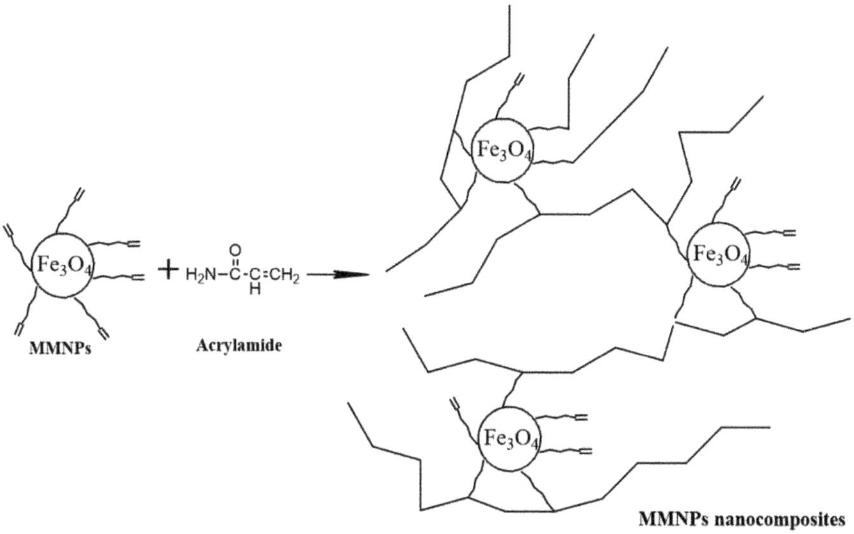

Figure 2. Preparation of polyacrylamide (PAA)/modified magnetic nanoparticles (MMNPs) nanocomposites.

2.5. Swelling Measurements

Swelling behavior of the hydrogels was performed in saline solution at 37 °C. The weight changes of the hydrogels were recorded at regular time intervals during swelling. The swelling degree of the hydrogels was determined according to the following equation [42,43]:

$$SD = \frac{W_t - W_o}{W_0} \cdot 100, \tag{1}$$

where W and W_0 denote the weight of the wet hydrogel at a predetermined time and the weight of the dry sample, respectively. The equilibrium swelling degrees (ESD) were measured until the weight of the swollen hydrogels was constant. At least three swelling measurements were performed for each hydrogel sample and the mean values were reported.

Swelling kinetics. The dynamics of the water sorption process was studied by monitoring the saline solution absorption by the hydrogels at different time intervals. For diffusion kinetic analysis, the swelling results were used only up to 60% of the swelling curves. Fick's equation was used [42–49]:

$$f = k \cdot t^n, \tag{2}$$

where f is the fractional water uptake, k is a constant, t is swelling time and n is the swelling coefficient that indicates whether diffusion or relaxation controls the swelling process. The fractional water content f is M_t/M_n where M_t is the mass of water in the hydrogel at time t, and M_n is the mass of the water at equilibrium.

2.6. Characterization Methods

FTIR analysis. FTIR spectra of native magnetite and 3-TPM modified magnetite were recorded on a Bruker Vertex 70 FT-IR spectrophotometer with attenuated total reflectance (ATR) accessory with 32 scans and 4 cm^{-1} resolution in mid-IR region.

XPS analysis. The X-ray photoelectron spectroscopy spectra for magnetite and modified magnetite were recorded to evaluate the covalent bonding of silane modifying agent. The spectra were recorded on a K-Alpha instrument from Thermo Scientific, using a monochromated Al Kα source (1486.6 eV), at a pressure of 2×10^{-9} mbar.

2.6.1. Evaluation of the Rheological Properties for the Nanocomposite Hydrogels

Rheological tests were performed with a rotational rheometer Kinexus Pro, Malvern Instruments, and a temperature control unit. In oscillating mode, a parallel plate and a geometric measuring system were used, and the gap was set according to the force value. The tests were performed on samples of 20 mm diameter with parallel plate geometry in a frequency range 1 to 30 Hz.

2.6.2. Magnetic Properties by Vibrating Sample Magnetometry (VSM)

Vibrating sample magnetometry (LakeShore 7404-s VSM) was used in order to investigate the magnetic behavior of the hydrogels. Hysteresis loops were recorded at room temperature with an applied field up to 15 kOe, increments of 200 Oe and ramp rate of 20 Oe/s.

2.6.3. Morphological Characterization by Scanning Electron Microscopy (SEM) and Transmission Electron Microscopy (TEM)

The microstructure of the samples was analyzed by Scanning Electron Microscopy (SEM) using a Quanta Inspect F50, with a field emission gun (FEG) having 1.2 nm resolution and an energy dispersive X-ray spectrometer (EDXS) having 133 eV resolution at MnKα. Morphology, geometrical evaluation (size and shape) of nanostructural characteristics and the crystalline structure of the samples were investigated by high-resolution transmission electron microscopy (HR-TEM) and selected area electron diffraction (SAED) using a TECNAI F30 G2 S-TWIN microscope operated at 300 kV with energy dispersive X-ray analysis (EDAX) facility.

3. Results and Discussion

3.1. Swelling Measurements

The most important property of a hydrogel is its ability to absorb and hold an amount of solvent in its network structure. The equilibrium swelling of a hydrogel is a result of the balance of osmotic forces determined by the affinity to the solvent and network elasticity. Hydrogel properties depend strongly on the degree of cross-linking, the chemical composition of the polymer chains, and the interactions of the network and surrounding liquid. Figure 3 shows the water swelling behavior of the PAA/MMNPs hydrogels. The swelling curves show a decreasing trend of swelling degree with the increase of the modified magnetite nanoparticles content (Figure 3). These results are sustained by the fact that a higher amount of MMNPs lead to a higher crosslinking density. The crosslinking of the hydrogel comes from the reaction between the double bonds from NPs surface and the double bonds of the acrylamide monomer without the adding of any other crosslinker.

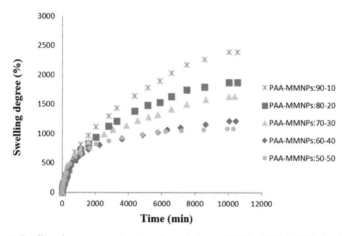

Figure 3. Swelling degree versus time in saline solution at 37 °C for PAA/MMNPs hydrogels.

Next, the swelling mechanism is evaluated by Equation (2). Here, by plotting ln f versus ln t, we may calculate the swelling coefficient n as the slope of the linear graph. It is known that the swelling process could be controlled by a Fickian-type mechanism, by relaxation of the chain or by both mechanisms depending on the composition. The values of n were below 0.5 for 2 samples (PAA/MMNPs 70/30, 60/40 ratio), which means a diffusion-controlled process (Fickian mechanism). The other three nanocomposite samples (PAA/MMNPs, 90/10, 80/20 and 50/50 ratio) are governed by a diffusion swelling coefficient with values above 0.5 and a water molecules transport model, done by chain relaxation [50,51]. These data are shown in Table 1.

Table 1. The swelling diffusion coefficient and the regression model-R^2.

Parameters/ Composition	PAA/MMNPs 90:10	PAA/MMNPs 80:20	PAA/MMNPs 70:30	PAA/MMNPs 60:40	PAA/MMNPs 50:50
n	0.6044	0.5816	0.4400	0.4818	0.5465
R^2	0.9980	0.9979	0.9985	0.9981	0.9939

3.2. FTIR Analysis

The modification of magnetite nanoparticles with 3-TPM was proved by FTIR investigation (Figure 4). FTIR spectrum of modified magnetite shows several new peaks specific to organic modifier 3-TPM. Therefore, the peak at 1170 cm^{-1} can be assigned to stretching vibration of ester bonds; peaks at 1299 cm^{-1} and 1325 cm^{-1} can be assigned to the stretching vibration of -Si-methylene- from the internal structure of modifier agent; peaks at 1454 cm^{-1} and 1412 cm^{-1} can be assigned to the bending vibration of methyl and methylene groups from the internal structure of the modifier agent; the peak at 1638 cm^{-1} is specific to the stretching vibration of –C=C– from the internal structure of the modifier agent; the peak at 1719 cm^{-1} is specific to the stretching vibration of carbonyl –C=O from the internal structure of the modifier agent [52]. Considering all of the attributed peaks, FTIR analysis was a very useful tool to evidence the modification of the magnetite nanoparticles with double bonds.

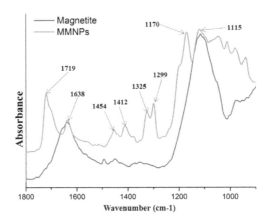

Figure 4. FTIR spectra for magnetite and double bond functionalized magnetite nanoparticles.

3.3. XPS Analysis

XPS analysis for both magnetite and double bond modified magnetite was carried out in order to reveal the interstitial organic/inorganic character of new generated magnetite lattice. The results for surface modification are well correlated with the reaction mechanism and morphological results. There is an increasing of C1s in the elemental composition up to the main elemental percent, due to the modification on the surface of magnetite nanoparticles. Figure 5 highlights the high resolution spectra of the O1s species from crude magnetite with two deconvoluted peaks, the first centered at 530.35 eV, which can be attributed to O-Fe in magnetite phase [53], and the second centered at 531.01 eV, probably corresponding to the hydroxyl bonding within magnetite lattice. Furthermore, Figure 5 reveals the high magnification spectra of O1s species for functionalized magnetite nanoparticles with three secondary deconvoluted peaks. The two O1s peaks at 529.67 eV and 531.13 eV can be attributed to the crude magnetite structure and the new peak centered at 533.01 eV can be attributed to a Si-O new formed species by covalent bonding of silane with magnetite hydroxyl groups [22].

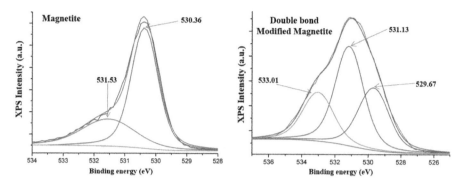

Figure 5. XPS spectra of magnetite and double bond modified magnetite.

3.4. Evaluation of the Rheological Properties for the Nanocomposite Hydrogels

Rheological behavior of novel nanocomposites was performed on swollen samples in aqueous NaCl 0.9 wt% solution at swelling equilibrium. The investigation involves the stress optimization in order to maintain a linear viscoelastic domain and samples to be dependent only on frequency and not on the applied stress. The elastic modulus for nanocomposite with 10% modified magnetite nanoparticles showed a unique behavior with significant differences, as compared to other samples. Figure 6 reveals a slow decreasing elastic of the modulus G′ up to 20 Hz, followed by a fast increasing until 30 Hz for the sample with 90% PAA and 10% modified magnetite nanoparticles. This behavior can be explained by a low amount of modified magnetite nanoparticles, which act as a crosslinking agent. The low amount of inorganic modified agent does not allow the specific elastic network to adapt to environmental mechanical changes [22]. The nanocomposite samples with a higher amount of modified magnetite nanoparticles (30%, 50%) showed a different specific elastic behavior with frequency variation, presenting a constant elastic modulus increasing from 1Hz up to 30 Hz. The specific elastic behavior allows for the environmental changes, due to the formation of elastically active chains by bridging multiple surrounding chains with various lengths. In the case of 30% modified magnetite nanoparticles, the elastic modulus exhibited higher values over the frequency range. This is probably due to the nanoparticles concentration that is optimal for a good dispersion into polymer matrix. In the case of the 50% modified magnetite nanoparticles, the elastic modulus showed lower values, probably due to a lower dispersion in the matrix, with significant influences on the segmental mobility of the 3D network.

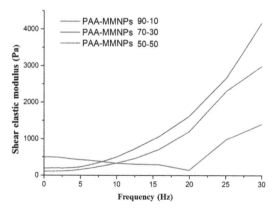

Figure 6. Elastic modulus G′ versus frequency.

3.5. Magnetic Properties by Vibrating Sample Magnetometry (VSM)

The magnetic properties of the magnetic iron oxide nanoparticles (Fe_3O_4 NPs) and of the hydrogels were investigated by vibrating sample magnetometry (VSM) at room temperature. In Figure 7, the magnetic hysteresis loops that are characteristic of superparamagnetic behavior can be observed for all of the samples, due to the presence of the magnetite nanoparticles. Superparamagnetism is the responsiveness to an applied magnetic field without retaining any magnetism after removal of the applied magnetic field. The measured saturation magnetization (Ms) of the Fe_3O_4 NPs is 63.128 emu/g. For PAA-MMNPs 90:10, the saturation magnetization was found at 9.74 emu/g, the lowest measured saturation of the hydrogels. The saturation magnetization for the PAA-MMNPs 70:30 was found at 26.73 emu/g and the highest saturation magnetization was at 31.88 emu/g, corresponding to the PAA-MMNPs 50:50, the hydrogel with the highest concentration (50%) of MMNPs. These results show that the magnetization of the hydrogels increases with the increase of the concentration of MMNPs present in the hydrogels.

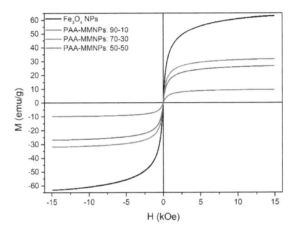

Figure 7. Vibrating sample magnetometry (VSM) magnetization curves of the Fe_3O_4 nanoparticles and the nanocomposites hydrogels.

3.6. Morphological Characterization by SEM and TEM

3.6.1. SEM Analysis

The microstructure of the PAA-MMNPs hydrogels was studied by SEM in cross-section and the results are shown in Figure 8 (PAA-MMNPs 90:10) and Figure 9 (PAA-MMNPs 50:50). The image in Figure 8A, (magnification ×2.000) shows submicronic areas of bright contrast (functionalized magnetite aggregates) evenly distributed in a dark contrast PAA matrix. At higher magnifications (×200.000, Figure 8B) it can be observed that the areas of bright contrast are aggregates of MMNPs. Also, the image shows that the modified Fe_3O_4 nanoparticles showed a good distribution in the polymer matrix by the presence of areas with high dispersed MMNPs and areas with local agglomeration of MMNPs. However, even the local agglomerations revealed that the modified magnetite nanoparticles (MMNPs) seem to be addressed by the polymer polyacrylamide matrix due to the effect of the crosslinking agent of the MMNPs (Figure 8A,B). Thus, the polymer matrix covering the MMNPs is chemically linked by the MMNPs and the whole ensemble displays a crosslinked network-like architecture.

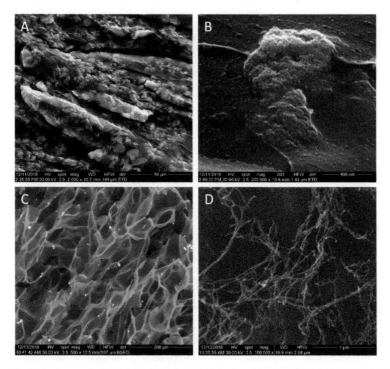

Figure 8. SEM micrographs of PAA-MMNPs 90:10 block hydrogel (**A,B**) and lyophilized PAA-MMNPs 90:10 hydrogel (**C,D**).

Figure 8C is a SEM backscattered electron image at a smaller magnification (×500), showing small MMNPs agglomerates (white spots) uniformly dispersed on a mesh of micro-pores. Figure 8D is a detail (×100.000 magnification) of a nano-size area from the central zone in Figure 8C, showing a nanostructure of the lyophilized hydrogel as fibrils having evenly incorporated MMNPs. The crosslinked network-like ensemble generated by the MMNPs is better highlighted by the lyophilized samples (Figure 8C,D). The fibrils revealed branches-like structures which are extending on the sample surface and evenly through the sample internal structure. The branched-like structures exhibited MMNPs linked to each other by the polymer matrix and serve as the basis of the crosslinked network-like ensemble.

The SEM image in Figure 9A (magnification ×2.000) shows a higher density in MMNPs clusters for the PAA-MMNPs 50:50 hydrogel due to the higher amount of modified magnetite, in comparison to the PAA-MMNPs 90:10 hydrogel from Figure 8A. Detail from Figure 9A is shown in Figure 9B (magnification ×200.000), proving that the clusters are made of nanoparticles. The polymeric matrix is not homogenous, due to the fact that it has smaller nanoparticle aggregates embedded. The cross-section of the lyophilized hydrogel shows microsize pores, with chains of MMNPs clusters, which seem to be located especially on the pore walls. At higher magnifications (Figure 8D), it can be observed that there are also nano-size areas having the same fibrils with branches-like structures with incorporated MMNPs. A very interesting result of the lyophilized sample of both PAA-MMNPs 50:50 and PAA-MMNPs 90:10 (Figure 8C,D and Figure 9C,D) showed less local MMNPs agglomeration with respect to un-lyophilized samples. This behavior can be explained by the lyophilization procedure. During the process, the polymer matrix between MMNPs swells and the space grows between them. Furthermore, the sublimation phenomenon leads to a rearrangement of the structure with the display of the MMNPs in the pore walls and fragmentation of the local agglomerates.

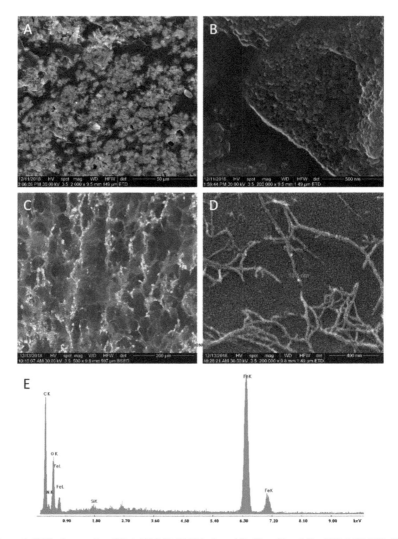

Figure 9. SEM micrographs of PAA-MMNPs 50:50 hydrogel (**A,B**).and lyophilized PAA-MMNPs 50:50 hydrogel (**C,D**); Energy dispersive X-ray (EDX) spectrum (**E**).

The EDXS spectrum (Figure 9E), acquired on a large area of the PAA-MMNPs 50:50 hydrogel surface and shows the presence in the sample of the elements Fe and O (from Fe_3O_4 NPs), C, N and Si (from 3-TPM and PAA).

3.6.2. TEM Analysis

The morphology and nanostructural characteristics of magnetic nanoparticles (MNPs), modified magnetic nanoparticles (MMNPs) and of polyacrylamide modified magnetic nanoparticles (PAA-MMNPs) hydrogels were analyzed by TEM, selected area electron diffraction (SAED) and high resolution electron microscopy (HR-TEM).

3.6.3. TEM Analysis for Magnetite Nanoparticles (MNPS) and Modified Magnetite Nanoparticles (MMNPs)

Figure 10A–C are TEM micrographs of the MNPs. The bright field TEM image (Figure 10A) shows that the magnetic Fe_3O_4 nanoparticles are nearly spherical with diameters between 5 and 12 nm. The SAED pattern (inset of Figure 10A) of MNPs exhibits a typical face centered cubic (fcc) crystalline structure. The lattice spacing measured based on the diffractions rings is in accordance with the standard lattice spacing of Fe_3O_4 from the Powder Diffraction File (PDF) database (ICCD file no. 04-002-5683). The HRTEM images of MNPs (Figure 10B,C) clearly show the single crystallinity of Fe_3O_4 nanoparticles. The interplanar distances measured from the adjacent lattice fringes with Fast Fourier Transform (FFT) (inset of Figure 10B) are 2.53 Å, 2.10 Å and 1.62 Å, corresponding to (311), (400) and (511) crystalline family planes of Fe_3O_4 with crystalline structure, according to the PDF database. Nanocrystalline particles with diameter size between 5.7 and 8.6 nm are highlighted in Figure 10B. In the HRTEM image from Figure 10C it clearly shows the crystalline planes with 2.97 Å and 2.53 Å measured interplanar distances corresponding to crystalline family planes with (220) and (311) Miller indices.

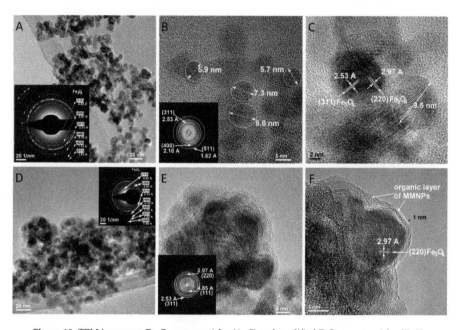

Figure 10. TEM images on Fe_3O_4 nanoparticles (**A–C**) and modified F_3O_4 nanoparticles (**D–F**).

The TEM results of MMNPs are presented in Figure 10D,E,F. According to Figure 10D, modified Fe_3O_4 nanoparticles still keep the morphological properties of Fe_3O_4 nanoparticles. According to HRTEM images (Figure 10E,F), morphological and nanocrystalline properties of Fe_3O_4 nanoparticles are maintained, but it is clearly shown that the nanoscale Fe_3O_4 nanoparticles are modified in the MMNPs sample, because of the organic layer surrounding the Fe_3O_4 nanoparticles (highlighted in Figure 10F). The tailoring of magnetite nanoparticles by chemically functionalization with silane 3-TPM revealed by physico-chemical X-photoelectron spectroscopy is also sustained by the morphological characterization by TEM. The high magnification Figure 10E,F exhibits a less ordered organic layer consisted by silane 3-TPM, which addresses the magnetite nanoparticles. However, the surrounding organic layer displayed a specific order and arrangement structure, which will be further discussed. The Figure 10D revealed an overview result with a considering functionalization of the whole nanoparticles and not as an isolated modification. Thus, the magnetic nanoparticles were tailored with double bonds by the presence of the silane structure (Figure 1).

3.6.4. TEM Analysis of Polyacrylamide-MMNPs Nanocomposite Hydrogels

The bright field TEM (BF-TEM) images from Figure 11A, 10D and 10G are results from PAA-MMNPs 90:10, PAA-MMNPs 70:30 and PAA-MMNPs 50:50 samples. These images show that all of the hydrogels have a similar morphology and nanostructure. The overview images (Figure 11A,D,G) revealed an expected decreasing of polymer matrix area with the increasing of MMNPs amount. All of the samples have isolated and local agglomerated magnetic Fe_3O_4 nanoparticles embedded within a polymer matrix. By comparing the BF-TEM images from PAA-MMNPs (Figure 11A,D,G) with the BF-TEM images from MNPs (Figure 10A) and MMNPs (Figure 10D), it can be concluded that the shape and the dimensions of the embedded nanoparticles are kept in the same range. The SAED image (inset of Figure 11G) shows that the PAA-MMNPs hydrogels contain similar Fe_3O_4 nanoparticles, well crystallized, with the same lattice spacing measured on the SAED image from MNPs (inset of Figure 10A). The diffraction of the matrix was not observed in the SAED image (inset of Figure 11G), which is probably because the organic layer and the PAA matrix are not highly ordered and are displaying short ordering range. In order to observe the detailed structure of PAA-MMNPs hydrogels, HRTEM was employed. Figure 11B,C (from PAA-MMNPs: 90-10), Figure 11E,F (From PAA-MMNPs: 70-30) and Figure 11H,I (from PAA-MMNPs: 50-50) show that the nanoparticles are embedded within a polymer matrix with amorphous structure. The nanoparticles have a round shape with diameters between 5 and 14 nm. The MMNPs are well integrated into polymer matrix revealing a clear interaction between the two phases. The nature of the interaction was revealed by the HR-TEM images, which rarely highlighted the specific order and arrangement structure of the organic layer from MMNPs. This result can be explained by the surrounding organic layer being in a chemical reaction with the acrylamide monomer by the consumption of the silane double bonds. Thus, the MMNPs act as an inorganic cross-linker by becoming generators of bridges between polymeric chains and development of a hybrid network (Figure 2). Also, the HRTEM results show that the nanoparticles are nanocrystals, disclosing the crystalline planes (220) and (311) of magnetite with 2.97 Å and 2.53 Å, respectively, which are characteristic interplanar distances. Furthermore, the HRTEM images also reveal a short ordering range in the matrix besides the amorphous phase, highlighted by squares (Figure 11F for PAA-MMNPs: 70-30 and Figure 11I for PAA-MMNPs: 50-50), which shows the structural arrangement of the polymer macromolecular chains compared with inorganic ordered magnetite nanoparticles.

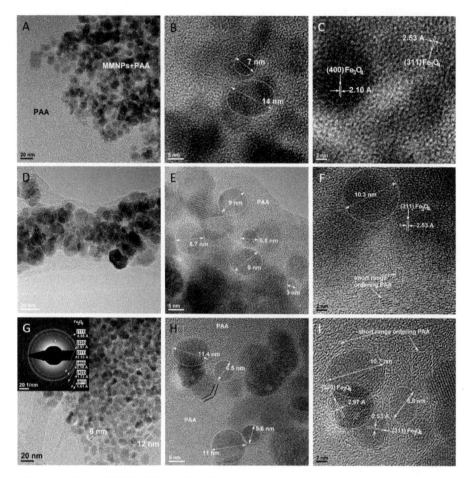

Figure 11. Bright field (BF)-TEM and HR-TEM images on PAA-MMNPs 90:10 (**A–C**); PAA-MMNPs 70:30 (**D–F**) and on PAA-MMNPs 50:50 (**G–I**).

4. Conclusions

This study provides a comprehensive approach in the wide field of polymer nanocomposite materials. A new hybrid polymer network was successfully developed by double bond modified magnetic nanoparticles, using polyacrylamide as the crosslinked network structure, thereby overcoming the limitation of traditional organically crosslinkers. Functionalization of magnetic nanoparticles with the double bond was monitored by physico-chemical investigations. The details of the microarchitecture were shown by modern morphological characterization techniques, highlighting the nature of the interaction between the organic and inorganic phases. Furthermore, the obtained nanocomposite hydrogels may have an efficient applicability in the soft tissue engineering field, in the form of repairing scaffolds, due to their mechanical and magnetization behavior that can stimulate tissue regeneration.

Author Contributions: Formal analysis, E.T., V.-A.S., B.S.V. and C.-M.D.; Investigation, E.T. and I.-C.R.; Methodology, C.Z.; Project administration, C.Z.; Validation, C.Z.; Writing—original draft, E.T. and I.-C.R.; Writing—review and editing, C.Z. and E.A.

Funding: This research received no external funding.

Conflicts of Interest: The authors declare no conflict of interest.

References

1. Itagaki, H.; Kurokawa, T.; Furukawa, H.; Nakajima, T.; Katsumoto, Y.; Gong, J.P. Water-Induced Brittle-Ductile Transition of Double Network Hydrogels. *Macromolecules* **2010**, *43*, 9495–9500. [CrossRef]
2. Nakajima, T.; Furukawa, H.; Tanaka, Y.; Kurokawa, T.; Osada, Y.; Gong, J.P. True Chemical Structure of Double Network Hydrogels. *Macromolecules* **2009**, *42*, 2184–2189. [CrossRef]
3. Okumura, Y.; Ito, K. The Polyrotaxane Gel: A Topological Gel by Figure-of-Eight Cross-links. *Adv. Mater.* **2001**, *13*, 485–487. [CrossRef]
4. Gong, J.P.; Katsuyama, Y.; Kurokawa, T.; Osada, Y. Double-Network Hydrogels with Extremely High Mechanical Strength. *Adv. Mater.* **2003**, *15*, 1155–1158. [CrossRef]
5. Hoare, T.R.; Kohane, D.S. Hydrogels in drug delivery: Progress and challenges. *Polymer* **2008**, *49*, 1993–2007. [CrossRef]
6. Lee, K.Y.; Mooney, D.J. Hydrogels for Tissue Engineering. *Chem. Rev.* **2001**, *101*, 1869–1880. [CrossRef]
7. Haraguchi, K.; Takehisa, T. Nanocomposite Hydrogels: A Unique Organic–Inorganic Network Structure with Extraordinary Mechanical, Optical, and Swelling/De-swelling Properties. *Adv. Mater.* **2002**, *14*, 1120–1124. [CrossRef]
8. Haraguchi, K.; Takehisa, T.; Fan, S. Effects of Clay Content on the Properties of Nanocomposite Hydrogels Composed of Poly(N-isopropylacrylamide) and Clay. *Macromolecules* **2002**, *35*, 10162–10171. [CrossRef]
9. Haraguchi, K.; Farnworth, R.; Ohbayashi, A.; Takehisa, T. Compositional Effects on Mechanical Properties of Nanocomposite Hydrogels Composed of Poly(N,N-dimethylacrylamide) and Clay. *Macromolecules* **2003**, *36*, 5732–5741. [CrossRef]
10. Shibayama, M.; Suda, J.; Karino, T.; Okabe, S.; Takehisa, T.; Haraguchi, K. Structure and Dynamics of Poly(N-isopropylacrylamide)–Clay Nanocomposite Gels. *Macromolecules* **2004**, *37*, 9606–9612. [CrossRef]
11. Haraguchi, K.; Li, H.-J. Mechanical Properties and Structure of Polymer–Clay Nanocomposite Gels with High Clay Content. *Macromolecules* **2006**, *39*, 1898–1905. [CrossRef]
12. Haraguchi, K.; Li, H.-J.; Matsuda, K.; Takehisa, T.; Elliott, E. Mechanism of Forming Organic/Inorganic Network Structures during In-situ Free-Radical Polymerization in PNIPA–Clay Nanocomposite Hydrogels. *Macromolecules* **2005**, *38*, 3482–3490. [CrossRef]
13. Haraguchi, K.; Li, H.-J. Control of the Coil-to-Globule Transition and Ultrahigh Mechanical Properties of PNIPA in Nanocomposite Hydrogels. *Angew. Chem. Int. Ed.* **2005**, *44*, 6500–6504. [CrossRef] [PubMed]
14. Haraguchi, K.; Matsuda, K. Spontaneous Formation of Characteristic Layered Morphologies in Porous Nanocomposites Prepared from Nanocomposite Hydrogels. *Chem. Mater.* **2005**, *17*, 931–934. [CrossRef]
15. Haraguchi, K.; Ebato, M.; Takehisa, T. Polymer–Clay Nanocomposites Exhibiting Abnormal Necking Phenomena Accompanied by Extremely Large Reversible Elongations and Excellent Transparency. *Adv. Mater.* **2006**, *18*, 2250–2254. [CrossRef]
16. Saegusa, T.; Chujo, Y. Organic-inorganic polymer hybrids. *Makromol. Chemie. Macromol. Symp.* **1992**, *64*, 1–9. [CrossRef]
17. Giannelis, E.P. Polymer Layered Silicate Nanocomposites. *Adv. Mater.* **1996**, *8*, 29–35. [CrossRef]
18. Mark, J.E. New developments and directions in the area of elastomers and rubberlike elasticity. *Macromol. Symp.* **2003**, *201*, 77–84. [CrossRef]
19. Okada, A.; Usuki, A. Twenty Years of Polymer-Clay Nanocomposites. *Macromol. Mater. Eng.* **2006**, *291*, 1449–1476. [CrossRef]
20. Usuki, A.; Kojima, Y.; Kawasumi, M.; Okada, A.; Fukushima, Y.; Kurauchi, T.; Kamigaito, O. Synthesis of nylon 6-clay hybrid. *J. Mater. Res.* **1993**, *8*, 1179–1184. [CrossRef]
21. Galateanu, B.R.; Radu, I.C.; Vasile, E.; Hudita, A.; Serban, M.V.; Costache, M.; Iovu, H.; Zaharia, C. Fabrication of novel silk fibroin-ldhs composite architectures for potential bone tissue engineering. *Mater. Plast.* **2017**, *54*, 659–665.
22. Cristianradu, I.; Vasile, E.; Damian, C.M.; Iovu, H.; Stanescu, P.O.; Zaharia, C. Influence of the double bond LDH clay on the exfoliation intercalation mechanism of polyacrylamide nanocomposite hydrogels. *Mater. Plast.* **2018**, *55*, 263–268.
23. Haraguchi, K. Synthesis and properties of soft nanocomposite materials with novel organic/inorganic network structures. *Polym. J.* **2011**, *43*, 223. [CrossRef]

24. Wang, Q.; Gao, Z. A constitutive model of nanocomposite hydrogels with nanoparticle crosslinkers. *J. Mech. Phys. Solids* **2016**, *94*, 127–147. [CrossRef]

25. Hu, Z.; Chen, G. Novel Nanocomposite Hydrogels Consisting of Layered Double Hydroxide with Ultrahigh Tensibility and Hierarchical Porous Structure at Low Inorganic Content. *Adv. Mater.* **2014**, *26*, 5950–5956. [CrossRef] [PubMed]

26. Huang, T.; Xu, H.G.; Jiao, K.X.; Zhu, L.P.; Brown, H.R.; Wang, H.L. A Novel Hydrogel with High Mechanical Strength: A Macromolecular Microsphere Composite Hydrogel. *Adv. Mater.* **2007**, *19*, 1622–1626. [CrossRef]

27. Wang, Q.; Mynar, J.L.; Yoshida, M.; Lee, E.; Lee, M.; Okuro, K.; Kinbara, K.; Aida, T. High-water-content mouldable hydrogels by mixing clay and a dendritic molecular binder. *Nature* **2010**, *463*, 339. [CrossRef] [PubMed]

28. Zhao, X. Multi-scale multi-mechanism design of tough hydrogels: Building dissipation into stretchy networks. *Soft Matter* **2014**, *10*, 672–687. [CrossRef] [PubMed]

29. Long, R.; Mayumi, K.; Creton, C.; Narita, T.; Hui, C.-Y. Time Dependent Behavior of a Dual Cross-Link Self-Healing Gel: Theory and Experiments. *Macromolecules* **2014**, *47*, 7243–7250. [CrossRef]

30. Na, Y.-H.; Kurokawa, T.; Katsuyama, Y.; Tsukeshiba, H.; Gong, J.P.; Osada, Y.; Okabe, S.; Karino, T.; Shibayama, M. Structural Characteristics of Double Network Gels with Extremely High Mechanical Strength. *Macromolecules* **2004**, *37*, 5370–5374. [CrossRef]

31. Na, Y.-H.; Tanaka, Y.; Kawauchi, Y.; Furukawa, H.; Sumiyoshi, T.; Gong, J.P.; Osada, Y. Necking Phenomenon of Double-Network Gels. *Macromolecules* **2006**, *39*, 4641–4645. [CrossRef]

32. Sharma, V.K.; McDonald, T.J.; Kim, H.; Garg, V.K. Magnetic graphene–carbon nanotube iron nanocomposites as adsorbents and antibacterial agents for water purification. *Adv. Colloid Interface Sci.* **2015**, *225*, 229–240. [CrossRef] [PubMed]

33. Lu, A.-H.; Salabas, E.L.; Schüth, F. Magnetic Nanoparticles: Synthesis, Protection, Functionalization, and Application. *Angew. Chemie Int. Ed.* **2007**, *46*, 1222–1244. [CrossRef] [PubMed]

34. Wu, W.; He, Q.; Jiang, C. Magnetic iron oxide nanoparticles: Synthesis and surface functionalization strategies. *Nanoscale Res. Lett.* **2008**, *3*, 397–415. [CrossRef]

35. Wu, W.; Wu, Z.; Yu, T.; Jiang, C.; Kim, W.-S. Recent progress on magnetic iron oxide nanoparticles: Synthesis, surface functional strategies and biomedical applications. *Sci. Technol. Adv. Mater.* **2015**, *16*, 023501. [CrossRef] [PubMed]

36. Sun, S.-N.; Wei, C.; Zhu, Z.-Z.; Hou, Y.-L.; Venkatraman, S.S.; Xu, Z.-C. Magnetic iron oxide nanoparticles: Synthesis and surface coating techniques for biomedical applications. *Chin. Phys. B* **2014**, *23*, 037503. [CrossRef]

37. Shi, D.; Sadat, M.E.; Dunn, A.W.; Mast, D.B. Photo-fluorescent and magnetic properties of iron oxide nanoparticles for biomedical applications. *Nanoscale* **2015**, *7*, 8209–8232. [CrossRef] [PubMed]

38. Sharifi, S.; Seyednejad, H.; Laurent, S.; Atyabi, F.; Saei, A.A.; Mahmoudi, M. Superparamagnetic iron oxide nanoparticles for in vivo molecular and cellular imaging. *Contrast Media Mol. Imaging* **2015**, *10*, 329–355. [CrossRef] [PubMed]

39. Revia, R.A.; Zhang, M. Magnetite nanoparticles for cancer diagnosis, treatment, and treatment monitoring: Recent advances. *Mater. Today* **2016**, *19*, 157–168. [CrossRef] [PubMed]

40. Tanasa, E.A.E.; Cernea, M.; Oprea, O.C. $Fe_3O_4/BaTiO_3$ composites with core-shell structures. *U.P.B. Sci. Bull. Ser. B* **2019**, *81*, 171–180.

41. Ahn, T.; Kim, J.H.; Yang, H.-M.; Lee, J.W.; Kim, J.-D. Formation Pathways of Magnetite Nanoparticles by Coprecipitation Method. *J. Phys. Chem. C* **2012**, *116*, 6069–6076. [CrossRef]

42. Zaharia, C.; Tudora, M.-R.; Stancu, I.-C.; Galateanu, B.; Lungu, A.; Cincu, C. Characterization and deposition behavior of silk hydrogels soaked in simulated body fluid. *Mater. Sci. Eng. C* **2012**, *32*, 945–952. [CrossRef]

43. Ganji, F.V.-F.E. Hydrogels in controlled drug delivery systems. *Iran. Polym. J.* **2009**, *18*, 63–88.

44. Afif, A.E.; Grmela, M. Non-Fickian mass transport in polymers. *J. Rheol.* **2002**, *46*, 591–628. [CrossRef]

45. Bajpai, A.K.; Shukla, S.K.; Bhanu, S.; Kankane, S. Responsive polymers in controlled drug delivery. *Prog. Polym. Sci.* **2008**, *33*, 1088–1118. [CrossRef]

46. Lee, H.; Zhang, J.; Lu, J.; Georgiadis, J.; Jiang, H.; Fang, N. *Coupled Non-Fickian Diffusion and Large Deformation of Hydrogels*; Springer: New York, NY, USA, 2011; pp. 25–28.

47. Liu, Q.; Wang, X.; De Kee, D. Mass transport through swelling membranes. *Int. J. Eng. Sci.* **2005**, *43*, 1464–1470. [CrossRef]

48. Rajagopal, K.R. Diffusion through polymeric solids undergoing large deformations. *Mater. Sci. Technol.* **2003**, *19*, 1175–1180. [CrossRef]

49. Vrentas, J.S.; Vrentas, C.M. Steady viscoelastic diffusion. *J. Appl. Polym. Sci.* **2003**, *88*, 3256–3263. [CrossRef]

50. Kim, S.J.; Lee, K.J.; Kim, I.Y.; Lee, Y.M.; Kim, S.I. Swelling kinetics of modified poly(vinyl alcohol) hydrogels. *J. Appl. Polym. Sci.* **2003**, *90*, 3310–3313. [CrossRef]

51. Zhao, Z.X.; Li, Z.; Xia, Q.B.; Bajalis, E.; Xi, H.X.; Lin, Y.S. Swelling/deswelling kinetics of PNIPAAm hydrogels synthesized by microwave irradiation. *Chem. Eng. J.* **2008**, *142*, 263–270. [CrossRef]

52. Zakirov, A.S.; Navamathavan, R.; Jang, Y.J.; Jung, A.S.; Lee, K.; Choi, C.K. Comparative study on the structural and electrical properties of low-bftextitk SiOC(-H) films deposited by using plasma enhanced chemical vapor deposition. *J. Korean Phys. Soc.* **2007**, *50*, 1809–1813. [CrossRef]

53. Márquez, F.; Herrera, G.M.; Campo, T.; Cotto, M.; Ducongé, J.; Sanz, J.M.; Elizalde, E.; Perales, Ó.; Morant, C. Preparation of hollow magnetite microspheres and their applications as drugs carriers. *Nanoscale Res. Lett.* **2012**, *7*, 210. [CrossRef] [PubMed]

Communication

Formation of Nanospikes on AISI 420 Martensitic Stainless Steel under Gallium Ion Bombardment

Zoran Cenev [1]**, Malte Bartenwerfer** [2,*]**, Waldemar Klauser** [2]**, Ville Jokinen** [3]**, Sergej Fatikow** [2] **and Quan Zhou** [1,*]

[1] Department of Electrical Engineering and Automation, School of Electrical Engineering, Aalto University, Maarantie 8, 02150 Espoo, Finland; zoran.cenev@aalto.fi
[2] Department of Computing Science, University of Oldenburg, Ammerländer Heerstraße 114-118, 26129 Oldenburg, Germany; waldemar.klauser@uni-oldenburg.de (W.K.); sergej.fatikow@uni-oldenburg.de (S.F.)
[3] Department of Chemistry and Materials Science, Aalto University, School of Chemical Engineering, Tietotie 3, 02150 Espoo, Finland; ville.p.jokinen@aalto.fi
* Correspondence: m.bartenwerfer@uni-oldenburg.de (M.B.); quan.zhou@aalto.fi (Q.Z.);
   Tel.: +49-179-682-1971 (M.B.); +358-40-855-0311 (Q.Z.)

Received: 11 September 2019; Accepted: 15 October 2019; Published: 19 October 2019

Abstract: The focused ion beam (FIB) has proven to be an extremely powerful tool for the nanometer-scale machining and patterning of nanostructures. In this work, we experimentally study the behavior of AISI 420 martensitic stainless steel when bombarded by Ga^+ ions in a FIB system. The results show the formation of nanometer sized spiky structures. Utilizing the nanospiking effect, we fabricated a single-tip needle with a measured 15.15 nanometer curvature radius and a microneedle with a nanometer sized spiky surface. The nanospikes can be made straight or angled, depending on the incident angle between the sample and the beam. We also show that the nanospiking effect is present in ferritic AISI 430 stainless steel. The weak occurrence of the nanospiking effect in between nano-rough regions (nano-cliffs) was also witnessed for austenitic AISI 316 and martensitic AISI 431 stainless steel samples.

Keywords: focused ion beam; nanospikes; martensite; stainless steel; gallium; bombardment; irradiation effects; sharp needle; incident angle

---

1. Introduction

The focused ion beam (FIB) technique has been established as a powerful tool for micro and nanoscale imaging [1], sputtering, deposition [2], 3D machining [3], and surface modifications [4]. When an incident ion comes into contact with a targeted material, the ion enters into a set of collisions (higher than normal thermal energies) with the target atoms, a process known as a collision cascade. Sputtering occurs when an incident ion comes into contact with a targeted surface and transfers its momentum to the host atoms. A host atom on the surface will absorb a part of the ion's kinetic energy. If the new energy state of the host surface atom is higher than the surface binding energy (SBE) of the targeted material, then the surface atom will be ejected as a sputtered particle [5]. A quantitative measure of sputtering is defined through sputtering yield, i.e., the number of atoms removed by an incident ion. The sputtering yield is affected by the material composition, angle of incidence, the crystal structure of the substrate, redeposition, scanning speed, temperature of the target, and surface contaminations [6].

The process which constrains the path of the ion in a crystalline solid is known as ion channeling [7]. Along low index directions in crystalline materials, ions may penetrate greater distances as compared to cascade collisions in amorphous materials. Since ion channeling has a direct impact on the ion penetration range, meaning the trajectory within the collision cascade, it also impacts the sputtering yield. Variations on sputtering yield within a sample target cause roughening of the surface, which has been observed for aluminum [8], tungsten [9], and polycrystalline gold [10]. A nanometer sized spiky structure, an extreme form of nano-roughening, with distinct and visibly pronounced spikes, occurs during the anisotropic etching of single crystal (100) copper [8], tungsten [11], and 18 Cr-ODS (Oxide Dispersive Strengthened) steel [12]. Pyramidal and conical (faceted pyramid) micro/nanometer-sized structures have been observed much earlier on tin crystals [13], and monocrystalline [14] and polycrystalline copper [15] when irradiated by argon, as well as krypton ions [16]. The origins and stability of ion-bombarded copper surfaces have been heavily analyzed and discussed by Auciello and Kelly [17,18].

Here, we experimentally demonstrate the formation of nanospikes occurring on a martensitic AISI 420 stainless steel surface when bombarded with gallium ions. We also show that nanospikes can be made straight or angled depending on the incident angle of the FIB. To demonstrate potential applications, we FIB-treated an electrochemically etched stainless steel (AISI 420) tip to induce nanospiking and thus obtain a single tip nano-needle with a measured diameter of 15.15 nanometers. Additionally, we FIB-treated an electrochemically etched stainless steel (AISI 420) tip with micrometer sharpness to induce nanospikes. Finally, we also show that the nanospiking effect is present in ferritic AISI 430 stainless steel. The weak occurrence of the nanospiking effect in between nano-rough regions (nano-cliffs) was also witnessed for austenitic AISI 316 and martensitic AISI 431 stainless steel samples.

Future research should focus on using the single sharp nano-needle for creating localized magnetic fields, as in [19], or laser-induced electron emission, as in [20,21]. Due to the soft magnetic properties of the martensitic ASI 420 stainless steel, nanometer sized spiky magnetic tips could be applied in producing magnetic nano-devices, for example, magneto-gravitational traps [22,23]. Another line of research can focus on producing superhydrophobic/hydrophobic microneedles by subsequent fluoropolymer deposition (low adhesive polymer) to the surface of the microneedle with the nanometer sized spiky surface.

2. Materials and Methods

2.1. Procedure of FIB Treatment of Martensitic, Austenitic and Ferritic Stainless Steel Plates

The treatment of the martensitic AISI 420 (Fe-86,7/Cr13,0/C0,3) stainless steel sample (Goodfellow, Cambridge, UK) was carried out with a dual-beam high-resolution scanning electron/focused ion beam microscope, namely, the Lyra FEG (TESCAN, Brno, Czech Republic). A 0.5 cm^2 piece was cut from the foil sheets. The piece was cleaned in an ultrasonic isopropanol bath, with a 10 min O_2 plasma treatment under 40 kHz at 100 W in the plasma system, using the Femto instrument (Diener electronic GmbH and Co KG, Ebhausen, Germany). An area of 10 μm^2 was exposed to an ion dose of 19.4 nC/μm^2 at a 30 keV beam energy and emission current of 2 μA. The same treatment was applied for the other stainless steel samples, i.e., AISI316 (Fe/Cr18/Ni10/Mo3), AISI430 (Fe81/Cr17/Mn/Si/C/S/P) and AISI431 (Fe82/Cr16/Ni2), as were received from the supplier (Goodfellow, UK).

2.2. Procedure of Fabrication of Martensitic Stainless Steel Needle with Nanometer Sharpness

A one millimeter thick stainless steel AISI 420 wire (Goodfellow, Cambridge, UK) was thinned with up to micrometer sharpness, as previously reported in [19]. The etched needle was installed into the FIB-SEM dual beam system (Lyra FEG) and is shown in Figure S1a–c. Prior to FIB exposure, the needle was cleaned in an ultrasonic acetone bath and rinsed with isopropanol. A series of FIB exposures with a total ion dose of roughly equal to 2000 nC/μm^2 at a 30 keV beam energy and emission current of 2 μA was applied in order to induce more spikes (Figure S1d). Once a prominent spike was obtained, it was isolated from further exposure, but the exposure was targeted towards removing the surrounding spikes and eventually providing the final result (Figure S1f).

2.3. Procedure of Fabrication of Stainless Steel Microneedle with Nanospikes

A one millimeter thick stainless steel AISI 420 wire (Goodfellow, USA) was installed into the collet of a milling 3-axis bridge router, as illustrated in Figure S2a. A face mill insert with four cutting edges was used for machining, where the wire would be thinned within a range of 0.4 and 0.7 mm thickness, with a length of about 3 mm. A thinned wire as such was mounted onto a holder of an in-house built electrochemical etching station, containing a 10% HCl bath, a computer-controlled voltage supply, and a motorized stage (further details can be found in [19]). The first step was electrochemical thinning, consisting of dipping the wire 3 mm into the HCl bath. The etching started when a voltage of 1V was supplied. Immediately after voltage application, the wire was pulled with a constant speed of 10 μm/s. The second etching step consisted of re-dipping the wire by 1 mm, with supply voltage of 1V and pulling the wire with a constant speed of 10 μm/s until the needle was completely out of the bath, as illustrated in Figure S2b. A sample micrograph of an etched needle can be seen in Figure S3a–c.

The electrochemically etched needle was installed into the FIB-SEM dual beam system, as illustrated in Figure S2c. Prior to FIB treatment, the machined/etched needles were cleaned in an ultrasonic acetone bath and rinsed with isopropanol. The needles were exposed to an ion dose of 10.6 nC/μm² at a 30 keV beam energy and emission current of 2 μA (Figure S3d,e).

3. Results

3.1. Nanospikes Formation on AISI 420 Martensitic Stainless Steels by FIB Treatment

Figure 1a shows the surface morphology of an FIB-irradiated AISI 420 sample with gallium ions. Details of the sample preparation and FIB treatment settings are provided in Section 2.1. From the figure, it can be seen that the sharpness of the nanospikes is in the sub-micron range. One can also see that the nanospikes on the edge feature higher aspect ratios than the nanospikes in the middle of the trench.

To demonstrate the potential usability of the nanospiking effect, we have fabricated two different types of needles, i.e., an extremely sharp needle with radius of 15.15 nanometers (Figure 1b) and a micrometer-sized needle, featuring a nanometer sized spiky topology (Figure 1c). The fabrication procedure of both needles is similar, and they are explained in detail in Sections 2.2 and 2.3, respectively. One should note that the fabrication procedure for both needles includes a certain level of randomness, however, the sharpness of the nanospikes is very often in the low nanometer range (from several up to tens of nanometers).

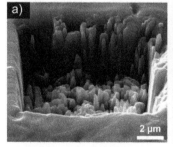  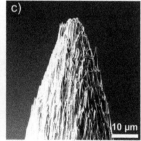

Figure 1. Nanospiking effects on martensitic AISI 420 stainless steel. (**a**) Nanometer sized spiky surface of the martensitic AISI 420 stainless steel sample plate after FIB treatment with gallium ions with a dose of 19.4 nC/μm². Fabrication results of a (**b**) sharp needle with nanometer resolution, the circle denotes fitting to the curvature of the tip. The original raw image without fitting is given in Figure S1f. (**c**) A micrometer-scale needle with nanospikes.

3.2. Morphological Evolution of AISI 420 during FIB Treatment

We also examined the morphological evolution of the martensitic AISI 420 in a step-by-step manner. Figure 2 shows the surface morphology evolution, and finally, the formation of the nanospikes. The samples were exposed to 30,000 scans overall (1000 scans correspond to an ion dose of 1435 nC/μm^2). The trench dimension was 10×10 μm^2. The AISI 420 surface was untreated at the beginning (0 scans), and after the first 500 scans, the appearance of a few pits on the surface was noted.

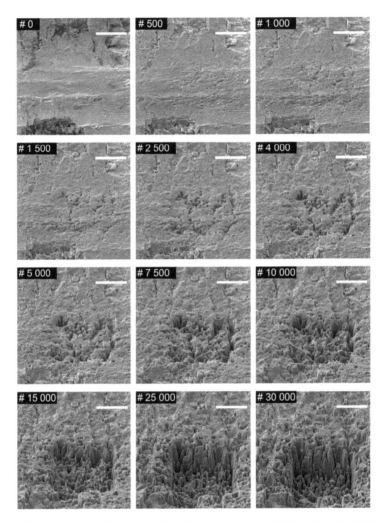

Figure 2. Evolution of nanospikes at normal incidence as a function of FIB dose from 0 to 30,000 scans (1000 scans correspond to a gallium ion dose of 1435 nC/μm^2). The spiky structure shift downwards along with increase of the FIB exposure. Scale bar in each image is 5 μm. Orange arrows denote the formation and size variation of the firstly formed nanospike.

The indentation of the initial pits increased with the increase of the number of scans (1000 to 5000 scans). At 7500 scans, the formation of the first nanospike (denoted with an orange arrow) was noticed. Further exposure of the earlier formed nanospikes causes their increase in sharpness, but also causes a decrease in height, as indicated by the orange arrows (10,000 to 30,000 scans). With further increase of the irradiation, new spikes started to form, and they could be found more within the central region of the trench, rather than on the edges. The nanospikes on the edges feature much higher aspect ratios than the ones in the central region. This difference in aspect ratios can be observed from scans 15,000 to 30,000.

3.3. Energy-Dispersive X-ray Spectroscopy (EDX) and X-ray Photoelectron Spectroscopy (XPS) Analysis of FIB Irradiated AISI 420 Stainless Steel Alloy with Gallium Ions

We have performed energy-dispersive X-ray spectroscopy (EDX) analysis on the whole gallium irradiated trench, a spot on a single nanospike, and a non-irradiated area (Figure S4). The only difference that the EDX results show is the presence of gallium in irradiated regions in comparison to non-irradiated regions. No significant change in the presence of the iron or chromium content within the AISI 420 sample before and after gallium irradiation was determined.

We also have performed X-ray photoelectron spectroscopy (XPS) measurements with the Kratos Axis Ultra ESCA system (Kratos Analytical Ltd., Manchester, UK), analyzing the gallium irradiated circular trench (diameter of ~35 µm) and the non-irradiated area (Figure S5). The XPS results show a reduction of iron (9.42% to 2.2% for XPS aperture of 27 µm) and chromium (0.97% to 0.44% for XPS aperture of 27 µm) between the non-irradiated and irradiated regions. Here, it could be that the gallium in the non-irradiated region was deposited during the gallium irradiation of the sample in the FIB system. The existence of high oxygen and carbon concentrations is due to exposure of the sample to ambient conditions.

3.4. Effect from Variation of Incident Angle

The effect from the variation of the incident angle has been studied by the different orientation of an AISI 420 stainless steel probes during gallium irradiation (Figure 3). At first, a probe was installed in a vertical position (Figure 3a) and it was subjected to gallium irradiation in FIB system. After a pre-defined FIB dose was delivered to the probe, the nanospikes formed in the direction of the beam. The same nanospikes formation occurred for a horizontally positioned probe (Figure 3b) and 40° inclined probe (Figure 3c). From these results, one can infer that nanospikes form regardless of the incident angle in this specific martensitic steel alloy.

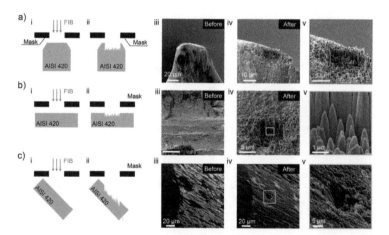

Figure 3. Spiking phenomena of AISI 420 stainless steel sample probes during gallium irradiation with different incident angles: (**a**) Vertical position of the probe; (**b**) Horizontal position of the probe; and (**c**) at an incident angle of 40° to the probe. Here, (**i**) and (**ii**) illustratively depict the orientation of the sample, the gallium irradiated regions and the spiking result, respectively. Here, (**iii**) and (**iv**) show the results obtained before and after the FIB irradiation. Here, (**v**) are close-ins of (**iv**).

3.5. Nanospiking Effect on Austenitic AISI 316, Ferritic AISI 430 and Martensitic AISI 431 during FIB Treatment

Figure S6 shows a comparison of FIB irradiated stainless steel plates with gallium ions of other three different stainless steel types, i.e., austenitic (AISI 316), ferritic (AISI 430) and martensitic (AISI 431) stainless steel plates. Details of the sample preparation and FIB treatment are provided in Section 2.1. All samples have an anisotropic etching behavior. The austenitic AISI 316 stainless steel sample (a and d) shows a mix of inhomogeneous nano-rough regions and regions with nanospikes (Figure S6d). Nano-rough regions look like mountain range or cliffs, therefore the notation "nano-cliffs", for instance, see the orange arrows in Figure S6. The ferritic AISI 431 (c and e) features only a region with nanospikes (Figure S6e). The AISI 431 displays a presence of nanospikes, however, these are seldom scattered in between the nano-roughed bottom (Figure S6f).

4. Discussion

As can be seen from Figures 2 and 3, the nanospikes on the edges have much higher aspect ratios than the ones in the central region of the trench. When sputtering occurs at the edges, it nucleates the edges (formation of nanospikes), due to the presence of the gallium ions outside of the beam spot (the beam power features Gauss distribution). The inner part of the trench continues to be sputtered, but the part outside of the trench is slightly affected by the satellite gallium ions. The satellite gallium ions also cause sputtering, but at significantly reduced rates than the inner part of the trench. The sputtering continuity of the inner part shifts the nano-spiky structure downwards into the bulk, but this shift barely occurs at the edges. This discrepancy in the structural shift can explain why the nanospikes in the edges feature higher aspect ratios than the nanospikes in the central region of the trench.

Polycrystalline alloys such as the martensitic AISI 420 stainless steel (and the other FIB treated stainless steels) investigated in this work, besides the difference in material content, feature domains with different crystallographic orientations. The sputtering rates of neighboring domains may vary greatly, depending on the structural configuration of the grains and the orientation of the lattices in the particular domain with respect to the incident ion beam. However, we have shown that nanospikes occurred in the gallium bombarded AISI 420 sample, but not as much in the AISI 431 sample, where nano-cliffs were more dominant, although both samples are martensitic stainless steels with very similar crystalline structure [24,25].

We have performed EDX and XPS analysis to investigate the material content within the non-irradiated and the irradiated regions of the martensitic AISI 420 stainless steel alloy. The EDX results (penetration depth up to 10 µm) show the presence of iron, chromium, and gallium in the gallium-irradiated regions. The XPS results show a decrease of the iron and the chromium in the irradiated trench with respect to non-irradiated surface. The XPS results do not indicate any saturation of a single element on the very surface in the irradiated regions.

Other studies have demonstrated that ion channeling affects the sputtering yield in polycrystalline materials such as [4,8,9], therefore inducing nano-roughening on the treated surface. However, we are not sure whether the same explanation can be attributed to the formation of the nanospikes. The nanometer sized spiky formations are special forms of the nano-roughed surface, and the exact mechanism has been recently discussed by Prenitzer et al. [8] and Ran et al. [12], but also heavily researched much earlier [13–18]. Auciello [18] claims that the micro/nanometer scaled pyramidal structures form due to sputtering differences in (1) the presence of intrinsic and/or bombardment-induced sub-surface defects, (2) the evolution of pre-existing and/or bombardment-induced asperities of convex-up curvatures, and (3) the erosion of nuclei formed by migration of sputter-deposited foreign atoms on the substrate of the surface. The observed nanospike formations by Prenitzer et al. [8] were attributed to a wide range of sputtering conditions, whereas the most likely one may be the quality of initial target surface. Ran et al. [12] imaged 18 Cr-ODS steel nanospikes with transmission electron microscopy (TEM), showing that two different crystal orientations do exist in one nanospike with distinct two grains and a clear grain boundary. The report claims that nanospike formation is not induced by grain recrystallization and regrowth during Ga$^+$ ion bombardment, but rather due to an interplay between a curvature-dependent sputtering and defect accumulation near the surface. Both reports address the importance of the initial surface topology. This interplay between a curvature-dependent sputtering and defect accumulation near the surface seems to be a valid argument and might be used to interpret our experimental observations, since the morphological variation of the targeted surfaces greatly impacts the dynamic competition of available atoms on the substrate, the atom evacuation due to sputtering, and the gathering of vacancies.

5. Summary and Conclusions

The nanospiking phenomenon has been previously reported for copper [8], tungsten [11], and 18Cr-ODS steel [12]. In this communication, we have shown that nanospikes are formed on martensitic AISI 420 stainless steel when treated with FIB. The nanospikes can be made straight or angled depending on the incident angle between the sample and the beam. We also showed fabrication of a <16 nanometer sharp single tip needle and a micrometer-sized sharp needle with nanospikes. The nanospiking effect occurs in ferritic AISI 430 stainless steel sample too. A weak occurrence of the nanospiking effect in between nano-rough regions (nano-cliffs) was also witnessed for the austenitic AISI 316 and martensitic AISI 431 stainless steel samples. Unlike the intermediate existence of the nano-pyramidal structures reported in [16], the nano-spiky structures reported here are stable and occur at different irradiation doses.

The nanospiking phenomenon in martensitic AISI 420 stainless steels has promising capacity for future research. The single sharp nano-needle has potential of being used for creating localizing magnetic fields, as in [19], or laser-induced electron emission as in [20,21]. A micrometer-scaled needle with nano-spiky topology could be utilized for making superhydrophobic needles, performing droplet manipulation on open hydrophobic and superhydrophobic surfaces, where needle-droplet adhesion is less than droplet-substrate adhesion, similar to as in [26,27]. Since the martensitic stainless steel has soft ferromagnetic properties, the Ga$^+$ ion bombardment process can be used for fabricating magnetic nanospikes, which might find application in the development of novel quantum devices, e.g., magneto-gravitational traps [22,23].

Supplementary Materials: The following are available online at http://www.mdpi.com/2079-4991/9/10/1492/s1, Figure S1: Intermediate steps of the fabrication process of martensitic stainless steel AISI420 needle with nanometer sharpness. Figure S2: Illustration of the fabrication procedure of microneedle with nanospikes. Figure S3: Intermediate steps of the fabrication process of martensitic stainless steel AISI420 microneedle with nanospikes. Figure S4: Energy-dispersive X-ray spectroscopy (EDX) results of (a) the completely irradiated trench; (b) a spot on a single nanospike; (c) non-irradiated area. Figure S5: X-ray Photoelectron Spectroscopy (XPS) results of (a) a ~35 μm in diameter gallium irradiated trench; (b) non-irradiated area. i and ii denote measurement with XPS aperture of 27 and 55 μm, respectively. Figure S6: Gallium irradiation of austenitic AISI 316 (a,d), ferritic AISI 430 (b,e), and martensitic AISI 431 (c,f) stainless steel plates with a dose of 19.4 C/μm^2. (a–c) Before and (d–f) after gallium irradiation.

Author Contributions: Z.C., W.K., and M.B. have jointly observed the nanospiking phenomena of AISI 420 martensitic stainless steel and fabricated the needles. Z.C. has electrochemically etched the martensitic AISI 420 stainless steel wires, performed the EDX and the XPS measurements and analysis. W.K. and M.B. have performed the FIB treatments and investigated the spiking effects on the other stainless steel types. Z.C. and V.J. have conceived the idea of fabrication of microscopic needles with nanospikes. Q.Z., S.F. and M.B. supervised the whole research throughout the whole duration and ensured credible conduction of experimental work. All authors wrote the paper.

Funding: This research work was supported by the Academy of Finland (projects: #304843, #295006, #297360) and German Academic Exchange Service (DAAD) (project: #57247327). The authors express their gratitude to Micronova Nanofabrication Center for providing laboratory facilities for microfabrication.

Conflicts of Interest: The authors declare no conflict of interest.

References

1. Gupta, J.; Harper, J.M.E.; IV, J.M.; Blauner, P.G.; Smith, D.A. Focused ion beam imaging of grain growth in copper thin films. *Appl. Phys. Lett.* **1992**, *61*, 663–665. [CrossRef]

2. Fujita, J.; Ishida, M.; Ichihashi, T.; Ochiai, Y.; Kaito, T.; Matsui, S. Growth of three-dimensional nano-structures using FIB-CVD and its mechanical properties. *Nucl. Instrum. Methods Phys. Res. Sect. B Beam Interact. Mater. At.* **2003**, *206*, 472–477. [CrossRef]

3. Young, R.J. Micro-machining using a focused ion beam. *Vacuum* **1993**, *44*, 353–356. [CrossRef]

4. Nagasaki, T.; Hirai, H.; Yoshino, M.; Yamada, T. Crystallographic orientation dependence of the sputtering yields of nickel and copper for 4-keV argon ions determined using polycrystalline targets. *Nucl. Instrum. Methods Phys. Res. Sect. B Beam Interact. Mater. At.* **2018**, *418*, 34–40. [CrossRef]

5. Lucille, A.; Giannuzzi, F.A.S. *Introduction to Focused Ion Beams*; Giannuzzi, L.A., Stevie, F.A., Eds.; Springer: Boston, MA, USA, 2005; ISBN 978-0-387-23116-7.

6. Guu, Y.H.; Hocheng, H. *Advanced Analysis of Nontraditional Machining*; Hocheng, H., Tsai, H.Y., Eds.; Springer: New York, NY, USA, 2013; ISBN 978-1-4614-4053-6.

7. Kempshall, B.W.; Schwarz, S.M.; Prenitzer, B.I.; Giannuzzi, L.A.; Irwin, R.B.; Stevie, F.A. Ion channeling effects on the focused ion beam milling of Cu. *J. Vac. Sci. Technol. B Microelectron. Nanometer Struct. Process. Meas. Phenom.* **2002**, *19*, 749. [CrossRef]

8. Prenitzer, B.I.; Urbanik-Shannon, C.A.; Giannuzzi, L.A.; Brown, S.R.; Irwin, R.B.; Shofner, T.L.; Stevie, F.A. The Correlation between Ion Beam/Material Interactions and Practical FIB Specimen Preparation. *Microsc. Microanal.* **2003**, *9*, 216–236. [CrossRef]

9. Ran, G.; Wu, S.; Liu, X.; Wu, J.; Li, N.; Zu, X.; Wang, L. The effect of crystal orientation on the behavior of a polycrystalline tungsten surface under focused Ga+ ion bombardment. *Nucl. Instrum. Methods Phys. Res. Sect. B Beam Interact. Mater. At.* **2012**, *289*, 39–42. [CrossRef]

10. Wagner, A. X-ray mask repair with focused ion beams. *J. Vac. Sci. Technol. B Microelectron. Nanometer Struct. Process. Meas. Phenom.* **1990**, *8*, 1557. [CrossRef]

11. Ran, G.; Liu, X.; Wu, J.; Li, N.; Zu, X.; Wang, L. In situ observation of surface morphology evolution in tungsten under focused Ga + ion irradiation. *J. Nucl. Mater.* **2012**, *424*, 146–152. [CrossRef]

12. Ran, G.; Chen, N.; Qiang, R.; Wang, L.; Li, N.; Lian, J. Surface morphological evolution and nanoneedle formation of 18Cr-ODS steel by focused ion beam bombardment. *Nucl. Instrum. Methods Phys. Res. Sect. B Beam Interact. Mater. At.* **2015**, *356–357*, 103–107. [CrossRef]

13. Stewart, A.D.G.; Thompson, M.W. Microtopography of surfaces eroded by ion-bombardment. *J. Mater. Sci.* **1969**, *4*, 56–60. [CrossRef]

14. Whitton, J.L.; Holck, O.; Carter, G.; Nobes, M.J. The crystallographic dependence of surface topographical features formed by energetic ion bombardment of copper. *Nucl. Instrum. Methods* **1980**, *170*, 371–375. [CrossRef]

15. Whitton, J.L.; Tanović, L.; Williams, J.S. The production of regular pyramids on argon ion bombarded surfaces of copper crystals. *Appl. Surf. Sci.* **1978**, *1*, 408–413. [CrossRef]

16. Auciello, O.; Kelly, R.; Iricibar, R. On the problem of the stability of pyramidal structures on bombarded copper surfaces. *Radiat. Eff.* **1979**, *43*, 37–42. [CrossRef]

17. Kelly, R.; Auciello, O. On the origin of pyramids and cones on ion-bombarded copper surfaces. *Surf. Sci.* **1980**, *100*, 135–153. [CrossRef]

18. Auciello, O. Critical Analysis on the Origin, Stability, Relative Sputtering Yield and Related Phenomena of Textured Surfaces Under Ion Bombardment. *Radiat. Eff.* **1982**, *60*, 1–26. [CrossRef]

19. Cenev, Z.; Zhang, H.; Sariola, V.; Rahikkala, A.; Liu, D.; Santos, H.A.; Zhou, Q. Manipulating Superparamagnetic Microparticles with an Electromagnetic Needle. *Adv. Mater. Technol.* **2018**, *3*, 1700177. [CrossRef]

20. Bionta, M.R.; Chalopin, B.; Champeaux, J.P.; Faure, S.; Masseboeuf, A.; Moretto-Capelle, P.; Chatel, B. Laser-induced electron emission from a tungsten nanotip: Identifying above threshold photoemission using energy-resolved laser power dependencies. *J. Mod. Opt.* **2014**, *61*, 833–838. [CrossRef]

21. Bionta, M.R.; Chalopin, B.; Masseboeuf, A.; Chatel, B. First results on laser-induced field emission from a CNT-based nanotip. *Ultramicroscopy* **2015**, *159*, 152–155. [CrossRef]

22. Slezak, B.R.; Lewandowski, C.W.; Hsu, J.F.; D'Urso, B. Cooling the motion of a silica microsphere in a magneto-gravitational trap in ultra-high vacuum. *New J. Phys.* **2018**, *20*, 063028. [CrossRef]

23. Houlton, J.P.; Chen, M.L.; Brubaker, M.D.; Bertness, K.A.; Rogers, C.T. Axisymmetric scalable magneto-gravitational trap for diamagnetic particle levitation. *Rev. Sci. Instrum.* **2018**, *89*, 125107. [CrossRef] [PubMed]

24. Baghjari, S.H.; AkbariMousavi, S.A.A. Experimental investigation on dissimilar pulsed Nd: YAG laser welding of AISI 420 stainless steel to kovar alloy. *Mater. Des.* **2014**, *57*, 128–134. [CrossRef]

25. Khorram, A.; Davoodi Jamaloei, A.; Jafari, A.; Moradi, M. Nd:YAG laser surface hardening of AISI 431 stainless steel; mechanical and metallurgical investigation. *Opt. Laser Technol.* **2019**, *119*, 105617. [CrossRef]

26. Long, Z.; Shetty, A.M.; Solomon, M.J.; Larson, R.G. Fundamentals of magnet-actuated droplet manipulation on an open hydrophobic surface. *Lab Chip* **2009**, *9*, 1567–1575. [CrossRef]

27. Gao, N.; Geyer, F.; Pilat, D.W.; Wooh, S.; Vollmer, D.; Butt, H.J.; Berger, R. How drops start sliding over solid surfaces. *Nat. Phys.* **2018**, *14*, 191–196. [CrossRef]

 *nanomaterials*

MDPI

Article

Enhanced Protective Coatings Based on Nanoparticle fullerene C60 for Oil & Gas Pipeline Corrosion Mitigation

Xingyu Wang [1], Fujian Tang [2], Xiaoning Qi [3], Zhibin Lin [1,*], Dante Battocchi [3] and Xi Chen [4]

[1]   Department of Civil and Environmental Engineering, North Dakota State University, Fargo, ND 58018, USA; Xingyu.wang@ndsu.edu
[2]   State Key Laboratory of Coastal and Offshore Engineering, School of Civil Engineering, Dalian University of Technology, Dalian 116024, China; ftang@dlut.edu.cn
[3]   Department of Coatings and Polymeric Materials, North Dakota State University, Fargo, ND 58018, USA; xiaoning.qi@ndsu.edu (X.Q.); bante.battocchi@ndsu.edu (D.B.)
[4]   Guilin University of Technology, Guilin Guangxi 541004, China; xi.chen_01@outlook.com
*   Correspondence: zhibin.lin@ndsu.edu; Tel.: +1-701-231-7204

Received: 23 September 2019; Accepted: 14 October 2019; Published: 17 October 2019

Abstract: Corrosion accounts for huge maintenance cost in the pipeline community. Promotion of protective coatings used for oil/gas pipeline corrosion control, in terms of high corrosion resistance as well as high damage tolerance, are still in high demand. This study was to explore the inclusion of nanoparticle fullerene-C60 in protective coatings for oil/gas pipeline corrosion control and mitigation. Fullerene-C60/epoxy nanocomposite coatings were fabricated using a solvent-free dispersion method through high-speed disk (HSD) and ultrasonication. The morphology of fullerene-C60 particles was characterized by transmission electron microscopy (TEM), and dynamic light scattering (DLS). The data analysis indicated that the nanoparticles were effectively dispersed in the matrix. The performance of the nanocomposites was investigated through their mechanical and electrochemical properties, including corrosion potential, tensile strength, strain at failure, adhesion to substrate, and durability performance. Dogbone shaped samples were fabricated to study the tensile properties of the nanocomposites, and improvement of strength, ultimate strain, and Young's modulus were observed in the C60/epoxy specimens. The results demonstrated that the C60/epoxy composite coatings also had improvements in adhesion strength, suggesting that they could provide high damage tolerance of coatings for engineering applications. Moreover, the electrochemical impedance spectroscopy (EIS) results generated from the accelerated durability test revealed that the developed fullerene-C60 loaded composite coatings exhibited significantly improved corrosion resistance. The nanocomposite with 0.5 and 1.0 wt.% of C60 particles behaved as an intact layer for corrosion protection, even after 200-h salt spray exposure, as compared to the control coating without nanofiller in which severe damage by over 50% reduction was observed.

Keywords: nano-modified high-performance coating; dispersion methods; fullerene-C60; corrosion mitigation; nanocomposite; gas and oil pipelines

1. Introduction

Corrosion has been a leading cause of metallic oil/gas pipeline failures in the United States and worldwide [1–4]. A report revealed that there are over 2.6 million miles of gas and oil pipeline in United States. Most of the pipelines are fabricated by low-carbon steel, and severe corrosion failures can be developed when they are exposed to corrosive media during transporting oil and gas [5]. As of 2014, the reported cost of corrosion in the oil and gas industry was more than $17 billion for the United

States [6]. Despite the fact that significant efforts have been targeted in corrosion control and mitigation, there are still challenges to develop effective corrosion prevention techniques [7–11].

In the last two decades, carbon-based nanoparticles have attracted great attention due to their ability to provide outstanding mechanical, tribological, and electrical properties with the various dimensions and geometrical shapes [12–19]. One particularly promising application of carbon-based nanoparticles is to assemble high-performance nanocomposites by incorporating nanofiller reinforcement into polymers [13] and coatings [18–20]. The enhancements of carbon-based nanocomposites are often considered to be associated with the shape and size of the nanofillers [14,15]. Nanofillers can be defined as zero, one, or two-dimensional materials [16]. Fullerene-C60 is a typical 0-dimensional spherical nanofiller that contains only carbon atoms, consisting of 12 pentagons and 20 hexagons arranged in a cage-like structure [17]. Their unique shapes provide a different combination of properties and assist the polymetric coating in overcoming their limitations [19,21–27]. In particular, fullerene-C60/polymer nanocomposites demonstrate enhanced mechanical and anti-corrosion properties, as compared to neat polymer materials [28,29].

Fullerene-C60 particles have been investigated as a nanofiller in polymer reinforcement due to their unique aromatic character and nanosized diameter [30]. Zuev et al. [29] suggested strong reinforcement could be obtained with a small amount of fullerene nanofillers in epoxy matrix, which should be less than 0.5 wt.%. Otherwise, agglomerates will be created with the excess amount of C60 nanoparticles, which leads to degradation on coating performance. Pikhurow et al. [28] reported that the addition of fullerene-C60 particles improved the Young's modulus and tensile strength of epoxy resin. The maximum reinforcement was observed in the sample containing 0.08 wt.% of fullerene-C60 particles. Differently, other authors [30] mentioned that the maximum reinforcement on mechanical properties and elastic resilience was observed with the addition of 2 wt.% of C60 in poly(styrene-b-butadiene-b-styrene (SBS)), and the tensile strength increased almost 13 times compared with base polymer. Meanwhile, Ogasawara et al. [31] also pointed out that the greatest improvement was obtained with 1.0 wt.% of fullerene-C60 nanoparticles in the epoxy, with both increased tensile strength and strain. In addition, compared with neat epoxy, researchers [17] suggested both enhanced anti-corrosion and tribological properties were observed after the incorporation of C60 particles. C60/epoxy nanocomposites containing C60 particles from 0.25 to 1.0 wt.% were fabricated. However, the performance of the coatings was weakened when the content of C60 was higher than 0.5 wt.%; hence, this observation may result from large number of aggregated nanoparticles. Clearly, fullerene-C60 nanoparticle showed its potential to the development of high-performance coatings, but the current findings as observed in the literature is still unclear, as there were still in high variances in many academic studies. Specifically, one may concern appropriate content of nanoparticles as demanded for a coating system; however, the suggestions of fullerene-C60 content in coating system were sources of conflict among different researchers. Moreover, most of existing studies on fullerene-C60 particles as nanofillers in polymer system have been directed to mechanical properties of the nanocomposites [28,29,31]. As compared, few studies [17] discussed corrosion protection behaviors of fullerene-C60 particle reinforced coatings, which are critical properties as required for metallic pipelines. Furthermore, to the best of authors' knowledge, evaluation on the long-term performance of C60/epoxy coatings with varied contents of C60 nanofillers has not been reported.

Until recently, some research efforts have been made on corrosion protection enhancement of modified epoxy [32–35]. Li et al. [32] developed an anti-corrosion epoxy with the addition of a-zirconium phosphate nanoplatelets (ZrP). The ZrP nanoplatelets were prepared with a refluxing method. The Zirconyl chloride was refluxed in H_3PO_4 solution for 24 h, and the solution was then dried for another 24 h. The obtained product was grounded into fine powders. The ZrP powders were dispersed in acetone before mixing with epoxy resin. Improved corrosion resistance was confirmed by the potentiodynamic polarization and EIS measurements. Zhang et al. [33] successfully used modified silicon nitride powders to enhance the corrosion performance of epoxy coatings. The silicon nitride powders were kept in a vacuum chamber at 80 °C for 24 h before the fabrication procedure. Then the powders were dispersed into methacryloxy propyl trimethoxyl silane with ultrasonication, and deionized water, ethyl alcohol, and acetic acid were added into the solution, while the PH value was adjusted between 3.5 to 5.0. The solution was mechanically stirred for 3 h in a water bath, and the modified silicon nitride powders were obtained after the solution was dried in an oven. Stronger barrier performance was observed in the epoxy coating with the addition of silane functionalized silicon nitride. Chhetri et al. [34] developed functionalized APTES–Mo–LDH reinforced epoxy for corrosion protection application. The $Mg(NO_3)_2 \cdot 6H_2O$ and $Al(NO_3)_3 \cdot 9H_2O$ salt were dissolved in distilled water, and the solution were stirred with $NaNO_3$ and $NaOH$ solutions, while the PH value was maintained around 10 by adding $NaOH$ solution. Then the obtained solution was dried inside a vacuum oven to obtained $MgAl–NO_3–LDH$ powder; the powder was mixed with $Na_2MoO_4 \cdot 2H_2O$ with mechanical stirring for 12 h, followed by a 6 h reflux. At this stage, modified Mo–LDH was obtained, then APTES, acetic acid, and anhydrous ethanol were used to functionalize modified Mo–LDH powder into APTES-modified Mo–LDH, as named as APTES–Mo–LDH. Improved corrosion resistance and adhesion were observed in the developed APTES–Mo–LDH/Epoxy coatings. The above studies have successfully increased the corrosion protection performance of epoxy coating with their developed modification techniques; however, the applied techniques were either time-consuming or required extensive chemical solvents in the fabrication process, which showed their limitations toward large scale commercial applications.

As such, this study aimed to investigate the inclusion of fullerene-C60 in epoxy for the development of nanocomposites for oil and gas pipeline corrosion mitigation, and the nanocomposite coatings were prepared by a solvent-free facile approach. The nanoparticle size distribution was investigated using dynamic light scattering (DLS), and the dispersion was inspected by scanning electron microscopy (SEM). The mechanical and electrochemical behavior of the developed nanocomposite coatings was systemically evaluated by electrochemical impedance spectroscopy (EIS), pull-off strength test, and tensile test. In addition, the salt fog test was employed to examine the durability of the nanofiller reinforced coatings.

2. Experiment

This section describes the detailed experimental methods for synthesizing and characterizing fullerene-C60/epoxy coatings, including material preparation and synthesis, the dispersion method, and characterization as shown below.

2.1. Material

Fullerene-C60 (Sigma-Aldrich Corp., St. Louis, MO, USA) nanoparticles were purchased and used without any modification. As shown in Figure 1, the size and shape of C60 nanoparticle were examined by transmission electron microscopy (TEM), and it was clear that the as-received C60 particles were initially in agglomerates with an average diameter of 20 nm. A two-component epoxy adhesive coating was used to mix with nanofillers in this study, and the coating was based on EPON™ Resin 828 resin and Epikure 3175 (Hexion Inc., Columbus, OH, USA). The EPON™ Resin 828 resin is a pure bisphenol A/epichlorohydrin derived liquid epoxy resin and able to provide good mechanical, adhesive, dielectric and chemical resistance properties when crosslinked with EPIKURE™ Curing Agent 3175.

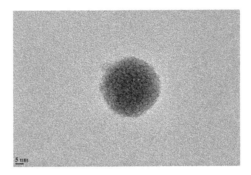

Figure 1. Transmission electron microscopy of a fullerene-C60 particle.

2.2. Fabrication of Nanofiller-Reinforced Epoxy Composites and Test Sample Preparation

The dispersion of the fullerene-C60 was carried out by integration of high-speed disk (HSD) and ultrasonication, as this method can effectively disperse nanoparticles into polymer matrix. The epoxy resin was first mixed with C60 particles by using HSD dispersers (high-speed impellers) at a speed of 4000 rpm for 30 min, and the shear stress during the high-speed rotation was able to break down large aggregated particles. After the first step, the solution was subjected to ultrasonication (Misonix S1805 sonicator) with a 19-mm probe at 100% amplitude, and the process used a 30 s on/off cycle for a total duration of 60 min. After this step, a proper dispersion of C60 nanoparticles was obtained in the epoxy resin. The solution was cooled to room temperature and was then mixed with EPIIKURE™ Curing Agent 3175 with a 1:1 mole ratio at 600 rpm for another 10 min.

Nanocomposites with varied C60 weight concentration were fabricated in this study, which included 0.1, 0.5, 1.0, 1.5 and 3.0 wt.%. The developed coatings were applied on standard Q-panel steel substrates, and the commercial panels were purchased from Q-lab. Inc, with a dimension of 76×152 mm and a thickness of 0.8 mm. The panels were cleaned with acetone before applying the developed coatings. The coatings were then allowed to completely dry for a few days. After that, the thickness of dried films was measured, and the average thickness of 110 ± 5 μm was obtained for all the tested specimens. For simplicity, the specimens were labeled based on the weight content of nanofillers, while 0.1% F-Epoxy denotes the coating mixed with 0.1 wt.% of fullerene-C60.

2.3. Characterization Method

2.3.1. Characterization

The characterization of nanoparticle and nanocomposite coatings was performed by several techniques, including dynamic light scattering (DLS), scanning electron microscopy (SEM) and transmission electron microscopy (TEM). A JEOL JEM-2100 (JEOL USA, Inc., Peabody, MA, USA) high-resolution analytical transmission electron microscopy (TEM) was employed to examine the shape and average size of fullerene-C60 nanoparticles (as typically shown in Figure 1). Moreover, the dispersion and stability of nanoparticles were characterized by dynamic light scattering, with a Particle Sizing Systems Nicomp 380 (PSS Nicomp, Santa Barbara, CA, USA). The average particle size distribution of fullerene-C60 was obtained for the specimens with varied weight content of nanoparticles, and the information was used to investigate the extent of agglomeration. In order to visually examine the effectiveness of employed dispersion method, scanning electron microscopy (SEM) technique was carried out to observe the distribution of nanoparticles in the epoxy matrix, with a JEOL JSM-7600F field-emission SEM (JEOL USA, Inc., Peabody, MA, USA). Meanwhile, after tension test, SEM images of fracture surface were used to analyze the impact resistance and fracture toughness for the dogbone shaped specimens.

2.3.2. Corrosion Resistance of the Composite Coating Using EIS

The corrosion protection performance of the developed coatings can be effectively evaluated by the electrochemical behavior obtained from electrochemical impedance spectroscopy (EIS) test [36]. The EIS test was employed by using Gamry equipment (Reference 600 potentio/Galvanostat/ZRA); a saturated calomel electrode reference electrode was used as the reference electrode, while a platinum mesh and the test panel were worked as the counter electrode and working electrode, respectively. A glass tube with a diameter of 30-mm was clamped on the panel and filled with 1.0% of NaCl solution during the test. The measurements were collected in the frequency range of 10^5 to 10^{-2} Hz, and the obtained data was described as impedance spectra. The impedance-frequency plot analysis was used to provide detailed information about the corrosion potential of coatings.

Salt spray test (ASTM B117) was applied as an accelerated durability test for the prepared coatings to examine their long-term performance. The specimens were exposed to salt fog spray with evaluated temperature in a Q-Fog CCT chamber (Q-Lab Corporation, Cleveland, OH, USA) for 200 h, and EIS measurements were conducted before, 100 h, and 200 h after the exposure.

For further investigation, the obtained data from EIS measurement were fitted into equivalent electrical circuit models (EEC). As described in Figure 2, the coating degradation process was characterized into four stages, and each stage could be represented by one EEC model. The EEC model consists of R_{sol} (solution resistance), R_c (coating resistance) and C_{po} (constant-phase element of the coating) is labeled as model A, indicates that coating behaves an intact layer to protect substrate. After that, once the coating was damaged and could not prevent electrodes from penetrating the coating layer to contact with substrate, that is, the corrosion reaction was initiated, then the coating reached the second stage. In this case, R_{ct}, charge transfer resistance, and C_{dl}, constant phase element of double-charge, were added into the EEC model, and the new model was named as model B. In the third stage, the model was described as model B with W, the included Warburg impedance element (W) indicating the diffusion effect dominated corrosion has occurred in the system. In the final stage, the coating suffered severe corrosion damage, and a thin corrosion product layer was accumulated by a large amount of corrosion products. The new parameters, including constant phase element of diffusion capacitance (C_{diff}) and diffusion resistance (R_{diff}) were joined to represent the corrosion product layer.

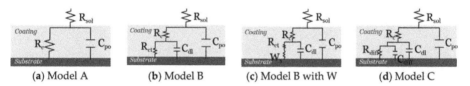

(a) Model A   (b) Model B   (c) Model B with W   (d) Model C

Figure 2. Equivalent electrical circuit models at four stages: (**a–d**). (**a**) Model A; (**b**) Model B; (**c**) Model B with the Warburg impedance element (W); (**d**) Model C.

As mentioned above, the degree of coating damage over exposure time can be evaluated using EIS data. In this study, a new corrosion protection index was introduced to effectively describe the coating corrosion resistance, and the formulation was modified from previous researchers' work [37,38]. Therefore, the coating corrosion protection index, CCPI, could be obtained through the EIS data at a wide range of frequencies, as shown below:

$$CCPI(\%) = \left(\frac{A_1 + A_2}{A_1 + A_2 + A_3} \right) \times 100 \tag{1}$$

where $(A_1 + A_2)$ is the area under the impedance curve in log-log axes for a damaged coating, and $(A_1 + A_2 + A_3)$ represents the impedance plot for an ideally intact coating (see Figure 3). In this case, a larger value of *CCPI* indicated less coating damage, and when the value reached up to 100%, the coating behaved as an intact layer for corrosion protection.

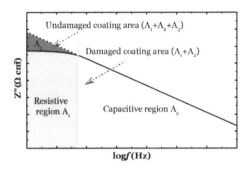

Figure 3. Corrosion protection index for the coating degradation assessment.

2.3.3. Adhesion of the Composite Coating Using Tensile Button Testing

The adhesive bonding strength between coating and substrate was evaluated by the pull-off tensile test. Before the test, dollies were glued to the surface of specimen. Then the dollies were pulled vertically away from substrate until completely detached, and the adhesion strength was measured. The test area was abraded with a 100-grit sandpaper to enhance the bonding between dollies and coating, and then the dollies were glued on the abraded surface. The samples were kept in room temperature for 24 h, which allowed the glue to dry. Before the adhesion test applied, the test area was isolated by using die cutting.

2.3.4. Tensile Strength, Ultimate Strain, and Young's Modulus

The coupon tensile test was conducted to characterize the tensile properties of nanofiller reinforced polymer composites. The test was carried out by following ASTM D638 standard with a Shimadzu's EZ-X tester (Shimadzu Scientific Instruments, Columbia, MD, USA). Tensile strength was applied to elongate the sample with a testing speed of 1mm/min until the sample broke in the narrow test section. In this test, maximum tensile strength, strain at failure, and Young's modulus of each specimen were calculated.

3. Results and Discussion

3.1. Particle Size Distribution and Dispersion

The DLS measurements confirmed that the developed dispersion method could effectively prevent the C60 particles from forming large agglomerates. The obtained results revealed that the C60 samples were dispersed into nano-sized particles with a diameter between 40 to 100 nm, and the average particle size did not significantly increase with the increasing weight concentration of nanofillers (Figure 4). A slight increase in particle size was observed, as the average particle diameter was around 65, 70, and 80 nm for the sample with 0.1, 1.0 and 3.0 wt.% C60, respectively.

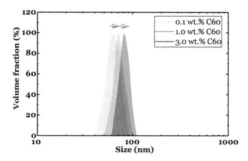

Figure 4. Particle size distribution of fullerene-C60 nanocomposites.

The cross-sectional SEM images of epoxy samples with varied content of C60 were presented in Figure 5a through to Figure 5c. The observation showed a strong agreement with results from the particle size distribution test, and the C60 particles behaved with good compatibility and a high dispersion level in the epoxy matrix, in which no large agglomerate was observed in all the tested samples. These results also indicated that fullerene particles exhibited a weak tendency to form agglomerates due to their unique spherical shape [17].

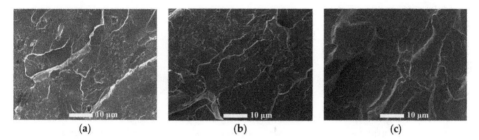

Figure 5. Cross-sectional SEM image of (**a**) 0.1 wt.%, (**b**) 1.0 wt.%, and (**c**) 3.0 wt.% C60/epoxy.

3.2. Barrier Performance of the New Composite Coatings

The corrosion barrier performance of the samples loaded with fullerene-C60 particles was evaluated by way of an electrochemical impedance spectroscopy (EIS) test. The EIS measurement was utilized before, and 100 h, and 200 h after exposure and the results were presented in terms of impedance and phase angle plots, as shown in Figure 6a–f. According to the collected data, the neat epoxy coating behaved fair in terms of corrosion protection for the substrate. However, as a clear bend was observed at the low-frequency region of impedance curve, it could be understood that the neat epoxy coating could not act as a solid barrier layer against corrosive media [39]. Researchers have suggested that micro-pores would be generated into a neat epoxy coating during the curing process, and these voids allowed corrosive media to penetrate through the coating and initiate a corrosion reaction at substrate, as confirmed elsewhere [40,41]. This was further confirmed by data in Table 1, where the neat epoxy coating behaved at the third stage of corrosion process, and the Warburg impedance element (W) indicates that the corrosion was diffusion dominant. As shown in Figure 6b,c, a clear degradation was observed in the neat epoxy coating during the exposure, and the degradation level was developed over time, suggesting that the coating failed to provide long-term corrosion protection for metallic substrate.

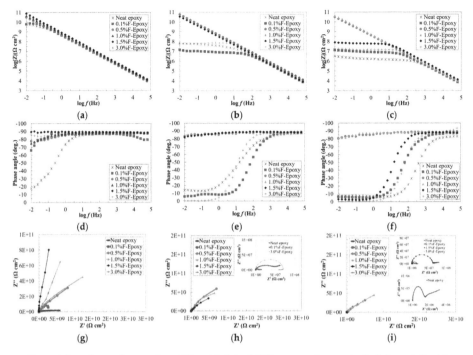

Figure 6. Impedance, phase angle and Nyquist plots of C60/epoxy coatings under (**a,d,g**) onset, (**b,e,h**) 100-h, and (**c,f,i**) 200-h exposure.

Obviously, higher content fullerene-C60 particles (from 0.5 to 3.0 wt.%) provided stronger corrosion resistance than the neat epoxy. Overall, it turned out that model A was suitable for the groups with 0.5 to 3.0 wt.% fullerene-C6o particles, which meant that the coatings were providing perfect corrosion resistance (Table 1). As shown in Figure 6a, similar to the neat epoxy, the group with 0.1 wt.% of fullerene-C60 exhibited low corrosion resistance and a clear bend was observed in the phase angle curve in the tested frequency region. The results suggested that there was no noticeable reinforcement on corrosion resistance when low content (0.1 wt.%) of fullerene-C60 was incorporated into epoxy resin. Differently, all the other fullerene-C60/epoxy groups exhibited higher impedance, suggesting improved corrosion barrier performance was obtained by the addition of C60 nanofillers.

Table 1. Electrochemical impedance spectroscopy (EIS) data associated with different stages of the equivalent electrical circuit models.

Label	Content of C60 (wt.%)	Exposure to Accelerated Environmental Stresses					
		Onset		100-h		200-h	
		Electrical Circuit Models (EEC)	$|Zmod|_{0.01Hz}$	EEC	$|Zmod|_{0.01Hz}$	EEC	$|Zmod|_{0.01Hz}$
Neat epoxy	/	Model B with W	6.10×10^9	Model B with W	6.46×10^7	Model B with W	3.29×10^6
0.1%F-Epoxy	0.1	Model B with W	6.39×10^9	Model B with W	1.45×10^7	Model B with W	1.31×10^7
0.5%F-Epoxy	0.5	Model A	3.21×10^{10}	Model A	6.03×10^{10}	Model A	3.22×10^{10}
1.0%F-Epoxy	1.0	Model A	4.69×10^{10}	Model A	4.87×10^{10}	Model A	4.64×10^{10}
1.5%F-Epoxy	1.5	Model A	8.04×10^{10}	Model A	3.47×10^{10}	Model B	7.86×10^7
3.0%F-Epoxy	3.0	Model A	6.51×10^{10}	Model B	5.85×10^7	Model B	1.83×10^7

After 200 h of exposure, similar to the fresh stage, the results from 0.5% and 1.0% F-Epoxy group showed their extraordinary anti-corrosion performance, regardless of exposure time. High impedance value was maintained during exposure, as the impedance value was over 10^{10} Ω/cm^2 during the exposure. Additionally, the 0.5% and 1.0% F-Epoxy groups remained in model A during the whole exposure, which confirmed that 0.5 and 1.0 wt.% of fullerene-C60 could dramatically improve the corrosion resistance of epoxy coatings. The 0.1%F-Epoxy exhibited the lowest corrosion resistance, and this result confirmed the conclusion above which low content (0.1 wt.%) of fullerene-C60 would not improve the corrosion resistance of epoxy. Additionally, it was clear to observe that the 3.0% F-Epoxy coating started to delaminate and damage due to the salt spray. This phenomenon indicated that the high content of fullerene-C60 particles (3.0 wt.%) provided a short-term reinforcement on corrosion resistance, but it was weak in durability for a long-term test.

Figure 7 was plotted for the results of coating corrosion protection index. Clearly, the results were in strong agreement with the previous observation from Bode plots. The neat epoxy demonstrated the weakest anti-corrosion performance, with significant development of coating damages, as compared to all the nanofiller/epoxy coatings during the exposure. The degree of coating damage at the initial moment and after exposure was reduced by the addition of C60. The samples that were reinforced by 0.5 and 1.0 wt.% C60 nanoparticles showed the highest corrosion protection and more stability against the severe environment. During the exposure, 0.5 and 1.0 wt.% C60-Epoxy remained 100% undamaged, indicating that the coating system was acting as an intact layer against the penetration of corrosive media.

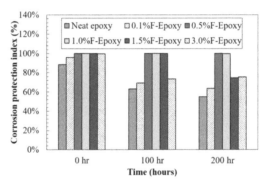

Figure 7. Coating corrosion protection index of C60/epoxy coatings.

3.3. Adhesive Bond Strength of Nano-Reinforced Composites to the Substrate

Understanding the bonding properties between the protective coating and the substrate could assist to characterize corrosion protection properties. The pull-off strength (adhesion) was measured by following ASTM D4541 to evaluate the tensile bond strength of nano-reinforced epoxy coatings with varied C60 concentrations. The performance of the neat epoxy was employed as a reference, and the adhesive strength was close to 3.05 MPa.

The pull-off bond strength over the nanofiller concentration is illustrated in Figure 8. Different to the corrosion barrier performance, the adhesive strength increased in the coating with 0.1 wt.% of fullerene-C60 and at maximum strength, reached 3.42 MPa. With 0.5 and 1.0 wt.% C60 particles, the adhesion strength reduced to 2.84 MPa but remained close to neat epoxy. Adhesion decreased when the coatings had a higher content of fullerene-C60 (1.5 to 3.0 wt.%). For instance, the adhesion dropped to 2.14 MPa in the sample with 3.0 wt.% C60 particles.

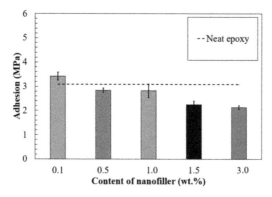

Figure 8. Pull-off strength of the nanocomposites with varied types of carbon nanofillers.

Different interfacial failure modes could reveal the level of bond strength between coating and substrate [42]. The failure mode of the coatings with and without C60 particles after the pull-off strength test as illustrated in Figure 9 demonstrated that the coatings were wholly detached from the substrate for both neat epoxy and C60/epoxy groups, indicating that an adhesive failure mode was obtained, as adhesive strength was less than their tensile strengths [43].

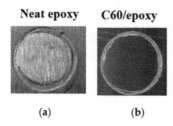

(a)   **(b)**

Figure 9. The adhesion failure mode of (**a**) neat epoxy and (**b**) epoxy with 1.0 wt.% C60 nanofillers.

3.4. Tensile Behavior of the Nanocomposite Coating

Tensile properties of the nanofiller reinforced epoxy account for whether the developed coating could provide high damage tolerance for real-world applications. The analysis of tensile properties of nanofiller reinforced epoxy was performed by a tensile test, following ASTM D638. The nanocomposites were evaluated by measuring the maximum tensile stress, strain at failure, and the Young's modulus during the test.

Figure 10a shows the maximum tensile stress of the tested nanocomposites at varied concentrations. The results indicated the C60 had a dramatic reinforcement in tensile strength; hence; the tensile stress of all the tested fullerene-C60/epoxy groups were higher than 45 MPa while of neat epoxy it was 24 MPa. Additionally, a gradual increase in the tensile stress was observed in 0.1 to 1.0 wt.% fullerene-C60 groups. The maximum stress was 56 MPa and found in the 1.0%C60-Epoxy, increased 130% compared to that of the pure epoxy group. Material degradation was found in the group with higher concentration (3.0 wt.%) of C60 nanofillers.

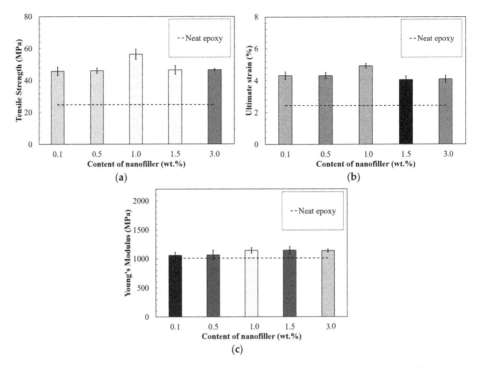

Figure 10. (a) Tensile strength, (b) ultimate strain, and (c) Young's modulus of nano filler/epoxy composites.

Furthermore, a similar tendency was observed in ultimate strains presented in Figure 10b. The ultimate strain of all the tested nanocomposites was increased by the addition of fullerene-C60 particles; thus, the ultimate strains of all F-Epoxy groups were larger than 4.0% while neat epoxy was 2.4%. The greatest improvement of ultimate strain was obtained by the addition of 1.0 wt.% of fullerene-C60, and the ultimate strain was 4.9%, which was as twice as neat epoxy samples.

The Young's modulus of the nanocomposites exhibited an identical trend as observed in their tensile strength and strain (Figure 10c). The value of Young's modulus for the fullerene-C60/epoxy groups was increased in all the tested nanocomposites. Similar to tensile strength and strain, the value of Young's modulus values was increased from 0.1 to 1.0 wt.% of fullerene-C60 groups. However, unlike the other properties, no degradation was observed in the nanocomposites with a higher amount of fullerene-C60 (1.0 to 3.0 wt.%).

The fracture surfaces for specimens fractured under tensile stress are shown in Figure 11. The relatively smooth surface was observed in pure epoxy comparing with nano-reinforced composites. This typical brittle fracture from pure epoxy represented the low impact resistance and fracture toughness of the non-reinforced composite. In contrast, the fracture surface was significantly rougher for the composite containing C60 than the pure epoxy (Figure 11b), where higher surface roughness and more compacted cleavages were observed, indicating signs of higher energy absorption and better fracture resistance. This observation is consistent with the experiment results that higher strain at failure was obtained from C60 reinforced epoxy.

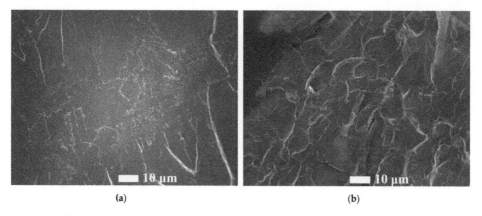

(a)   (b)

Figure 11. The fracture surface of (**a**) neat epoxy, (**b**) C60/epoxy with 1.0 wt.% of C60.

4. Conclusions

Nano reinforced composites were fabricated by incorporating fullerene-C60 into the epoxy matrix, potentially for high-performance coatings in pipelines. Highly dispersed nanoparticles were achieved using a solvent-free dispersion method through the integration of ultrasonication and high-speed disperser. The level of particle dispersion was examined via DLS measurements and cross-sectional SEM images. Major properties affecting the performance of corrosion protection coatings, including corrosion barrier property, mechanical strength, and long-term durability, were systemically characterized and the findings are summarized as follows:

(a) The fullerene-C60/epoxy coatings exhibited improved electrochemical, mechanical properties with excellent durability, indicating the coatings enabled protection of the substrate against a harsh environment with corrosive media that oil/gas pipelines often experience.

(b) Particle distribution results from DLS measurements revealed the developed dispersion method effectively overcome agglomeration, and no large particles were observed in all the tested samples.

(c) The incorporation of fullerene-C60 as a coating additive led to dramatically improved corrosion resistance, as suggested by EIS results. Excellent barrier performance was observed in the samples with higher content fullerene-C60 particles (from 0.5 to 3.0 wt.%).

(d) EIS results after salt fog exposure confirmed that nanofiller coatings could provide a much longer life as compared with the neat epoxy. Particularly, as compared to an over 50% reduction in the control samples, the coatings with 0.5 and 1.0 wt.% of fullerene-C60 particles remained intact even after 200-h exposure to salt spray, as identified on their impedance values in Bode plots.

(e) Enhancement in mechanical properties was observed in all the coatings with fullerene-C60 particles. The 1.0% F-Epoxy group exhibited the highest increase in tensile properties, including increased strength, strain, and Young's modulus. In addition, improvement on adhesion was observed in the coating with low content of fullerene-C60 particles (0.1 wt.%).

Author Contributions: X.W. designed, conducted this research and wrote the paper under the supervision of Z.L., F.T., X.Q., D.B. and X.C. assisted the experiments and edited the paper.

Funding: This research was funded by ND DOC Venture Grant, USDOTs (DTPH5616HCAP03, 693JK318500010CAAP and 693JK31850009CAAP), North Dakota State University and the Mountain-Plains Consortium, a University Transportation Center funded by the U.S. Department of Transportation.

Acknowledgments: The authors gratefully acknowledge the financial support provided by the ND DOC Venture Grant, USDOTs (DTPH5616HCAP03, 693JK318500010CAAP, and 693JK31850009CAAP). The work presented in this paper was also conducted with partial support from North Dakota State University and the Mountain-Plains Consortium, a University Transportation Center funded by the U.S. Department of Transportation. The results, discussion, and opinions reflected in this paper are those of the authors only and do not necessarily represent those of the sponsors.

Conflicts of Interest: The authors declare no conflict of interest.

References

1. Wang, X.; Qi, X.; Li, M.; Lin, Z.; Battocchi, D. Characterization of Graphene Reinforced Epoxy Coatings for Internal Surface of Oil and Gas Pipelines. In Proceedings of the ASCE Pipelines 2019 Conference, Nashville, TN, USA, 21–24 July 2019.
2. Zi, Z.; Wang, X.; Pan, H.; Lin, Z. Corrosion-induced damage identification in metallic structures using machine learning approaches. In Proceedings of the 2019 Defense TechConnect Innovation Summit, National Harbor, MD, USA, 7–10 October 2019.
3. Zi, Z.; Pan, H.; Lin, Z. Data-Driven Identification for Early-Age Corrosion-Induced Damage in Metallic Structures. In Proceedings of the Bridge Engineering Institute Conference 2019, Honolulu, HI, USA, 22–25 July 2019.
4. Pan, H.; Azimi, M.; Yan, F.; Lin, Z. Time-frequency-based data-driven structural diagnosis and damage detection for cable-stayed bridges. *J. Bridge Eng.* **2018**, *23*, 04018033. [CrossRef]
5. Sampath, S.; Bhattacharya, B.; Aryan, P.; Sohn, H. A Real-Time, Non-Contact Method for In-Line Inspection of Oil and Gas Pipelines Using Optical Sensor Array. *Sensors* **2019**, *19*, 3615. [CrossRef] [PubMed]
6. Abbas, M.H.; Norman, R.; Charles, A. Neural network modelling of high pressure CO2 corrosion in pipeline steels. *Process Saf. Environ. Prot.* **2018**, *119*, 36–45. [CrossRef]
7. Pan, H.; Gui, G.; Lin, Z.; Yan, C. Deep BBN Learning for Health Assessment toward Decision-Making on Structures under Uncertainties. *KSCE J. Civ. Eng.* **2018**, *22*, 928–940. [CrossRef]
8. Lin, Z.; Pan, H.; Wang, X.; Li, M. Data-driven structural diagnosis and conditional assessment: From shallow to deep learning. In Proceedings of the International Society for Optics and Photonics, Denver, CO, USA, 27 March 2018; Volume 10598, p. 1059814.
9. Gui, G.; Pan, H.; Lin, Z.; Li, Y.; Yuan, Z. Data-driven support vector machine with optimization techniques for structural health monitoring and damage detection. *KSCE J. Civ. Eng.* **2017**, *21*, 523–534. [CrossRef]
10. Pan, H.; Ge, R.; Xingyu, W.; Jinhui, W.; Na, G.; Zhibin, L. Embedded Wireless Passive Sensor Networks for Health Monitoring of Welded Joints in Onshore Metallic Pipelines. In Proceedings of the ASCE 2017 Pipelines, Phoenix, AZ, USA, 6–9 August 2017.
11. Wang, X.; Qi, X.; Pearson, M.; LI, M.; Lin, Z.; Battocchi, D. Design and Characterization of Functional Nanoengineered Epoxy-Resin Coatings for Pipeline Corrosion Control. In Proceedings of the Coating Trends and Technologies 2019, Rosemont, IL, USA, 10–11 September 2019.
12. Varga, M.; Izak, T.; Vretenar, V.; Kozak, H.; Holovsky, J.; Artemenko, A.; Hulman, M.; Skalakova, V.; Lee, D.S.; Kromka, A. Diamond/carbon nanotube composites: Raman, FTIR and XPS spectroscopic studies. *Carbon* **2017**, *111*, 54–61. [CrossRef]
13. Wang, X.; Tang, F.; Qi, X.; Lin, Z. Mechanical, electrochemical, and durability behavior of graphene nano-platelet loaded epoxy-resin composite coatings. *Compos. Part B Eng.* **2019**, 107103. [CrossRef]
14. Shadlou, S.; Alishahi, E.; Ayatollahi, M. Fracture behavior of epoxy nanocomposites reinforced with different carbon nano-reinforcements. *Compos. Struct.* **2013**, *95*, 577–581. [CrossRef]
15. Kuilla, T.; Bhadra, S.; Yao, D.; Kim, N.H.; Bose, S.; Lee, J.H. Recent advances in graphene based polymer composites. *Prog. Polym. Sci.* **2010**, *35*, 1350–1375. [CrossRef]
16. Roy, S.; Mitra, K.; Desai, C.; Petrova, R.; Mitra, S. Detonation nanodiamonds and carbon nanotubes as reinforcements in epoxy composites—A comparative study. *J. Nanotechnol. Eng. Med.* **2013**, *4*, 011008. [CrossRef]
17. Liu, D.; Zhao, W.; Liu, S.; Cen, Q.; Xue, Q. Comparative tribological and corrosion resistance properties of epoxy composite coatings reinforced with functionalized fullerene C60 and graphene. *Surf. Coat. Technol.* **2016**, *286*, 354–364. [CrossRef]
18. Bhattacharya, M. Polymer nanocomposites—A comparison between carbon nanotubes, graphene, and clay as nanofillers. *Materials* **2016**, *9*, 262. [CrossRef] [PubMed]
19. Abbasi, H.; Antunes, M.; Velasco, J.I. Recent advances in carbon-based polymer nanocomposites for electromagnetic interference shielding. *Prog. Mater. Sci.* **2019**, *103*, 319–373. [CrossRef]

20. Wang, X.; Qi, X.; Pearson, M.; Lin, Z.; Battocchi, D. NanoModified Protective Coatings for Pipeline Corrosion Control and Mitigation. In Proceedings of the 2019 TechConnect World Innovation Conference, Boston, MA, USA, 17–19 June 2019.

21. Cui, L.-J.; Wang, Y.-B.; Xiu, W.-J.; Wang, W.-Y.; Xu, L.-H.; Xu, X.-B.; Meng, Y.; Li, L.-Y.; Gao, J.; Chen, L.-T.; et al. Effect of functionalization of multi-walled carbon nanotube on the curing behavior and mechanical property of multi-walled carbon nanotube/epoxy composites. *Mater. Des.* **2013**, *49*, 279–284. [CrossRef]

22. Zabet, M.; Moradian, S.; Ranjbar, Z.; Zanganeh, N. Effect of carbon nanotubes on electrical and mechanical properties of multiwalled carbon nanotubes/epoxy coatings. *J. Coat. Technol. Res.* **2016**, *13*, 191–200. [CrossRef]

23. Jiang, J.; Xu, C.; Su, Y.; Guo, Q.; Liu, F.; Deng, C.; Yao, X.; Zhou, L. Influence of carbon nanotube coatings on carbon fiber by ultrasonically assisted electrophoretic deposition on its composite interfacial property. *Polymers* **2016**, *8*, 302. [CrossRef]

24. Shimamura, Y.; Oshima, K.; Tohgo, K.; Fujii, T.; Shirasu, K.; Yamamoto, G.; Hashida, T.; Goto, K.; Ogasawara, T.; Naito, K.; et al. Tensile mechanical properties of carbon nanotube/epoxy composite fabricated by pultrusion of carbon nanotube spun yarn preform. *Compos. Part A Appl. Sci. Manuf.* **2014**, *62*, 32–38. [CrossRef]

25. Wang, X.; Qi, X.; Pearson, M.; Li, M.; Lin, Z.; Battocchi, D. Characterization of Nano-Particle Reinforced Epoxy Coatings for Structural Corrosion Mitigation. In Proceedings of the Bridge Engineering Institute Conference 2019, Honolulu, HI, USA, 22–25 July 2019.

26. Yu, Y.-H.; Lin, Y.-Y.; Lin, C.-H.; Chan, C.-C.; Huang, Y.-C. High-performance polystyrene/graphene-based nanocomposites with excellent anti-corrosion properties. *Polym. Chem.* **2014**, *5*, 535–550. [CrossRef]

27. Chang, C.-H.; Huang, T.-C.; Peng, C.-W.; Yeh, T.-C.; Lu, H.-I.; Hung, W.-I.; Weng, C.-J.; Yang, T.-I.; Yeh, J.-M. Novel anticorrosion coatings prepared from polyaniline/graphene composites. *Carbon* **2012**, *50*, 5044–5051. [CrossRef]

28. Pikhurov, D.V.; Zuev, V.V. The effect of fullerene C60 on the dielectric behaviour of epoxy resin at low nanofiller loading. *Chem. Phys. Lett.* **2014**, *601*, 13–15. [CrossRef]

29. Zuev, V.V. The mechanisms and mechanics of the toughening of epoxy polymers modified with fullerene C60. *Polym. Eng. Sci.* **2012**, *52*, 2518–2522. [CrossRef]

30. Bai, J.; He, Q.; Shi, Z.; Tian, M.; Xu, H.; Ma, X.; Yin, J. Self-assembled elastomer nanocomposites utilizing C60 and poly (styrene-b-butadiene-b-styrene) via thermally reversible Diels-Alder reaction with self-healing and remolding abilities. *Polymer* **2017**, *116*, 268–277. [CrossRef]

31. Ogasawara, T.; Ishida, Y.; Kasai, T. Mechanical properties of carbon fiber/fullerene-dispersed epoxy composites. *Compos. Sci. Technol.* **2009**, *69*, 2002–2007. [CrossRef]

32. Li, P.; He, X.; Huang, T.-C.; White, K.L.; Zhang, X.; Liang, H.; Nishimura, R.; Sue, H.-J. Highly effective anti-corrosion epoxy spray coatings containing self-assembled clay in smectic order. *J. Mater. Chem. A* **2015**, *3*, 2669–2676. [CrossRef]

33. Zhang, Y.; Zhao, M.; Zhang, J.; Shao, Q.; Li, J.; Li, H.; Lin, B.; Yu, M.; Chen, S.; Guo, Z. Excellent corrosion protection performance of epoxy composite coatings filled with silane functionalized silicon nitride. *J. Polym. Res.* **2018**, *25*, 130. [CrossRef]

34. Chhetri, S.; Samanta, P.; Murmu, N.C.; Kuila, T. Anticorrosion Properties of Epoxy Composite Coating Reinforced by Molybdate-Intercalated Functionalized Layered Double Hydroxide. *J. Compos. Sci.* **2019**, *3*, 11. [CrossRef]

35. Yang, F.; Liu, T.; Li, J.; Zhao, H. Long Term Corrosion Protection of Epoxy Coating Containing Tetraaniline Nanofiber. *Int. J. Electrochem. Sci.* **2018**, *13*, 6843–6857. [CrossRef]

36. Wang, X.; Qi, X.; Lin, Z.; Wang, J.; Gong, N. Electrochemical Characterization of the Soils Surrounding Buried or Embedded Steel Elements. In Proceedings of the ASCE 2016 Pipelines, Kansas City, MO, USA, 17–20 August 2016.

37. Ammar, S.; Ramesh, K.; Ma, I.; Farah, Z.; Vengadaesvaran, B.; Ramesh, S.; Arof, A.K. Studies on SiO2-hybrid polymeric nanocomposite coatings with superior corrosion protection and hydrophobicity. *Surf. Coat. Technol.* **2017**, *324*, 536–545. [CrossRef]

38. Ramezanzadeh, B.; Niroumandrad, S.; Ahmadi, A.; Mahdavian, M.; Moghadam, M.M. Enhancement of barrier and corrosion protection performance of an epoxy coating through wet transfer of amino functionalized graphene oxide. *Corros. Sci.* **2016**, *103*, 283–304. [CrossRef]

39. Wang, X.; Qi, X.; Lin, Z.; Battocchi, D. Graphene Reinforced Composites as Protective Coatings for Oil and Gas Pipelines. *Nanomaterials* **2018**, *8*, 1005. [CrossRef]

40. Ammar, S.; Ramesh, K.; Vengadaesvaran, B.; Ramesh, S.; Arof, A.K. A novel coating material that uses nano-sized SiO2 particles to intensify hydrophobicity and corrosion protection properties. *Electrochim. Acta* **2016**, *220*, 417–426. [CrossRef]

41. Ramezanzadeh, B.; Haeri, Z.; Ramezanzadeh, M. A facile route of making silica nanoparticles-covered graphene oxide nanohybrids (SiO2-GO); fabrication of SiO2-GO/epoxy composite coating with superior barrier and corrosion protection performance. *Chem. Eng. J.* **2016**, *303*, 511–528. [CrossRef]

42. Zhai, L.; Ling, G.; Wang, Y. Effect of nano-Al2O3 on adhesion strength of epoxy adhesive and steel. *Int. J. Adhes. Adhes.* **2008**, *28*, 23–28. [CrossRef]

43. May, M.; Wang, H.; Akid, R. Effects of the addition of inorganic nanoparticles on the adhesive strength of a hybrid sol–gel epoxy system. *Int. J. Adhes. Adhes.* **2010**, *30*, 505–512. [CrossRef]

Review

Fly Ash, from Recycling to Potential Raw Material for Mesoporous Silica Synthesis

Marius Gheorghe Miricioiu and Violeta-Carolina Niculescu *

National Research and Development Institute for Cryogenics and Isotopic Technologies-ICSI Ramnicu Valcea, 4th Uzinei Street, 240050 Ramnicu Valcea, Romania; marius.miricioiu@icsi.ro
* Correspondence: violeta.niculescu@icsi.ro; Tel.: +40-250-732-744

Received: 17 February 2020; Accepted: 29 February 2020; Published: 5 March 2020

Abstract: In order to meet the increasing energy demand and to decrease the dependency on coal, environmentally friendly methods for fly ash utilization are required. In this respect, the priority is to identify the fly ash properties and to consider its potential as raw material in the obtaining of high-value materials. The physico-chemical and structural characteristics of the fly ash coming from various worldwide power plants are briefly presented. The fly ash was sampled from power plants where the combustion of lignite and hard coal in pulverized-fuel boilers (PC) and circulating fluidized bed (CFB) boilers was applied. The fly ash has high silica content. Due to this, the fly ash can be considered a potential raw material for the synthesis of nanoporous materials, such as zeolites or mesoporous silica. The samples with the highest content of SiO_2 can be used to obtain mesoporous silica materials, such as MCM-41 or SBA-15. The resulting mesoporous silica can be used for removing/capture of CO_2 from emissions or for wastewater treatment. The synthesis of various porous materials using wastes would allow a high level of recycling for a sustainable society with low environmental impact.

Keywords: coal combustion; fly ash; mesoporous silica; recycling

---

1. Introduction

Global coal consumption rose with 0.9% in 2018, the main contributors being India and China, followed by Turkey and Russia [1] (Figure 1). Due to power plants and some industrial sectors (such as steel, chemicals and cement) China is responsible for approximately half of global coal consumption [1], this value reaching about 3770 Mt, which represented 55% of the worldwide consumption in 2018, approximately four times higher than in 1990 [1].

By contrast, the United States reached the lowest level in the last 40 years, the coal consumption decreasing 4% in 2018 due to the availability of natural gas at lower prices and to stronger emission legislation (Figure 2).

CO_2 emissions are strongly related to coal consumption, and consequently the United States was expected to reduce CO_2 emissions by 2.2% in 2019 by further reducing coal consumption, and by 3.6% in 2020 [2].

Climate policies, renewable gas and CO_2 emissions costs have been responsible for a decrease of coal consumption in the last 6 years in Europe (Figure 3). Only Turkey was an exception, due to coal consumption that increased by 11% in 2018 [1].

Regarding the statistics for the last 30 years (Figure 3), it can be observe that the coal consumption in Asia slightly increased during 1990–1995, reaching approximately 2000 Mt/year, and became constant until 1999, while in Europe consumption continuously and slightly decreased, reaching approximately half of the consumption value recorded in Asia. From 2000 to 2013, coal consumption has shown a significant increase in Asia, becoming constant until 2018 (more than 3 times higher than in the 1990s), meanwhile, in Europe, it varied slightly.

Due to coal consumption decreasing, the capture and storage of CO_2 from emissions released into the atmosphere using new and cheap materials with high properties can be considered a reliable solution for greenhouse gases reduction. The most efficient adsorbents for CO_2 reported in literature are zeolites, porous silica and active carbons [3].

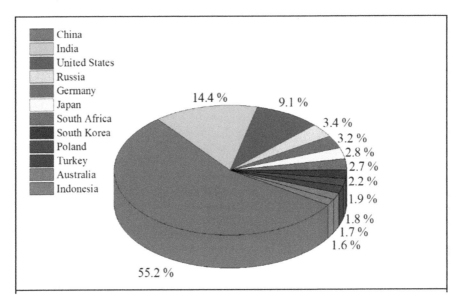

Figure 1. Global coal consumption (Mt) in 2018.

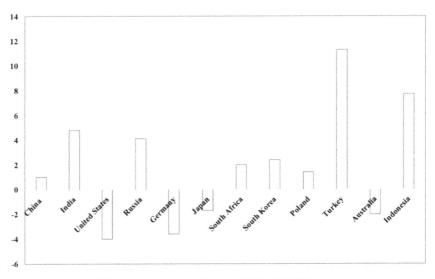

Figure 2. Coal consumption trend (%) in 2017–2018.

Along with the increase of CO_2 resulting from coal consumption, a considerable amount of fly ash is obtained as by-product of coal combustion. The fly ash is widely used as raw material in the cement industry. However, if the coal fly ash exceeds the worldwide demand it could be a problem due to storage spaces. For example, in China the utilization rate increases directly with the increase of the amount of fly ash obtained, reaching 70% 2015 [4]. In India, the utilization rate registered a significant increase in the period 2001–2011, reaching 62% [4].

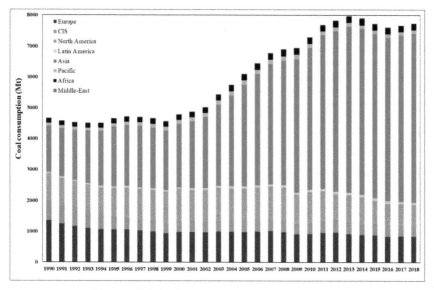

Figure 3. Global coal consumption trend (Mt) in 1990–2018.

The same tendency was observed in the USA, where the utilization rate reached about 50% in 2015 [4]. According to the American Coal Ash Association, the percent of used fly ash increased from 40% in 2000 to 60% in 2018, this being normal due to the coal consumption decrease [5]. In Russia, the utilization of fly ash in the period 1990–2005 was approximately 19%, the main application being in the cement industry [6]. Also, the consumption trend could increase due to the attention given to geo-polymer materials production [7]. Therefore, it is necessary to find some applications for coal fly ash use.

In the last decade, fly ash was used as raw materials for obtaining zeolites X (FAU framework type) [8,9], Y (FAU) [10], A (LTA) [8,9,11] or ZSM-5 (MFI) [12]. Furthermore, the synthesis of mesoporous silica from fly ash has also attracted interest due to the resulting material characteristics [13–18]. Mesoporous silica materials obtained from fly ash are considered to surpass the limitations of the microporous zeolites in the removal of macromolecule pollutants by adsorption [13].

The mesoporous silica materials, known since 1992 as M41S, present great potential in worldwide applications, such as catalysis or wastewater treatment, due to their properties, namely, uniformity of pore distribution (with size between 2 and 50 nm), high surface area, (around 1000 m^2/g) and good stability in thermal conditions [19,20].

A close survey of the data highlights the fact that the chemical composition of fly ash (mainly silica and alumina compounds) is significantly different. Moreover, the synthesis conditions, including the pre-treatment step, are often not completely described or some inconsistencies among the literature sources were observed.

This overview systematically explores the synthesis of various mesoporous silica materials derived from fly ash by taking into consideration the waste properties (fly ash), pre-treatment procedures or the hydrothermal treatment parameters (temperature, time and substrate concentrations).

2. Fly Ash Properties

Fly ash is a complex material, being a by-product resulting from the combustion of various coals with high contents of minerals [4,21–23]. Consequently, fly ash is rich in metallic oxides, in the order

$SiO_2 > Al_2O_3 > Fe_2O_3 > CaO > MgO > K_2O$ and large amount of unburned carbon. Furthermore, fly ash contains trace elements that can have a negative impact on the environment [24–26]. These can easily migrate from fly ash, through interaction with water, conducting to the soil and ground water contamination with heavy metals such as Cr, V, Ni, Cd and Pb [27]. Also, wind action contributes to the environmental pollution, by spreading the ash particles in the air.

As it is shown in Figure 4, the content of metallic oxides in fly ash is depended by the coal type. Thus, the SiO_2 contents were higher in the fly ash derived from sub-bituminous (40%–60%) and bituminous (20%–60%) samples than in lignite samples (15%–45%). The same trend was seen for Al_2O_3. In the case of CaO and MgO higher contents were observed in fly ash from lignite samples (15%–40% and 3%–10%, respectively), followed by sub-bituminous samples.

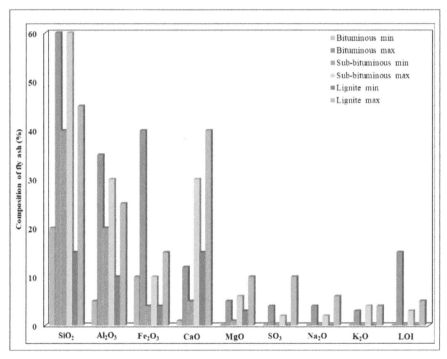

Figure 4. Different coal fly ash composition.

Small contents of SO_3, Na_2O and K_2O were observed in all the fly ash samples. The highest percent value for unburned carbon, determined by its loss-on-ignition (LOI), was found in bituminous fly ash (15%).

Comparing the sample contents, it can be highlighted that all types of fly ash are rich sources of SiO_2 and Al_2O_3, their recovery being an issue raised by the waste management. Also, the high content of CaO can be used for CO_2 capture and permanent sequestration, resulting in $CaCO_3$ [27]. Bituminous fly ash could be a precursor for activated carbon sorbent.

The intensive investigations carried out for fly ash reuse have resulted in the development of techniques for producing glass ceramics, ceramic wares, silicon carbide, silicon nitride, hollow/masonry/concrete blocks, cordierite or mullite. Recently, fly ash was applied in the development of mesoporous silica for CO_2 capture [28].

3. Environmental Risk Assessment

The reuse of fly ash must be encouraged for many reasons. For example, the disposal costs would be minimized; also, less landscape would be reserved for its disposal; and the by-products may be used as raw materials.

Fly ash may have metal concentrations up to 10 times higher than coal [4]. As it was already mentioned, there are many natural factors (e.g., rain, wind) that contribute to interaction of the metals with humans, reaching significant concentration in soil and water and finally in crops. This process is directly dependent on several parameters, such as particle size, pH, interaction time, trace elements concentration in fly ash [4].

A typical metallic composition in fly ash collected from different power plants in India is presented in Table 1. Furthermore, the effect of fly ash on soil quality for maize and rice crops was investigated, in two different cultivation areas from the eastern part of India and the results are presented in Figures 5–7 [29,30]. The fly ash was dried and mixed with cellulose to obtain pellets and was applied to the soil (about 200 t/ha).

Table 1. The elemental composition of the fly ash.

Element	Fe	Mn	Cu	Zn	As	Se	Mo	Pb
Unit	%	ppm	ppm	ppm	ppm	ppm	ppm	ppm
Concentration Range	2–3	148–261	64–83	103–150	3–6	2–3	3–4	15–40

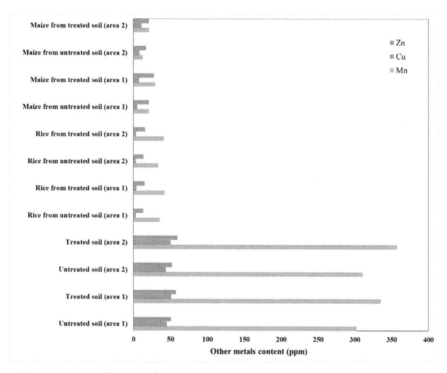

Figure 5. Heavy metal contents in soil, rice and maize from two areas, treated and untreated with fly ash.

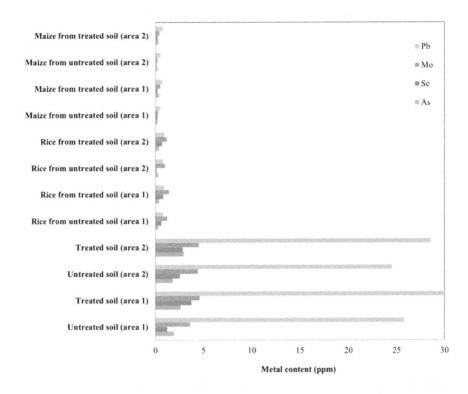

Figure 6. Toxic elements contents in soil, rice and maize from two areas, treated and untreated with fly ash.

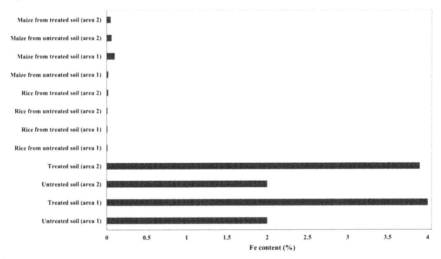

Figure 7. Fe content in soil, rice and maize from two areas, treated and untreated with fly ash.

As it can be seen from Figure 5, the soil enrichment with heavy metals in rice crops did not exceed the upper control limits (in soil: 1500 ppm for Mn, 100 ppm for Cu and 300 ppm for Zn; in plant: 300 ppm for Mn, 50 ppm for Cu and 100 ppm for Zn) [31–33]. Thus, the soils treated with fly ash presented a slight enrichment with microelements, such as Zn, Mn and Cu, which are essential for plant growth. Also, the enrichment of rice and maize crops with microelements was observed. In the case of rice crops, the enrichment with Cu was lower than with Zn and Mn.

Regarding the concentration of toxic elements in soil and crops, Figure 6 shows that the Mo concentration reached almost the critical level, being a concern for the environment. The differences between the rest of the toxic element concentrations (Pb, Se, As) and their upper control limits are higher (in soil: 200 ppm for As, 10 ppm for Se, 5 ppm for Mo and 100 ppm for Pb; in plant: 30 ppm for As, 0.69 ppm for Se, 1.2 ppm for Mo and 0.87 ppm for Pb), without significant risk to the crops.

It was observed that Fe was the major element. As it can be seen in Figure 7, there are significant differences between the enrichment of Fe in crops from untreated and treated soils (upper control limit for Fe being 5.6% in soil and 0.2% in plant).

It was also demonstrated that the soil treated with fly ash contributed to the increase of crop yields, by the action of essential plant nutrients, such as Mg, S, K, Ca, Mn, Fe and Zn [29]. The fly ash can modify the physical properties of the soil, such as porosity and density, increasing the water retaining capacity [29,34]. Due to low concentration of toxic elements in fly ash, the critical levels were not reached and the grains were considered safe for consumers.

As it can be seen, the use of coal fly ash meets technical, economic and legal barriers. The main technical milestones refer to coal fly ash production, specifications and standards, product demonstration and commercialization.

For coal-using plants, the incomes from the sale of fly ash are often negligible, the most important economic barriers coming from the increased cost of transportation of fly ash and competition from locally available natural sources. Legal barriers resulting from the lack of knowledge regarding the potential ash application, insufficient data on environmental and health risks, lack of regulations and procurement guidelines.

Worried industry and government representatives, scientists, and engineers have created national and international organizations to overcome the milestones of coal fly ash reuse. An integrated approach must be designed in order to produce a superior quality of fly ash-based materials, which can satisfy the consumer's expectations. Furthermore, high quantities of fly ash are often followed by emission of greenhouse gases, significantly intruding global warming. As a consequence, to conform to the environmental requirements, major efforts must be achieved in fly ash management, reducing the negative effect on the environment.

A deep analysis on the cost of fly ash use versus conventional building materials is required, being necessary to apply the best engineering practices in order to minimize as much as possible the environmental risk.

4. Fly Ash as Raw Material for Mesoporous Silica Synthesis

In order to increase the recycling of the coal fly ash and to reduce the environmental risk of contamination with heavy metals, several applications for recovery of valuable components from fly ash were developed.

Thus, among the fly ash used in conventional applications (cements obtaining), various mesoporous silica materials with improved properties were obtained, able to be applied in environmental depollution, namely CO_2 reduction and wastewater treatment.

The methods for obtaining these materials from fly ash were basically similar, the differences consisting in varying different parameters such as: pH, the temperature and the time of hydrothermal or aging treatment and also the quantities of raw materials or reagents. These variations influenced more or less the obtained material properties.

The used coal fly ash for the silica synthesis had various particle sizes (2.0–30 μm) and chemical compositions (31.6–60.1 wt.% Si, 10.2–40.8 wt.% Al, 0.8–8.4 wt.% Fe, 0.4–24.8 wt.% Ca) [35–38].

Porous materials, such as Al-MCM-41 and SBA-15, were obtained from coal fly ash, and they were used as catalysts in the cumene cracking reaction [35]. In this respect, the fly ash was treated with an aqueous solution of NaOH under thermal condition; the supernatant obtained after the resulting suspension filtration was used as a silica source for MCM-41. Its composition, beside 11,000 ppm Si, was rich in Na and Al, with 35,000 and 380 ppm, respectively. It was demonstrated that the pH adjustment has directly influenced the Al-MCM-41 synthesis. The Al incorporation was poor in the case of SBA-15, compared with the MCM-41, remaining in the dissolved state in the supernatant due to the high acidic medium. Furthermore, the Al-MCM-41, derived from both fly ash and pure chemicals (obtained by a conventional method), was tested for a cumene cracking reaction involving a high acid medium [35]. Despite the small differences of physical properties of the two mesoporous materials (surface area: 842 and 940 m^2/g, pore volume: 0.75 and 0.85 cm^3/g and pore diameter: 3.7 and 2.7 nm) it was observed that the Al-MCM-41 from the conventional method was more efficient in the cumene conversion because not all of the Al presented in the fly ash-derived porous material had catalytic activity. Thus, the cumene conversion was threefold higher, ~22%, at the beginning of the reaction for the Al-MCM-41 prepared from pure reagents and decreased drastically during the first hour, slowly decreasing the next hour for the two materials, reaching 6%, two times higher than Al-MCM-41 derived coal fly ash.

MCM-41 was also obtained by adapting the mesoporous synthesis from the previous study [35], with moderate modification of the parameters, such as temperature and time [36]. Also, the catalytic performance of the MCM-41 synthesized from fly ash and from conventional reagents for Mannich reaction application were compared [36]. It was supposed from the beginning that there would be superior catalytic activity of the obtained material, due to the lower particle size and higher surface area of MCM-41 than the fly ash. The materials were tested as catalysts in the Mannich reaction of acetophenone, benzaldehyde and aniline by varying the catalyst quantity (0.1 g, 0.2 g and 0.3 g) and the solvent nature (EtOH, CH_3CN, Toluene, Tetrahydrofuran (THF), CH_2Cl_2 and H_2O). A 90% yield was obtained in 4 h for 0.2 g MCM-41 in ethanol, providing a good catalytic active area for the reaction. The yield remained constant with the increase of catalyst quantity at 0.3 g. Instead, a lower efficiency, down to 50%, was observed with the decrease of catalyst quantity, to 0.1 g, even if the reaction time has been extended to 8 h [36]. The MCM-41 can be an optimal replacement for environment-unfriendly solvents used in catalysis.

A supernatant with higher contents of Si, Al and Na (15.37, 499.00 and 48.70 ppm, respectively) was obtained starting also from the method previously described [35], the fly ash (raw material) being rich in these elements [37]. The mesoporous Al-MCM-41 adsorbent obtained for the removal of methylene blue (MB) was finally synthesized by varying the ethyl acetate amount [37]. This had a direct influence on specific surface area and pore volume of the silica material, thus the Al-MCM-41 obtained by using 10 mL of ethyl acetate shown improved properties [37]. Thus, when 10 mL of ethyl acetate was used, the surface area increased from 25 to 525 m^2/g, the pore diameter decreased from 7.54 to 5.13 nm and the pore volume increased from 0.13 to 0.71 cm^3/g, although not reaching the properties of the material obtained through conventional methods [38]. The Al-MCM-41 capacity to adsorb the methylene blue was influenced by the pH, contact time, temperature and concentration. Regarding the contact time parameter, a drastic increase of the MB adsorption was observed, followed by a slower increase in intervals 0–10 min and 10–120 min, respectively. After 120 min, the MB reached the equilibrium adsorption level. The MB adsorption had a continuous increasing trend with the increase of pH from 3 to 10, with a higher increment in pH interval 3–7. The highest adsorption capacity was up to 277.78 mg/g and it was reached at room temperature and pH 10.

MCM-41 was also synthesized from brown and hard coal fly ash (10 samples) using two procedures, namely pulverized-fuel (PC) and fluidized-bed boilers (CFB) [39]. The X-ray fluorescence spectroscopy (XRF) revealed different chemical composition of fly ashes, the SiO_2 and Al_2O_3 ranging from 46.15 wt% to 56.52 wt% and from 18.48 wt% to 31.06 wt%, respectively. The lowest and the highest value of SiO_2 were obtained from PC combustion. Compared to other studies [35,37], where the silica content resulted in the filtrate was directly proportional with the percentage in the coal fly ash, this research highlighted a silica concentration significant lower compared to its content in the coal fly ash. In consequence, this sample could not be used for further MCM-41 synthesis because the properties of the resulting material are strongly dependent of the Al/Si ratio in the filtrate.

It was observed that the impurities presented in the coal fly ash composition, such as Fe, Ca, K, S and P, were also present in the channels of the obtained mesoporous material, conducting to poor properties compared to the commercial material [39].

For comparison, the ash resulting from the rice husk combustion was tested, using similar extraction steps as in the case of fly ash [40]. Even if the silica content was 33% higher in the rice husk ash, the MCM-41 obtained from the two raw materials had the same silica content.

It was concluded that, despite the similar chemical composition, the textural properties of the rice husk and fly ash-derived materials were superior to the commercial MCM-41, presenting higher pore volume for all synthesized materials, higher specific surface area and similar morphologies.

Also, the reaction time influenced the obtained suspension containing the surfactant (hexadecyltrimethylammonium bromide-CTAB) [40]. A longer time (96 h) was required to conduct the decrease of the surface area and pore volume. After the polyethyleneimine (PEI) impregnation, the obtained MCM-41 and a MCM-41 commercial sample were compared for CO_2 capture. The results shown higher CO_2 uptakes (with 2 wt.%) for the synthesized materials containing 60 wt.% PEI, the materials having faster kinetics due to larger pore volume comparing to the commercial materials.

It was established that the fly ash desilication rate can reach 46.3%, by using sodium hydroxide at a mass ratio of 1:6.4, during 4 h under thermal condition (95 °C). This solution was used as silica source to obtain SBA-15 mesoporous silica. The method involved further hydrothermal treatment for two days at 110 °C and a triblock copolymer as template-poly(ethylene oxide)/poly(propylene oxide)/poly(ethylene oxide) [41].

The functionalization of SBA-15 with amino groups involved, as in case of MCM-41, aminopropyltriethoxysilane (APTES), as source of NH_2- groups, and toluene as solvent [41–43]. An application of SBA-15 silica type derived from fly ash was Pb^{2+} adsorption [41] (see Table 2).

Thus, the effect of time, temperature and Pb^{2+} initial concentration on amino-SBA-15 adsorbent was studied and it was remarked that the adsorption equilibrium time was achieved after 1 h, reaching 131 mg/g (Table 2). Also, the temperature influenced the adsorption rate, increasing drastically between 20 and 30 °C (98.40% removal efficiency) becoming constant up to 40 °C. The increase of the Pb^{2+} concentration led to a higher removal efficiency, up to 93% for a concentration of 100 mg/g.

It was demonstrated that the aluminosilicate with mesoporous structure has great potential for the removal of organic dyes (such as methylene blue or crystal violet) from wastewater. In this respect, mesoporous aluminosilicate was synthesized from fly ash by varying the Si/Al molar ratio [44]. X-ray diffraction analysis of the obtained materials revealed a well-ordered hexagonal structure, as in the case of MCM-41, direct proportionally with the increase of Si/Al ratio. Thus, the surface area (~1080 m²/g) and the pore volume (~0.96 cm³/g) increased with the increase of Si/Al molar ratio, up to 25. The efficiency of the aluminosilicate on the methylene blue and crystal violet removal from wastewater was evaluated, revealing its high adsorption capacity (Table 2). Higher adsorption capacity was obtained for mesoporous materials with a Si/Al ratio equal to 10, due to the ordered mesoporous structure and pore size distribution with narrow pores. Also, as it was already mentioned [37], the pH has a direct influence on the adsorption capacity of the mesoporous materials, the highest value being obtained for the material resulted from a solution with a pH value of 11 (1100 mg/g for methylene blue and 1400 mg/g for crystal violet) [44].

Table 2. Applications of the synthesized mesoporous silica.

No.	Mesoporous Silica	Application	Reaction Conditions	Efficiency/Yield	Ref.
1	Al-MCM-41	Catalyst-cumene cracking reaction	T = 623 K W/F = 0.2 h (contact time) carrier gas flow (N_2) = 40 mL/min cumene partial pressure = 7.9 kPa	~ 3%–6 %	[35]
2	MCM-41	Catalyst—Mannich reaction	acetophenone = 5 mmol benzaldehyde = 5 mmol aniline = 5 mmol MCM-41 = 0.2 g ethanol = 20 mL t = 4 h (reaction time)	90 %	[36]
3	Al-MCM-41-10	MB (methylene blue) adsorbent	T = 298 K T = 308 K T = 318 K pH = 3 (at 293 K) pH = 5 (at 293 K) pH = 7 (at 293 K) pH = 10 (at 293 K) t = 2 min (adsorption time) t = 10 min (adsorption time) t = 120 min (adsorption time) t = 480 min (adsorption time)	278 mg/g 276 mg/g 272 mg/g ~137 mg/g ~156 mg/g ~175 mg/g ~181 mg/g ~223 mg/g ~255 mg/g ~266 mg/g ~267 mg/g	[37]
4	MCM-41 Commercial + 50% PEI	CO₂ adsorption	T = 348 Kgas concentration = 15 % CO_2	8.37 wt.% CO_2 uptake	[40]
	MCM-41 Commercial + 60% PEI			11.17 wt.% CO_2 uptake	
	MCM-41 PFA-1 + 50 % PEI loading			10.64 wt.% CO_2 uptake	
	MCM-41 PFA-1 + 60 % PEI loading			13.08 wt.% CO_2 uptake	
	MCM-41 PFA-2 + 50 % PEI loading			8.79 wt.% CO_2 uptake	
	MCM-41 PFA-2 + 60 % PEI loading			12.91 wt.% CO_2 uptake	
	MCM-41 RHA + 50 % PEI loading			10.13 wt.% CO_2 uptake	
	MCM-41 RHA + 60 % PEI loading			13.31 wt.% CO_2 uptake	
5	NH₂-SBA-15	Pb^{2+} adsorption	30 min (adsorption time)	~100 mg/g (adsorption)	[41]
			60 min (adsorption time)	~131.00 mg/g (adsorption)	
			90 min (adsorption time)	~130.00 mg/g (adsorption)	
			120 min (adsorption time)	~131.00 mg/g (adsorption)	
			180 min (adsorption time)	~131.00 mg/g (adsorption)	
			20 °C (adsorption time)	~96.51% (removal efficiency)	
			30 °C (adsorption time)	~98.40% (removal efficiency)	
			40 °C (adsorption time)	~98.50% (removal efficiency)	
			Pb^{2+} concentration = 20 mg/g	~87.5% (removal efficiency)	
			Pb^{2+} concentration = 60 mg/g	~91.5% (removal efficiency)	
			Pb^{2+} concentration = 100 mg/g	~93% (removal efficiency)	
6	Aluminosilicate (SA)	Methylene blue (MB) adsorption	475 mg/L (initial concentration) adsorbent loading = 0.2 mg/L	2003 mg/g	[44]
		Crystal violet (CV) adsorption	200 mg/L (initial concentration) adsorbent loading = 0.2 mg/L	458 mg/g	

5. Economic Assessment

Commercial MCM-41 production is expensive, due to intense energy consumption or utilization of inorganic and organic silicate reagents, the identification of new cheap silica sources being a viable future solution [40,45]. Beside the coal fly ash, these sources can be found in various wastes, such as agriculture slag [46], electronic components [47] or rice and wheat husk [48,49].

The synthesis of silica from waste is not free of cost, because the procedure is similar to the conventional methods, involving hydrothermal treatment with intensive energy consumption. The difference between conventional and unconventional synthesis is the source of silica, which leads to extra synthesis steps. The significant reduction of the synthesis time, from 72 h (time used in conventional methods) to 2 h, could result in the cost decrease, by saving energy [50]. The estimated cost for the synthesis of MCM-41 from fly ash was about 1200 euro/kg, representing half the price of a similar commercial material [40,51]. The higher purity of this type of material could reach even a price of 5000 euro/kg [51].

Taking into account the depollution and storage costs for fly ash or other silica source wastes, it could lead to a fair price and a sustainable process.

Nanomaterials **2020**, *10*, 474

6. Conclusions

In the last few decades, worldwide coal consumption has been monitored. High quantities of fly ash are discharged as a by-product from the coal power plants. The influence of fly ash on soil-crops chain (maize and rice) in different areas was demonstrated. In this respect, an effective process to recycle this waste is mandatory.

Fly ash proved to be suitable to use in environmental applications, for example by replacing activated carbon and zeolites as adsorbents for air pollutants or wastewater treatment. Its adsorption capacity mainly depends on the origin and the activation method. To date, no industrial set-up has been developed. Economic milestones must be surpassed. Taking into consideration the amorphous alumina-silicate nature of fly ash, it can be applied as raw material in various industrial reactions, such as obtaining ultramarine blue. Intensive harvesting depletes trace elements in the soil. Despite the fact that many studies were achieved for the use of fly ash as soil amendment, full-scale application has not been accomplished. In the near future, farmers may consider fly ash to be a substitute for lime to enrich the soil.

Fly ash has high silica content, which makes it a potential source for obtaining nanoporous materials, such as mesoporous silica. In this overview, various methods for obtaining mesoporous silica materials from fly ash were examined by considering the ash composition, pretreatment procedures, and synthesis conditions, in order to identify the optimal parameters for synthesis. The physical properties of the materials derived from fly ash confirmed their mesoporous nature, similar to the mesoporous silica materials obtained by conventional methods. The mesoporous silica materials with a high surface area and large pore volume derived from fly ash can be applied as a support or as a surface-functionalized host for capturing CO_2 from gaseous emissions or for various water pollutants. The synthesis of different nanoporous materials using as raw materials various wastes would enable a high level of recycling for a sustainable society with a low environmental impact.

Author Contributions: M.G.M. contributed to the designed the study, the collection of data and interpretation, as well as to the manuscript writing. V.-C.N. contributed to the design of the study, supervised the project, contributed to the data interpretation and to the manuscript writing. All authors have read and agreed to the published version of the manuscript.

Funding: This research and the APC were funded by the Romanian Ministry of Education and Research under 19N/2019 Financing Contract, NUCLEU Program, Project PN 19 11 03 01—"Studies on the obtaining and improvement of the acido-basic properties of the nanoporous catalytic materials for application in wastes valorization".

Acknowledgments: We thank the support offered by the Romanian Ministry of Education and Research through 19N/2019 Financing Contract, NUCLEU Program, Project PN 19 11 03 01.

Conflicts of Interest: The authors declare that the research was conducted in the absence of any commercial or financial relationships that could be construed as a potential conflict of interest.

References

1. Global Energy Statistical Yearbook 2019. Available online: https://yearbook.enerdata.net/coal-lignite/coal-world-consumption-data.html (accessed on 22 August 2019).
2. Enerdata Intelligence + Consluting. Available online: https://www.enerdata.net/publications/daily-energy-news/us-eia-expects-us-energy-related-co2-emissions-dip-22-2019.html (accessed on 23 August 2019).
3. Modak, A.; Jana, S. Advances in porous adsorbents for CO_2 capture and storage. In *Carbon Dioxide Chemistry, Capture and Oil Recovery*; IntechOpen: London, UK, 2018.
4. Yao, Z.T.; Ji, X.S.; Sarker, P.K.; Tang, J.H.; Ge, L.Q.; Xia, M.S.; Xi, Y.Q. A comprehensive review on the applications of coal fly ash. *Earth-Sci. Rev.* **2015**, *141*, 105–121. [CrossRef]
5. ACAA. Production & Use Reports. Available online: https://www.acaa-usa.org/publications/productionusereports.aspx (accessed on 16 December 2019).
6. Putilov, V.Y.; Putilova, I.V. Modern approach to the problem of utilization of fly ash and bottom ash from power plants in Russia. In Proceedings of the World of Coal Ash (WOCA), Lexington, Kentucky, USA, 11–15 April 2005.

7. Kargin, A.; Baev, V.; Mashkin, N.; Uglyanica, A. Fly ash: Perspective resource for geo-polymer materials production. In Proceedings of the Advanced Materials in Technology and Construction, AIP Publishing, Tomsk, Russia, 6–9 October 2015.

8. Soe, J.T.; Kim, S.S.; Lee, Y.R.; Ahn, J.W.; Ahn, W.S. CO2 capture and Ca2+ exchange using Zeolite A and 13X prepared from power plant fly ash. *Bull. Korean Chem. Soc.* **2016**, *37*, 490–493. [CrossRef]

9. Belviso, C.; Cavalcante, F.; Huertas, F.J.; Lettino, A.; Ragone, P. The crystallisation of zeolite (X- and A-type) from fly ash at 25 °C in artificial sea water. *Micropor. Mesopor. Mat.* **2012**, *162*, 115–121. [CrossRef]

10. Fotovat, F.; Kazemian, H.; Kazemini, M. Synthesis of Na-A and faujasitic zeolites from high silicon fly ash. *Mater. Res. Bull.* **2009**, *44*, 913–917. [CrossRef]

11. Belviso, C.; Agostinelli, E.; Belviso, S.; Cavalcante, F.; Pascucci, S.; Peddis, D.; Varvaro, G.; Fiore, S. Synthesis of magnetic zeolite at low temperature using a waste material mixture: Fly ash and red mud. *Micropor. Mesopor. Mat.* **2015**, *202*, 208–216. [CrossRef]

12. Vichaphund, S.; Ong, D.A.; Sricharoenchaikul, V.; Atong, D. Characteristic of fly ash derived-zeolite and its catalytic performance for fast pyrolysis of Jatropha waste. *Environ. Technol.* **2014**, *35*, 2254–2261. [CrossRef]

13. Li, D.; Min, H.; Jiang, X.; Ran, X.; Zou, L.; Fan, J. One-pot synthesis of aluminium containing ordered mesoporous silica MCM-41 using coal fly ash for phosphate adsorption. *J. Colloid Interface Sci.* **2013**, *404*, 42–48. [CrossRef]

14. Chen, C.; Kim, J.; Ahn, W.S. CO2 capture by amine-functionalized nanoporous materials: A review. *Korean J. Chem. Eng.* **2014**, *31*, 1919–1934. [CrossRef]

15. Watermann, A.; Brieger, J. Mesoporous Silica Nanoparticles as Drug Delivery Vehicles in Cancer. *Nanomaterials* **2017**, *7*, 189. [CrossRef]

16. Manzano, M.; Vallet-Regi, M. Mesoporous silica nanoparticles in nanomedicine applications. *J. Mater. Sci.-Mater.* **2018**, *29*, 65. [CrossRef]

17. Sannino, F.; Costantini, A.; Ruffo, F.; Aronne, A.; Venezia, V.; Califano, V. Covalent Immobilization of β-Glucosidase into Mesoporous Silica Nanoparticles from Anhydrous Acetone Enhances Its Catalytic Performance. *Nanomaterials* **2020**, *10*, 108. [CrossRef] [PubMed]

18. Califano, V.; Sannino, F.; Costantini, A.; Avossa, J.; Cimino, S.; Aronne, A. Wrinkled Silica Nanoparticles: Efficient Matrix for β-Glucosidase Immobilization. *J. Phys. Chem. C* **2018**, *122*, 8373–8379. [CrossRef]

19. Kresge, C.T.; Leonowicz, M.E.; Roth, W.J.; Vartuli, J.C.; Beck, J.S. Ordered mesoporous molecular sieves synthesized by a liquid-crystal template mechanism. *Nature* **1992**, *359*, 710–712. [CrossRef]

20. Beck, J.S.; Vartuli, J.C. Recent advances in the synthesis, characterization and applications of mesoporous molecular sieves. *Curr. Opin. Solid State Mater. Sci.* **1996**, *1*, 76–87. [CrossRef]

21. Vassilev, S.V.; Vassileva, C.G. Geochemistry of coals, coal ashes and combustion wastes from coal-fired power stations. *Fuel Process. Technol.* **1997**, *51*, 19–45. [CrossRef]

22. Vassilev, S.V.; Menendez, R.; Alvarez, D.; Diaz-Somoano, M.; Matrtinez-Tarazona, M.R. Phase-mineral and chemical composition of coal fly ashes as a basis for their multicomponent utilization. 1. Charaterization of feed coals and fly ashes. *Fuel* **2003**, *82*, 1793–1811. [CrossRef]

23. Vassilev, S.V.; Vassileva, C.G. Methods for characterization of composition of fly ashes from coal-fired power stations: A critical overview. *Energy fuel* **2005**, *19*, 1084–1098. [CrossRef]

24. Fernández-Turiel, J.L.; de Carvalho, W.; Cabañas, M.; Querol, X.; López-Soler, A. Mobility of heavy metals from coal fly ash. *Environ. Geol.* **1994**, *23*, 264–270.

25. Ram, L.C.; Srivastava, N.K.; Tripathi, R.C.; Thakur, S.K.; Sinha, A.K.; Jha, S.K.; Masto, R.E.; Mitra, S. Leaching behavior of lignite fly ash with shake and column tests. *Environ. Geol.* **2007**, *51*, 1119–1132. [CrossRef]

26. Sarode, D.B.; Jadhav, R.N.; Khatik, V.A.; Ingle, S.T.; Attarde, S.B. Extraction and leaching of heavy metals from thermal power plant fly ash and its admixtures. *Pol. J. Environ. Stud.* **2010**, *19*, 1325–1330.

27. Dindi, A.; Quang, D.V.; Vega, L.F.; Nashef, E.; Abu-Zahra, M.R.M. Applications of fly ash for CO$_2$ capture, utilization, and storage. *J. CO$_2$ Util.* **2019**, *29*, 82–102. [CrossRef]

28. Ahmaruzzaman, M. A review on the utilization of fly ash. *Progr. Energy Combust. Sci.* **2010**, *36*, 327–363. [CrossRef]

29. Patra, K.C.; Rautray Tapash, R.; Nayak, P. Analysis of grains grown on fly ash treated soils. *Appl. Radiat. Isotopes* **2012**, *70*, 1797–1802. [CrossRef] [PubMed]

30. Pendias, A.K.; Pendias, H. *Trace Elements in Soils and Plants*, 3rd ed.; CRC Press: Boca Raton, FL, USA, 2000.

31. Onyedikachi, U.B.; Belonwu, D.C.; Wegwu, M.O. Human health risk assessment of heavy metals in soils and commonly consumed food crops from quarry sites located at Isiagwu, Ebonyi State. *Ovidius Univ. Ann. Chem.* **2018**, *29*, 8–24. [CrossRef]

32. ECDGE, European Commission Director General Environment. Heavy Metals and Organic Compounds from Wastes Used as Organic Fertilizers. 2010, pp. 73–74. Available online: http://ec.europa.eu/environment/waste/compost/pdf/hm_finalreport.pdf (accessed on 15 October 2019).

33. WHO. *Permissible Limits of Heavy Metals in Soil and Plants*; World Health Organization: Geneva, Switzerland, 1996.

34. Page, A.L.; Elseewi, A.A.; Straughan, I.R. Physical and chemical properties of fly ash from coal-fired power plants with special reference to environmental impacts. *Residue Rev.* **1979**, *71*, 83–120.

35. Kumar, P.; Mal, N.; Oumi, Y.; Yamana, K.; Sano, T. Mesoporous materials prepared using coal fly ash as the silicon and aluminium source. *J. Mater. Chem.* **2001**, *11*, 3285–3290. [CrossRef]

36. Dhokte, A.O.; Khillare, S.L.; Lande, M.K.; Arbad, B.R. Synthesis, characterization of mesoporous silica materials from waste coal fly ash for the classical Mannich reaction. *J. Ind. Eng. Chem.* **2011**, *17*, 742–746. [CrossRef]

37. Zhou, C.; Gao, Q.; Luo, W.; Zhou, Q.; Wang, H.; Yan, C. Preparation, characterization and adsorption evaluation of spherical mesoporous Al-MCM-41 from coal fly ash. *J. Taiwan Inst. Chem. Eng.* **2015**, *52*, 147–157. [CrossRef]

38. Eimer, A.G.; Pirella, L.B.; Monti, G.A.; Anunziata, O.A. Synthesis and characterization of Al-MCM-41 and Al-MCM-48 mesoporous materials. *Catal. Lett.* **2002**, *78*, 1–4. [CrossRef]

39. Majchrzak-Kuceba, I.; Nowak, W. Characterization of MCM-41 mesoporous materials derived from polish fly ashes. *Int. J. Miner. Process.* **2011**, *101*, 100–111. [CrossRef]

40. Panek, R.; Wdowin, M.; Franus, W.; Czarna, D.; Stevens, L.A.; Deng, H.; Liu, J.; Sun, C.; Liu, H. Fly ash-derived MCM-41 as a low-cost silica support for polyethyleneimine in post-combustion CO_2 capture. *J. CO_2 Util.* **2017**, *22*, 81–90. [CrossRef]

41. Li, G.; Wang, B.; Sun, Q.; Xu, W.Q.; Han, Y. Adsorption of lead ion on amino-functionalized fly-ash-based SBA-15 mesoporous molecular sieves prepared via two-step hydrothermal method. *Micropor. Mesopor. Mat.* **2017**, *252*, 105–115. [CrossRef]

42. Miricioiu, M.G.; Iacob, C.; Nechifor, G.; Niculescu, V.C. High Selective Mixed Membranes Based on Mesoporous MCM-41 and MCM-41-NH_2 Particles in a Polysulfone Matrix. *Front. Chem.* **2019**, *7*, 332. [CrossRef] [PubMed]

43. Niculescu, V.; Miricioiu, M.; Geana, I.; Ionete, R.E.; Paun, N.; Parvulescu, V. Silica mesoporous materials—An efficient sorbent for wine polyphenols separation. *Rev. Chim.* **2019**, *70*, 1513–1517. [CrossRef]

44. Du, T.; Zhou, L.F.; Zhang, Q.; Liu, L.Y.; Li, G.; Luo, W.B.; Liu, H.K. Mesoporous structured aluminaosilicate with excellent adsorption performances for water purification. *Sustain. Mater. Technol.* **2018**, *17*, e00080. [CrossRef]

45. Niculescu, V.C.; Miricioiu, M.; Enache, S.; Constantinescu, M.; Bucura, F.; David, E. Optimized method for producing mesoporous silica from incineration ash. *Prog. Cryog. Isot. Sep.* **2019**, *22*, 65–76.

46. Ghorbani, F.; Younesi, V.; Mehraban, Z.; Celik, M.S.; Ghoreyshi, A.A.; Anbia, M. Preparation and characterization of highly pure silica from sedge as agricultural waste and its utilization in the synthesis of mesoporous silica MCM-41. *J. Taiwan Inst. Chem. Eng.* **2013**, *44*, 821–828. [CrossRef]

47. Liou, T.H. A green route to preparation of MCM-41 silicas with well-ordered mesostructure controlled in acidic and alkaline environments. *Chem. Eng. J.* **2011**, *171*, 1458–1468. [CrossRef]

48. Ma, Y.; Chen, H.; Shi, Y.; Yuan, S. Low cost synthesis of mesoporous molecular sieve MCM-41 from wheat straw ash using CTAB as surfactant. *Mater. Res. Bull.* **2016**, *77*, 258–264. [CrossRef]

49. Bhagiyalakshmi, M.; Yunn, L.J.; Anuradha, R.; Jang, H.T. Utilization of rice husk ash as silica source for the synthesis of mesoporous silicas and their application to CO_2 adsorption through TREN/TEPA grafting. *J. Hazard. Mater.* **2010**, *175*, 928–938. [CrossRef]

50. Lawrence, G.; Baskar, A.V.; El-Newehy, M.H.; Cha, W.S.; Al-Deyab, S.S.; Vinu, A. Quick high-temperature hydrothermal synthesis of mesoporous materials with 3D cubic structure for the adsorption of lysozyme. *Sci. Technol. Adv. Mater.* **2015**, *16*, 1–11. [CrossRef] [PubMed]

51. ACS Material, Advanced Chemical Supplier. Available online: https://www.acsmaterial.com/mcm-41.html (accessed on 18 October 2019).

Communication

Stretchable and Low-Haze Ag-Nanowire-Network 2-D Films Embedded into a Cross-linked Polydimethylsiloxane Elastomer

Ki-Wook Lee, Yong-Hoe Kim, Wen Xuan Du and Jin-Yeol Kim *

School of Advanced Materials Engineering, Kookmin University, Seoul 136-702, Korea;
Kwlee7908@kookmin.ac.kr (K.-W.L.); 20091304@kookmin.ac.kr (Y.-H.K.); duwenxuan1314@naver.com (W.X.D.)
* Correspondence: jinyeol@kookmin.ac.kr; Tel.: +82-2-910-4663

Received: 13 March 2019; Accepted: 6 April 2019; Published: 9 April 2019

Abstract: We report the fabrication of stretchable transparent electrode films (STEF) using 15-nm-diameter Ag nanowires networks embedded into a cross-linked polydimethylsiloxane elastomer. 15-nm-diameter Ag NWs with a high aspect ratio (>1000) were synthesized through pressure-induced polyol synthesis in the presence of AgCl particles with KBr. These Ag NW network-based STEF exhibited considerably low haze values (<1.5%) with a transparency of 90% despite the low sheet resistance of 20 Ω/sq. The STEF exhibited an outstanding mechanical elasticity of up to 20% and no visible change occurred in the sheet resistance after 100 cycles at a stretching-release test of 20%.

Keywords: 15-nm silver nanowire; Pressure-induced polyol method; stretchable transparent electrode 2-D films; Low-haze; Embedded electrode film

1. Introduction

Transparent conductive electrode films have increasingly attracted attention owing to their potential applications in optoelectronic fields, including touch screens, organic light emitting diodes and organic solar cells [1–4]. In particular, functionalized stretchable films, which have recently begun to be widely used as stretchable transparent electrode films (STEF), respond to mechanical deformations by the changes in electrical characteristics, such as resistance, owing to their stretchability and reproducibility. In this regard, nanomaterials, such as silver nanowires (Ag NWs) [5], single-walled carbon nanotubes (SWCNTs) [6–8], and graphene sheets [9–11], and their hybrid structures have been reported for use as sensitive strain sensors, which make them ideal for use as a transparent conductor in flexible or stretchable devices [12–15].

Among them, silver nanowires (Ag NW) have been gaining interest as a promising transparent conductive electrode material because of its simple synthesis and the possibility of large-area coating film fabrication via solution processes [16–21]. Particular attention has been focused on random network films of Ag NWs because such films can be easily fabricated in solutions and exhibit enhanced optoelectronic properties. The intrinsic properties of NWs mainly depend on the diameter and length of NWs. Many recent studies have also focused on the synthesis of Ag NWs with small diameters and large aspect ratios, which possess a low haze value due to low light scattering and good plasmonic properties. Polyol synthesis is known to be the most widely used and versatile method for the preparation of Ag NWs. To thin down the diameters, various polyol processes are being developed. In this regard, Wiley and a co-worker [3] recently reported the synthesis of Ag NWs with diameters of ~20 nm by controlling the bromide ion concentration in the conventional polyol method. Our group also have recently reported the synthesis of 20-nm-diameter Ag NWs under a pressure-induced polyol method in the presence of NaCl–KBr co-salts [22], but their synthesis mechanisms did not fully account for the synthesis. However, despite the progress, the synthesis of thin Ag NWs less than 20 nm has had limited success, meaning more research is required to synthesize wires with diameters below that amount.

Herein, we report a novel pressure-induced polyol method for synthesizing ultra-thin Ag NWs with a diameter of 15-nm-diameter or less and a high aspect ratio (>1000), a relatively unreported area so far. In particular, we have investigated the growth of Ag NWs and seed crystals in the presence of the AgCl-KBr co-salts instead of the NaCl salt used in the previous work [22] under a pressure of 1000 psi, and found that the K^+ ions cause a remarkable pressure effect.

For a two-dimensional (2-D) film consisting of an Ag NW networks, in particular, it has excellent transmittance and sheet resistance, yet its optical haze still needs to be improved in order for it to be suitable for display applications. Therefore, ultra-thin Ag NWs can be a good candidate for low-haze transparent electrodes. In particular, in order to obtain low-haze Ag NW network conductive films superior to indium tin oxide (ITO, up to 90% transmittance and ~1% haze at the low sheet resistance of 60 ohm/sq) in terms of opto-electrical performance, a diameter of at least 20 nm Ag NWs is required. However, to achieve the required optical characteristics, more effective processes that can control the shapes and sizes of the synthesized Ag NWs are required. In this work, 2-D films based on 15-nm-diameter Ag NW networks embedded into cross-linked polydimethylsiloxane (PDMS) elastomer, STEF, were formed via a conventional wet-coating technique that adhered the NWs to a PDMS substrate film, for flexible display applications, as shown in Figure 1. In particular, the conductor comprising an Ag NW network embedded into PDMS exhibited high elasticity, cycling stability, transparency, and excellent electrical conductivity. In addition, these films were also confirmed to exhibit good responses to the stretch/release for ≥100 cycles, while hysteresis tests without the loss of conductivity under stretching conditions of 20% were also conducted.

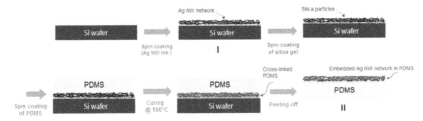

Figure 1. Fabrication of highly strain electrode films based on a Ag NW networks embedded into the cross-linked PDMS elastomer.

2. Results and Discussion

Herein, we newly synthesized the ultra-thin Ag NWs with 15 nm or less in diameter and aspect ratio to as high as 1000 using a pressure-induced polyol method via the chemical reduction of $AgNO_3$ in the presence of an AgCl crystal and KBr (molar ratio = 2:1), according to a previous report [22]. Figure 2 is a plot of the change in diameter of Ag NWs synthesized at various pressure conditions (the four pressure values; 0, 110, 250, and 1000 psi). As shown in Figure 2(I), the diameter of the Ag NWs decreased with increasing pressure in the presence of KBr supplemented with AgCl, including NaCl and $FeCl_3$ salts. In particular, at the highest reaction pressure (1000 psi (69 bar)), the Ag NWs that formed in the presence of AgCl with KBr were ultrathin with a mean diameter of 15 nm and a narrow size distribution (within ±5 nm). In any case, Ag NWs synthesized under the pressure-induced conditions of the present experiment were noticeably smaller and more evenly dispersed than those produced at atmospheric pressure. In contrast, in NaCl, AgCl, and $FeCl_3$ supplemented with NaBr, the NW diameter was independent of pressure, as shown in Figure 2(II). These results suggest that in the presence of KBr, particularly in the presence of K^+ ions, the pressure controls the rate of the formation of Ag^+ ions, thereby suppressing the growth in the thickness direction of the wire. As a result, in the pressure-induced polyol reaction, the reduction in diameter of Ag NW was observed to be affected by pressure only in the presence of K$^+$ ions. This suggests that K$^+$ ion acts as an effect of pressure on the growth of Ag NW, but it was difficult to describe the kinetics of K$^+$ ions involved in the formation and growth of Ag NWs in this work. However, in the process of synthesizing the Ag NWs, K$^+$ ions greatly acted on the pressure, and ultra-fine Ag NWs with a diameter of 15 nm could be successfully synthesized. Liao et al. [23] explained that increasing the reaction pressure lowers the energy barrier of nucleation and accelerates nucleation, resulting in a controlled rate of metal nanostructure formation when pressure is applied. Here, the nucleation rate of Ag ions is also closely related to the wire size. Figure 2(III) shows the SEM images of the produced Ag NWs synthesized in the presence of AgCl–KBr salts; (a) 19–25, (b) 17–18, and (c) 15–16 nm, respectively. These NWs correspond with 0, 250, and 1000 psi (69 bar), respectively.

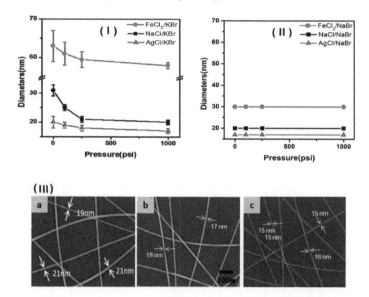

Figure 2. Ag NW diameter vs. pressure in the presence of various salts: [I] AgCl–KBr, NaCl–KBr, and FeCl$_3$–KBr and [II] AgCl–NaBr, NaCl–NaBr, and FeCl$_3$–NaBr (the error range is observed within the range of 2~3 nm, respectively). [III] SEM images of the Ag NWs synthesized in the presence of AgCl-KBr salts; (a) 19–25, (b) 17–18, and (c) 15–16 nm, respectively. These NWs correspond with 0, 250, and 1000 psi (69 bar), respectively.

Figure 3 displays SEM images at low magnification of the 15-nm-diameter Ag NWs synthesized at 1000 psi. Subsequently, the small-size Ag seed particles grew into Ag NWs with a mean diameter of 15 nm (range: 6–20 nm; aspect ratio: ~800). The diameter distribution of the synthesized wires is plotted in Figure 3(II). The mean diameter is at least 5 nm smaller than that of NWs formed at 0 psi (mean diameter = 22 nm; distribution = 14–28 nm). The surface plasmon resonance (SPR) signals have inherent characteristics depending on the size and structure of the nanomaterials [4,24,25]. Therefore, the size of Ag NWs can be predicted from the absorption bands appearing at different frequencies critically in the SPR data. In this regard, the SPR characteristic peak of 15-nm-diameter Ag NW synthesized at 1000 psi pressure with AgCl-KBr present shows at 354 and 362 nm, as shown in Figure 3(V). The SPR peak in Figure 3(V) appeared at 362 nm, which was significantly shorter than those in wires with diameters of 20–22 nm [366-nm peak; see Figure 3(IV)] and 30–32 nm [372 nm peak in Figure 3(III)]. This indicates that the transverse modes appeared at significantly shorter wavelengths in NWs with pentagonal cross sections than in the abovementioned wires. Besides causing a blue shift in the peaks, reducing the NW diameter reduces the amount of scattered light.

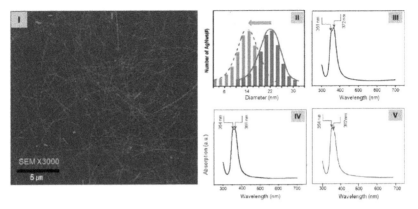

Figure 3. [**I**] SEM images of Ag NWs at low magnification (3000×). The average diameter of the NWs is 15 nm. [**II**] Diameter distribution of the Ag NWs synthesized at 1000 and 0 psi. Surface Plasmon resonance (SPR) absorption characteristics of the synthesized Ag NWs with diameters of [**III**] 30–32 nm, [**IV**] 20–22 nm, and [**V**] 15 nm.

STEF films based on a 15-nm-diameter Ag NW networks embedded into cross-linked PDMS elastomer were formed via a conventional spin-coating technique that adhered the NWs to a substrate, as shown in Figure 1. Ag NW were dispersed in DI water at a density of 0.2 mg/mL and directly coated onto the Si wafer substrate, which was previously cleaned with acetone, following by drying at 80 °C for 5 min. Second, 0.01 wt% silica gel dispersed in ethanol was spin-coated at 1000 rpm, and post drying, liquid PDMS with a thickness of ~50 μm was coated on the upper surface of silica and the Ag NWs network layer, followed by curing and crosslinking. Afterwards, we peeled the cured PDMS from the Si wafer. Here, when liquid PDMS covers the Ag NW network layer, it penetrates into the interconnected pores of the Ag NW network because of its low viscosity and low surface energy. After curing, all Ag NW networks are buried on the cross-linked PDMS surface (crosslinking between PDMS and silica gel) without considerable voids, indicative of the successful transfer of Ag NW networks from Si wafers to PDMS and excellent adhesion between Ag NW and PDMS. The Ag NW network is embedded into the surface of ~50-μm-thick-PDMS films.

Figure 4(I) shows a photograph of the finally produced Ag NW network embedded into the PDMS film sample, which is STEF, and Figure 4(II) and (III) shows the SEM and AFM surface images of the Ag NW conductive network layer, respectively. In particular, the SEM image of Figure 4(II) shows a highly transparent, extensible, and reliable "STEF" based on a 15-nm-diameter Ag NW network layer embedded in the surface layer of the cross-linked PDMS elastomer film. Here, PDMS completely penetrated into the Ag NW network and filled the gaps between Ag NWs, as shown in the SEM surface image of Figure 4(II), affording an Ag NW network and PDMS. The Ag NW network structure embedded in the surface layer of the cross-linked PDMS elastomer film was clearly observed as the current map image of AFM in Figure 4(III). The sheet resistance and optical value of the Ag NW conductive network layer was determined as a function of density of the Ag NWs in the network. That is, the change in the sheet resistance with increasing density (the content of the Ag NW networks in the layer is described by areal density, namely the Ag NW weight per unit area of the films) of the Ag NW network layer is obtained. However, the sheet resistance of the Ag NW network layer significantly decreased with increasing Ag NW density. Their results showed a low sheet resistance of 20, 40, 50, and 85 Ω/sq at transmittances of 90%, 95%, 96%, and 97% (based on PDMS), respectively. In particular, these 15-nm-diameter Ag NW network embedded PDMS films were exhibited to have low haze values of less than 1.5% (net Haze) with a transparency of 90% despite the low sheet resistance of 20 Ω/sq (up to 90% transmittance and ~1% haze at the sheet resistance of 60 Ω/sq). These haze values shown above were approximately 0.2–0.3 lower at the same sheet resistance condition than that of the 20-nm-diameter Ag NWs reported in the previous work [23]. However, as the diameter is decreased, the optical haze parameter improved; thus, the scattered light can be reduced and the haze value is greatly decreased. As a result, it has been suggested that a 2-D percolating network film constructed using at least 15-nm-diameter Ag NWs is needed to satisfy the electrical and optical properties of crystalline ITO glass.

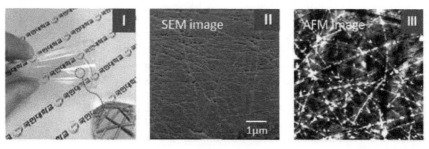

Figure 4. [I] Photograph of a STEF sample and [II] surface SEM image and [III] AFM current image of the Ag NW network-embedded PDMS.

Elastic behavior was observed for the sample under dynamic loading. In Figure 5(I), at tensile strains of 10%, 20% and 30%, the change in R/R0 was observed at strain restoration for the tensile strain. The initial sheet resistance (R_0) was almost completely recovered for a stretch/release cycle test with strains ε of 10% and 20%, revealing the outstanding stretchable property of film. Nevertheless, at a strain of greater than or equal to 30%, the sheet resistance of the film was not restored to its original position. Given that the flexible and stretchable characteristics of Ag NW network-embedded PDMS film can obtain highly reliable mechanical performance under continuous strain deformation, repeated stretch/release tests were conducted on the films. An automated testing tool was utilized, which enabled the electrode to exhibit repeated alternate stretch and release. This repeated stretch and release led to cyclic fatigue failure. Thus, the resistance of the Ag NW network-embedded PDMS film sharply increases at the very first stretching and then returns to its initial value. In this test, elongation values of 10% and 20% were utilized. With the repetition of the test for ≥100 cycles under stretching conditions of 10% and 20%, the change in the resistance was restored to its original position without any change in the resistance (Figure 5(II)). However, highly stretchable films based on the 15-nm Ag NW networks embedded into the cross-linked PDMS elastomer were simply fabricated using a spin-coating method.

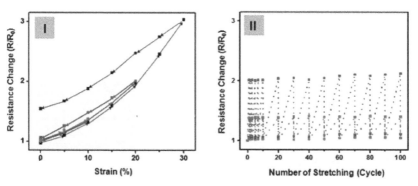

Figure 5. (I) Hysteresis curve of the film comprising a Ag NW-network-embedded PDMS film (at tensile strains (ε) of 10%, 20%, and 30%). (II) Effect of repeated stretching on the resistance change (R/R0) at strain recovery (stretch/release cycles of ε = 10% and 20%).

3. Conclusions

In conclusion, we demonstrated for the first time that ultra-fine Ag NWs with 15-nm-diameter that could not be realized in previous work [23] via a pressure-induced polyol process and in the presence of AgCl with KBr and K^+ ions induced a notable pressure effect. The characteristic SPR of these 15-nm-diameter NWs appeared at 362 nm. This is a novel finding for Ag NWs and provides evidence of their high optical performances. Furthermore, we fabricated the stretchable transparent electrode films (STEF) based on a 15-nm-diameter Ag nanowires networks embedded into a cross-linked polydimethylsiloxane elastomer. These 2-D embedded Ag NW network film with a 15-nm-diameter Ag NW showed a low sheet resistance of 20 Ω/sq. at 90% transparency with haze values (<1.5%). The electrode films also exhibited a high elasticity of 20%, and the strain films exhibited a good response to the stretch/release of 100 cycles and hysteresis tests. However, these 2-D STEF exhibit good flexibility, making them promising candidates for use as a transparent electrode in flexible electronics. In particular, in the case of the Ag NW embedded elastomer films having a high stretchability and a high electric conductivity, as in the present study, it is expected that these films will provide a much higher performance material in many areas for flexible transparent devices that can replace ITO.

Author Contributions: Y.-H.K. and K.-W.L. participated in the experiment design, carried out the synthesis of silver nanowires, tested the films, and helped draft the manuscript. W.X.D. supported experimentation and data analysis. J.-Y.K. wrote the paper and supervised the work. All authors read and approved the final manuscript.

Funding: This work was financially supported in part by the Korea Foundation Grant funded by the Korean Government (KRF-2017R1D1A1B03031246), and the fusion and complex R&D program (S2448726) of the SMBA.

Conflicts of Interest: The authors declare that there is no conflict of interest in the results and content of the study.

References

1. Wu, Y.; Xiang, J.; Yang, C.; Lu, W.; Lieber, C.M. Single-crystal metallic nanowires and metal/semiconductor nanowire heterostructures. *Nature* **2004**, *430*, 61–65. [CrossRef]
2. Loh, K.P.; Tong, S.W.; Wu, J. Graphene and Graphene-like Molecules: Prospects in Solar Cells. *J. Am. Chem. Soc.* **2016**, *138*, 1095–1102. [PubMed]
3. Li, B.; Ye, S.; Stewart, I.E.; Alvarez, S.; Wily, B.J. Synthesis and Purification of Silver Nanowires To Make Conducting Films with a Transmittance of 99%. *Wily Nano Lett.* **2015**, *15*, 6722. [CrossRef]
4. Cho, S.; Kang, S.; Pandya, A.; Shanker, R.; Khan, Z.; Lee, Y.; Park, J.; Craig, S.L.; Ko, H. Large-Area Cross-Aligned Silver Nanowire Electrodes for Flexible, Transparent, and Force-Sensitive Mechanochromic Touch Screens. *ACS Nano* **2017**, *11*, 4346–4357.
5. Amjadi, M.; Pichitpajongkit, A.; Lee, S.; Ryu, S.; Park, I. Highly stretchable and sensitive strain sensor based on silver nanowire-elastomer nanocomposite. *ACS Nano* **2014**, *8*, 5154–5163.
6. Fan, Q.; Qin, Z.; Gao, S.; Wu, Y.; Pionteck, J.; Mäder, E.; Zhu, M. The use of a carbon nanotube layer on a polyurethane multifilament substrate for monitoring strains as large as 400%. *Carbon* **2012**, *50*, 4085–4092. [CrossRef]
7. Luo, S.; Liu, T. Structure-property-processing relationships of SWCNT thin film piezosensitive sensors. *Carbon* **2013**, *59*, 315–324.
8. Zhang, R.; Deng, H.; Valenca, R.; Jin, J.; Fu, Q.; Bilotti, E.; Peijs, T. Strain sensing behaviour of elastomeric composite films containing CNT under cyclic loading. *Compos. Sci. Technol.* **2013**, *74*, 1–5. [CrossRef]
9. Li, X.; Zhang, W.; Wang, K.; Wei, J.; Wu, D.; Cao, A.; Li, Z.; Cheng, Y.; Zheng, Q. Stretchable and highly sensitive graphene-on-polymer strain sensors. *Sci. Rep.* **2012**, *2*, 870. [CrossRef] [PubMed]
10. Hempel, M.; Nezich, D.; Kong, J.; Hofmann, M. A novel class of strain gauges based on layered percolative films of 2D materials. *Nano Lett.* **2012**, *12*, 5714. [CrossRef] [PubMed]
11. Bae, S.H.; Lee, Y.; Sharma, B.K.; Lee, H.J.; Kim, H.J.; Ahn, J.H. Graphene-based transparent strain sensor. *Carbon* **2013**, *51*, 236–242. [CrossRef]
12. Strevens, A.E.; Drury, A.; Lipson, S.; Kröll, M.; Blau, W.; Hörhold, H. Hybrid light-emitting polymer device fabricated on a metallic nanowire array. *Appl. Phys. Lett.* **2005**, *86*, 143503–143507. [CrossRef]

13. Murphy, C.J.; Sau, T.K.; Gole, A.; Orendorff, C.J. Surfactant-directed synthesis and optical properties of one-dimensional plasmonic metallic nanostructures. *MRS Bull.* **2005**, *30*, 349–355. [CrossRef]

14. Favier, F.; Walter, E.C.; Zach, M.P.; Benter, T.; Penner, R.M. Hydrogen sensors and switches from electrodeposited palladium mesowire arrays. *Science* **2001**, *293*, 2227–2231. [CrossRef] [PubMed]

15. Zimmermann, E.; Ehrenreich, P.; Pfadler, T.; Dorman, J.A.; Weickert, J.; Schmidt-Mende, L. Erroneous efficiency reports harm organic solar cell research. *Nat. Photonic* **2014**, *8*, 669–672. [CrossRef]

16. Mutiso, R.M.; Sherrott, M.C.; Rathmell, A.R.; Wiley, B.J.; Winey, K.I. Integrating Simulations and Experiments to Predict Sheet Resistance and Optical Transmittance in Nanowire Films for Transparent Conductors. *ACS Nano* **2013**, *7*, 7654–7663. [PubMed]

17. Bari, B.; Lee, J.; Jang, T.; Won, P.; Ko, S.H.; Alamgir, K.; Arshad, M.; Guo, L.J. Simple hydrothermal synthesis of very-long and thin silver nanowires and their application in high quality transparent electrodes. *J. Mater. Chem. A* **2016**, *4*, 11365–11371. [CrossRef]

18. Kim, T.; Canlier, A.; Kim, G.H.; Choi, J.; Park, M.; Han, S.M. Electrostatic Spray Deposition of Highly Transparent Silver Nanowire Electrode on Flexible Substrate. *ACS Appl. Mater. Interfaces* **2013**, *5*, 788–794. [CrossRef] [PubMed]

19. Araki, T.; Jiu, J.; Nogi, M.; Koga, H.; Nagao, S.; Sugahara, T.; Suganuma, K. Low haze transparent electrodes and highly conducting air dried films with ultra-long silver nanowires synthesized by one-step polyol method. *Nano Res.* **2014**, *7*, 236–245. [CrossRef]

20. Kim, D.H.; Yu, K.C.; Kim, Y.; Kim, J.W. Highly stretchable and mechanically stable transparentn composite of silver nanowires and polyurethane-urea. *ACS Appl. Mater. Interfaces* **2015**, *7*, 15214–15222. [CrossRef]

21. Park, K.H.; Im, S.H.; Park, O.O. The size control of silver nanocrystals with different polyols and its application to low-reflection coating materials. *Nanotechnology* **2011**, *22*, 045602. [CrossRef]

22. Lee, E.J.; Kim, Y.H.; Hwang, D.K.; Choi, W.K.; Kim, J.Y. Synthesis of small diameter silver nanowires *via* a magnetic-ionic-liquid-assisted polyol process. *RSC Adv.* **2016**, *6*, 11702–11709. [CrossRef]

23. Liao, S.C.; Mayo, W.E.; Pae, K.D. Theory of high pressure/low temperature sintering of bulk nanocrystalline TiO$_2$. *Acta Mater.* **1997**, *45*, 4027–4040. [CrossRef]

24. Sun, Y.; Xia, Y. Gold and silver nanoparticles: A class of chromophores with colors tunable in the range from 400 to 750 nm. *Analyst* **2003**, *128*, 686–691. [CrossRef]

25. Wiley, B.J.; Im, S.H.; Li, Z.Y.; McLellan, J.; Siekkinen, A.; Xia, Y. Maneuvering the surface plasmon resonance of silver nanostructures through shape-controlled synthesis. *J. Phys. Chem. B* **2006**, *110*, 15666–15675.

Review

Ceramic Composite Materials Obtained by Electron-Beam Physical Vapor Deposition Used as Thermal Barriers in the Aerospace Industry

Bogdan Stefan Vasile [1,2,3,*], **Alexandra Catalina Birca** [1,2,3], **Vasile Adrian Surdu** [1,2,3], **Ionela Andreea Neacsu** [1,2,3] and **Adrian Ionut Nicoară** [1,2,3]

[1] National Research Center for Micro and Nanomaterials, University Politehnica of Bucharest, 010164 Bucharest, Romania; alexandra.birca@upb.ro (A.C.B.); adrian.surdu@upb.ro (V.A.S.); ionela.neacsu@upb.ro (I.A.N.); adrian.nicoara@upb.ro (A.I.N.)

[2] Department of Science and Engineering of Oxide Materials and nanomaterials, Faculty of Applied Chemistry and Materials Science, University Politehnica of Bucharest, 010164 Bucharest, Romania

[3] National Research Centre for Food Safety, University Politehnica of Bucharest, 010164 Bucharest, Romania

* Correspondence: bogdan.vasile@upb.ro; Tel.: +40-727-589-960

Received: 20 January 2020; Accepted: 17 February 2020; Published: 20 February 2020

Abstract: This paper is focused on the basic properties of ceramic composite materials used as thermal barrier coatings in the aerospace industry like SiC, ZrC, ZrB$_2$ etc., and summarizes some principal properties for thermal barrier coatings. Although the aerospace industry is mainly based on metallic materials, a more attractive approach is represented by ceramic materials that are often more resistant to corrosion, oxidation and wear having at the same time suitable thermal properties. It is known that the space environment presents extreme conditions that challenge aerospace scientists, but simultaneously, presents opportunities to produce materials that behave almost ideally in this environment. Used even today, metal-matrix composites (MMCs) have been developed since the beginning of the space era due to their high specific stiffness and low thermal expansion coefficient. These types of composites possess properties such as high-temperature resistance and high strength, and those potential benefits led to the use of MMCs for supreme space system requirements in the late 1980s. Electron beam physical vapor deposition (EB-PVD) is the technology that helps to obtain the composite materials that ultimately have optimal properties for the space environment, and ceramics that broadly meet the requirements for the space industry can be silicon carbide that has been developed as a standard material very quickly, possessing many advantages. One of the most promising ceramics for ultrahigh temperature applications could be zirconium carbide (ZrC) because of its remarkable properties and the competence to form unwilling oxide scales at high temperatures, but at the same time it is known that no material can have all the ideal properties. Another promising material in coating for components used for ultra-high temperature applications as thermal protection systems is zirconium diboride (ZrB$_2$), due to its high melting point, high thermal conductivities, and relatively low density. Some composite ceramic materials like carbon–carbon fiber reinforced SiC, SiC-SiC, ZrC-SiC, ZrB$_2$-SiC, etc., possessing low thermal conductivities have been used as thermal barrier coating (TBC) materials to increase turbine inlet temperatures since the 1960s. With increasing engine efficiency, they can reduce metal surface temperatures and prolong the lifetime of the hot sections of aero-engines and land-based turbines.

Keywords: thermal protection systems; ultrahigh temperature applications; EB-PVD

---

1. Introduction

One branch of engineering that deals with the maintenance, development and study of airplanes and spacecraft is aerospace engineering, where research into materials for the construction of aerospace components is in continuous development. Although metals are the most widely used materials in aircraft components, discoveries in materials science, particularly in composite science and technology, have allowed the development of new materials for aerospace engineering [1,2]. Lightweight design of aircraft frames and engines with materials of improved mechanical properties can improve fuel efficiency, increase payload, and flight range, which directly reduce the aircraft operating cost [3–5].

The aerospace industry is based on the use of composite materials for both primary and secondary constitutional components such as engine nacelles, rocket motor castings, aircraft wings, antenna dishes, landing gear doors, centre wing boxes, tall cones, engine cowls and others [6,7].

At the present time, the use of composite materials in the aerospace industry inspire in a positive way the development and outline of modern and complex aero vehicles. In this sense, the properties like high specific strength and individual stiffness together with other unique properties makes this type of materials very attractive and suitable for this kind of applications. A class of composite materials is classified as advanced composites which is defined by metal matrix composites, high-performance fibre-reinforced polymers, and those most used in high-performance aerospace vehicles, and their properties are the ceramic matrix composites. This class of composite materials provide supplementary functional advantages, the most highlighted being the temperature resistance [8,9]. Using composite materials in developing parts of aero vehicles implies more than just replacing the metals or other regular materials, it is about the introduction of advanced materials which have a role in a multitude of features starting from new designs in morphological structures, which were initially not possible with traditional materials [10].

One of the problems in the development of some aero vehicles consists in obtaining parts that must have specific properties for the field of use. The most attractive characteristic of advanced composite materials is based on the high ratio between strength, which is a basic feature when speaking about aerospace, and weight which is another goal in this industry, compared to the metals frequently used in aerospace. Moreover, the production techniques are a very important subject in this field. Manufacturing components by using composite materials favors the production of numerous distinct structures [11,12].

When it comes to temperatures that can be reached in this field, the aerospace industry has an ultrahigh temperature class that is generally placed from 1600 °C and can reach up to 2200 °C [13]. These temperatures require the use of materials that can withstand very high temperatures and also have exclusive mechanical properties [14,15].

Considering that a single material cannot have as many properties as are needed for aerospace applications, there has been a need to study and develop composite materials that have advantages that situate them in an advantageous position when it comes to their use in the aerospace industry. The latest air vehicles models contain more than 50% of their weight in terms of composite materials. However, there is still a lack of information regarding mechanical behavior, which leads to stricter regulations to guarantee safety standards [3,16]. This has led to the impossibility of reaching the full potential of the composites in the aerospace industry and, of course, to the need for further studies [11].

Ceramic composites are obtained by linking ceramics using continuous fibers, particles or whiskers. The literature data provide information about the conventional types of reinforcement for ceramic matrix composites which include silicon carbide, titanium carbide and boron carbide, silicon nitride and boron nitride, alumina and zirconia, carbon and boron. Below are presented the advantageous characteristics of ceramic composites (Figure 1) [1,17]:

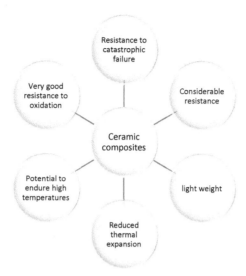

Figure 1. Properties of ceramic composites [18].

Metallic composites are manufactured by reinforcing various types of metal matrices, such as titanium, aluminum, copper, magnesium, etc. [19]. Typical blends for metal composites are ceramic particle or fiber in particular, but carbon fiber or metallic fiber can also be used. When it comes to processing techniques, metal composites can be obtained by diverse methods such as casting and powder metallurgy, but with specific limitations because of the metallic use [20]. This is despite the fact that There are limitations in the aerospace industry for metallic composites, the properties of which are presented in the Figure 2 [1]:

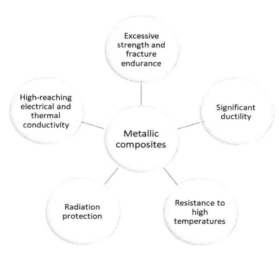

Figure 2. Properties of metallic composites [18].

Another class of materials that have applications in the aerospace industry is represented by the ultra-high temperature ceramics. These materials are described as possessing a blend of properties that are characterized by very good and suitable mechanical properties and at the same time a significant meting point, which can reach up to 3000 °C and even exceed this value [13,21].

In this sense, the materials that possess specific thermo-mechanical and thermo-chemical properties are required for aerospace applications, especially in ultra-high temperature area [22,23]. The ultra-high temperature class includes several applications like the manufacturing of solid rocket motors which need a very high temperature that are increased starting from room temperature to approximatively 3000 °C. Because of the fact that the application takes place at high temperatures, it is necessary that the materials for the components such as rocket combustion chambers to have properties that are dependent on each other such as high melting temperature, high-reach strength and of course significant resistance to environmental factors. Hypersonic vehicles also require components part manufactured from materials that reflect properties that ensure a specific action at temperatures beyond 1600 °C [24].

Some of the most notable properties of the materials that have applications at high temperatures are good oxidation resistance, high melting point, high hardness, and thermal shock and ablation endurance [14].

Demonstrating the behavior of different materials under high temperature applications, it was concluded that these materials should present a layer or more that covers the surface of the materials. At this point in time, thermal barrier coatings represent a subject that involves numerous and modern deposition techniques to increase the properties of the usual materials that are used in developing the component parts of aero vehicles. Surfaces of engines and gas turbine blades are the most covered components for the reason that at high temperature there is a need for thermal barrier behavior, considering the action of this as thermal insulation to the high temperature gas that flows within the turbine blades [25]. By covering the surface of aero vehicle components with materials that act as a thermal barrier, this also leads to a reduction in the thermal stresses. Criteria of thermal barrier coatings are to present low weight and low thermal conductivity, but there is still an issue because of the fact that after the heat-treatment processes, thermal conductivity of the coatings may increase [26].

2. Thermal Barrier Coating

Thermal-barrier coatings are defined as ceramic materials that present suitable resistance at high temperatures. Components like metal turbine blades used in aircraft engines need to be covered by depositing thermal barrier which allow these engines to perform at high temperatures [27]. The activity of these coatings is based on protecting from oxidation or melting because of that fact that hot gases from the engine core may affect the metal that is used at manufacturing these components for aero vehicles [25].

One essential role of thermal-barrier coatings components is to present various properties against the harsh environment such as corrosive atmosphere, high temperature and variation of this and complex stress conditions. It is well understood that it is complicated for a single coating component to possess all these conditions. At this level of depositing the coatings, the thermal barrier layers are planned to last for thousands of landings and take-offs in aero engines. When speaking about the complexity and diversity of thermal barrier coatings structures, there is an impediment of premature failure that can appear during operating conditions [28]. At one point in time, the use of thermal barrier coatings decreased and moreover, their full characteristics were discredited. In order to avoid and eliminate these impediments, more detailed analyses were considered regarding materials, processing principles, performance and, not least, failure mechanisms were enhanced, in order to better understand how to respond beneficially. This research field presents associative subjects of materials science, chemistry, physics, mechanics and thermodynamics [29].

At the same time, the advantageous development of thermal barrier coatings are essential to bring improvements in the case of inlet gas temperature which leads to a boost of the performance of gas turbines. Hence, to develop thermal barrier coatings with interdependent features such as high resistance to sintering, low thermal conductivity and also phase stability, it is necessary to highlight the increased demands in order to obtain a proper final material [30]. Commonly, thermal barrier coatings include a ceramic top coat and a metallic bond coat. The utilization of a bond coat is required to secure the metal substrate in the case of oxidation and corrosion because of the high temperature and also for coupling the ceramic top coat and the metallic substrate, being located between the substrate and the ceramic top coat [31].

However, work has been published on the conventional thermal-barrier coatings system, which in fact contain three layers, covering the substrate. The first layer is the metallic bond coat, the second layer is the middle thermally grown oxide and the third layer is the ceramic top coat. Separately, these layers cannot provide the thermal and mechanical properties necessary for their use under special conditions, but which are directly proportional to the processing conditions that may impose modifications [32,33].

The first layer seems to possess critical characteristics, due to the fact that this layer performs two fundamental roles. The certainty of the coatings system starts with the first layer, in this sense, one role is to ensure a very good adhesion between the substrate and the ceramic top layer. The second function is to act in the case of severe oxidation, because the oxygen ions from the environmental conditions may pass through the ceramic layer, due to the porosity and high diffusivity. The top coat requires high thermal stability and low thermal conductivity [34].

For these layers to act as demanded under special conditions at high temperatures, it is necessary that them to become common parts with the metal substrate that need coatings. For this reason, diverse physical methods were developed to deposit the ceramic top coat as a thermal barrier coating to the metallic substrate. The following methods are electron beam physical vapor deposition (EB–PVD), laser chemical vapor deposition, and atmospheric plasma sprayed, high-velocity oxy-fuel, sol-gel, plasma spray physical vapor deposition [1,34,35]. One of the most used of these kinds of application is electron beam physical vapor deposition (EB-PVD) and the second is atmospheric plasma spray (APS) [27,31,36].

Over time, measures have been taken to improve the competency of a gas turbine. These actions leaded to operating temperatures exceeding 1300 °C, which require thicker thermal barrier coating which influence the chemistry together with an additional cooling system. As a result, the top coat layer, present an increase of thickness which manage the surface temperature of the thermal barrier coating to a faster cooling components system with a rate of temperatures of 4–9 °C along with 25 µm [32].

The research in this domain surrounded by experimental activity and the implication of numerous people concluded that that thermal barrier coatings must meet a number of well-defined and interdependent conditions. The first condition speaking about aero vehicles is to present low weight, also low thermal conductivity is required. Because of the fact that the environmental medium may suffer drastic thermic changes, the coatings should resist variation from heating to cooling and vice versa and indeed to thermal shock. In order not to encounter problems that can have a significant impact later, the coatings must be chemically compatible with the substrate and resist oxidation process [32,37]. Thermal insulation is another mandatory condition for thermal-barrier coatings to the elemental superalloy engine components. The compliance of the superalloy parts with the thermal expansion is another necessity to minimize the discrepancy stresses. Moreover, thermal-barrier coatings must reverse as much as possible of the radiant heat produced by hot gas and is mandatory to prevent the contact of the heat with the substrate. It is desired for the thermal barrier coatings to ensure thermal protection for the coated substrate and to be capable of resisting for prolonged service times [31].

How to improve the protection of the components that are in contact with high temperatures, which use thermal-barrier coatings, has attracted the attention of researchers for many years. The coatings are deposited on the substrate using, in general, EB-PVD methods. This advanced technique involves high electron beam heating of rough materials which subsequently generate steam. The produced steam will be subjected to the substrate surface which is deposited as a coating [38]. It is understood that the coating is formed as a layer of vertical column grains that are standing upright on the substrate. Between the columns there are consecutive gaps, that separate pores in the structure of the grains which can be open pores or closed pores. Due to these structures of the coatings, the characteristics of the thermal barrier will be improved [39].

3. Electron Beam Physical Vapour Deposition (EB-PVD) Technology

The EB-PVD method is based mostly on the activity of the electron beam, which is considered the most important part having a role as thermal source in this deposition technique. One of the best and most attractive features of EB-PVD is the capability of depositing all types of material. The deposition procedure is based on the action of an electron beam established at 2000 °C within an electron gun, acting in accordance with the acceleration of thermal electrons supported by high voltage. The equipment includes a target of the material of interest, which is subsequently hit by high-speed electrons. Due to the energy generated by the electrons, the target material is melted and after that the material is transformed into vapor and deposited on the surface of the substrate as a coating. The highlighted advantage of this technique is the high deposition rate compared to other coating technique. The parameters applied for specific materials can be managed more easily and the surface also can be controlled when speaking about the dimension of the deposition. One mandatory property in obtaining the deposition materials is to present a strong adhesion between the coating and the substrate, which in the case of use of the EB-PVD technique, is fulfilled [29,31,40].

Depending on the needs of the final material, the coatings can be deposited differently from ceramic to ceramic, metallic to metallic, ceramic to metallic, or metallic to ceramic. Moreover, the best of the characteristic of this deposition technique is the multi material that can be used. In this sense, multilayer coatings can be deposited and also may be disposed of like alternative layers of distinct composition comprising ceramics, metals and polymers. All of these materials can be arranged as different and various layers on the substrate. Pointing to time efficiency, in this technique the deposition rate is high, and also in a short period the coating presents a dense structure. The microstructure may be controlled surrounded by a managed composition, trying to erase every possibility to be contaminated, and all of these properties are obtained finally regarding easily controlled parameters and flexible deposition. There are only minor exceptions where the deposited layers do not have a homogeneous microstructure, but generally the finished materials possess a good surface and uniform microstructure. Therefore, there is a fine relationship between the manipulating the process parameters and the final microstructure of the materials and also uniformity [39,41,42]. Below are showed the schematic illustrations of electron beam physical vapor deposition (EB-PVD) equipment (Stage 1) and the generation of the film for coating (Figure 3).

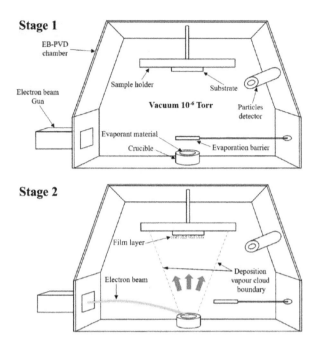

Figure 3. Schematic of electron beam physical vapor deposition (EB-PVD) equipment (Stage 1) and the generation of the film for coating [43].

4. Ultra-High Temperature Ceramics

Over time there has been new materials and modifications of the materials in the aerospace field have been developed. A new generation of aero vehicles are based on the incorporation of components that are composed from a special class of materials known as ultra-high temperature ceramics. These kind of materials are used as thermal protection and in the engine parts of the space vehicles, and in fact ultra-high temperature ceramics can be also used in critical applications on the ground where is a need of resistance to high temperature [44].

Ultra-high temperature ceramics appear in the periodic table in the groups IVB and VB transition metals, and are based especially on carbides along with nitrides and borides. These ceramics exhibit a superior combination of properties characterized by high melting points together with mechanical properties. In this sense, the use of ultra-high temperature ceramics in extreme environments make them excellent potential candidate for these applications [13].

Extreme applications require the use of materials that are not susceptible to oxidation attack in particular, and by using single-phase materials excluding secondary phases materials is not enough. The single phase materials own all the undesired properties for the use in extreme environment such as low thermal shock resistance, low fracture toughness which make these kinds of material unacceptable for aero vehicle applications and also for engineering parts of the vehicles. To erase all the possibilities of failure, the best way is to use a combination of at least two secondary phase of ultrahigh temperature ceramics. One of the most used composites contain silicon carbides (SiC) or other ceramics that involve silicon in different microstructures such as particles, whiskers or fibers. By using composite materials with the required special properties for aerospace application a better thermal shock resistance will be displayed in aggressive environments [45,46].

Ultra-high temperature applications proposed after years of testing and research, and most used with high potential materials in extreme environments, are fundamentally substances such as C (carbon), Ta (tantalum), W (wolfram), Os (osmium), Re (rhenium) and non-oxide compounds such as monocarbides, diborides and mononitrides of transition metals of IVB and VB groups in the periodic table, highlighted as Ti (titanium), Hf (hafnium), Zr (zirconium), Nb (niobium), and Ta (tantalum) [47].

Research interest in aluminum matrix composites has also increased in the last few years, referring to aerospace industries based on the properties of these, such as low density and high strength. From the various types of materials, Al_2O_3 is the most usual ceramic, forming a composite matrix by reinforcing with others materials [48,49]. Alumina have constantly been considered proper for aerospace applications both at ambient and at elevated temperatures. Even if the Al_2O_3 possess polymorphs character, the corundum α-Al_2O_3 is found to be the most suitable form for applications which include a medium temperature. However, oxide ceramics are ideal candidates when speaking about the high-temperature applications due to the fact that these ceramics possess proper behavior in oxidative environments and characteristic high melting point, but especially in combination with a material that supports these properties [50].

In this sense, a material which provides a better view in the aerospace application by forming a composite matrix with Al_2O_3 is tantalum carbide TaC. Its melting point is 3997 °C and it possesses the greatest chemical stability among other carbide [51]. Moreover, the properties of the TaC such as low thermal expansion and high electrical conductivity stimulate the use of this attractive candidate to establish a composite material with aluminum matrix. The literature data provide information about the difficulty developing a composite material by reinforcing TaC particles, but at the smallest possible size. Also the distribution of these particles represents an issue in the development process, because of the fact that the normal distribution of the smallest TaC particles in the alumina matrix is very hard to obtain. The agglomeration process occurs when it is desire to distribute the smallest particles in alumina matrix [52]. Off all the reinforcement particles in the matrix processes, the powder metallurgy process supports in the best way the uniform distribution of the particles in the matrix, however, there are also some problems with the agglomeration mechanism in this process. In this case, when the particles present agglomeration and the composite materials are assesed for sintering, there are possibilities to appear and to retain porosities, which can lead to unappropriated mechanical properties [53,54]. From the sintering method point of view, a spark plasma sintering process based on using aluminum matrix composites has proper behavior when developing adequate dense composites which also possess suitable mechanical properties, in comparison to conventional sintering methods [55,56].

As a basic idea, from the chemical point of view, all ultra-high temperature ceramics are compounds of carbon, boron, or nitrogen in combination with at least one of the early transition metals of IVB and VB in the periodic table. The binary compounds (transition metals and carbon, boron or nitrogen) finally present strong covalent bonds leading to properties of a composite material such as high melting temperature, high stiffness and high hardness. Moreover, all of the characteristics of the ultra-high temperature ceramics are increased compared to oxide ceramics. Due to the fact that ultra-high temperature ceramics involve the action of a mix of ceramics and metals, the final features make the materials suitable and attractive for extreme temperatures and other aggressive conditions of the application environment highlghting capabilities that are beyond other materials [57].

5. Ceramic Matrix Composites

Aerospace engineering includes an important part which is based on the choice of the materials for aero vehicles components. The requirements for a material vary simultaneously and in direct correlation with the specific component that possesses a suitable property for the aerospace industry. Some particular behaviors are being in consideration in materials selections when the design of a vehicles is desired. Each component is analyzed for design requirements which consist in manufacturability, loading conditions, maintainability and geometric limits. Aircraft engines are a point of interest in engineering this component. The most important aspects are the weight reduction and thrust improvement, which mandatory implicate materials with superior properties. The engine materials should present some specific features such as low densities which leads to weight reduction, and it is very important to possess essential mechanical properties under high-temperature conditions and an aggressive oxidative environment. Speaking about the design of an aircraft turbine engines, two divisions are described. The cold sections consist of the compressor, fan and casting, and the hot sections consisting of chamber, combustion and turbine. The category of cold and hot suggest that the sections present different temperatures, which affect the material selection where temperature is a crucial condition for aircraft engine materials. Corrosion resistant and high specific strength materials are suitable for use in the cold section. Composites that include titanium or aluminum and polymers are optimal materials for the cold section. The temperature that is reached in this section is usually in the range of 500–600 °C. On the other hand, for the hot section the materials should present high temperature resistance, hot corrosion resistance and high specific strength. In this section, the temperature is usually between 1400–1500 °C. Titanium composite in this section can not be used, in this case the suitable materials are nickel superalloys, due to their significant high temperature resistance strength [3,7,10,58,59].

The use of composite materials in aerospace vehicle engineering is about more than putting together the individual properties and increasing the final composite material characteristics and behavior. By means of using composite materials, the weight is reduced and the assembly is less complex. Moreover, the use of composite materials involves reducing fuel burn which is a major problem, and also reducing greenhouse gas emissions. Two methods can help to accomplish reduction in fuel burn. The first is about rediching the weight of gas turbine engines, and the second is about raising the thermal performance of the engines. As a matter of fact, composite materials are involved in both situations [2].

Even if the developing stages of composite materials compared to the developing stages of metal production seem to be identical, at the final stage the properties will be specific and beyond the classic metal product and manipulate the design procedures for composite systems. During the process of designing a composite material, at each step various options are available, making the design process persistent and interactive. The design of a component part from an aero vehicle, such as an airframe or a wing involve a considerable number of design variables. These variables need to accomplish various constraints from particular disciplines and also diverse targets have to be performed. Relevant models are used to correlate the constraints and targets to the design variables. At this point in time, aircraft designers possess the ability to use new techniques consisting in multidisciplinary design optimization. Due to the reason that high-performance computational tools are now available, changes and modifications can be correlated at every step of the design process under desired conditions [10,60,61].

The very varied options accessible in the nature and category of matrix and reinforcing components generate composite materials with a broad variety of pattern and characteristics which are an interdependent association of the particular constituent features [1,62].

5.1. Carbon–Carbon Composites

Carbon–carbon composite materials are part of a category of materials that are called advanced composite materials, due to their properties. A large variety of shapes are characteristic to this type of materials starting from one-dimensional to n-dimensional (usually $n = 1,2$ or maximum 3), conditioned by the raw utilized material. By taking into account this benefit, the performance of the materials can be customized in direct contact with the applications. The first use of carbon–carbon composites was in the aerospace domain of applications; at the present time, this type of composites possesses various properties with applications in numerous sectors that brings them to the fore of research into ceramic composite materials [59,63].

For aerospace applications, carbon-based ceramic composites possess attractive properties, such as remarkable thermal stability and also low weight, making them the most favorable materials. Carbon fibers and the carbon matrix are basically components of engineered carbon–carbon composite materials, occasionally improved with different components. One attractive characteristic is the selection of the constituent materials and fiber orientations, which highlight the possibility to manage the properties of the final carbon–carbon composites. Generally, carbon–carbon materials and components are created at the same time, so that the final composite properties can be directed to increase the component capabilities. Resistance to oxidation at high temperatures, fracture toughness, strength and stiffness are principal characteristics of this composite carbon materials [59,64].

This blend of characteristics, leads to their use as preferred materials for manufacturing numerous aero vehicles components parts such as landing gear door, flaps, ailerons and others. Still, the deficiency of stability above 500°C in aggressive environments has placed them in the category of materials that require enhancing. Because of this major drawback, only for short duration can they be used in a harsh environment. However, these composites can endure very high heat fluxes, but only for limited durations, which makes them appropriate for parts of the vehicles that not require continuous withstand for long durations such as re-entry nose tips. Furthermore, the carbon–carbon composites can be improved by extending the application duration and multiple consecutive use. There are some methods to improve the oxidation resistance such as coatings with a material exhibiting oxidation resistance. The second method is to enhance the composite matrix by supplementing with a third phase or to modify the carbon matrix to carbides such as silicon carbide (SiC). By improving the oxidation resistance with the addition of Si, carbon fibre-reinforced SiC matrix composites, are termed C/SiC composites. The oxidation and erosion resistance is enhanced due to the properties of the C/SiC composites. Additionally, the C/SiC composites can be used for lightweight and harsh applications, due to the fact that the density of the carbon is below the density of numerous metallic materials [65–67].

5.2. Hafnium Carbide (HfC) Composites

Pointing to one of the most important properties in the aerospace applications, hafnium carbide (HfC) present the maximal melting point (~3950 °C) among the transition metal carbides. Another attractive feature is low vapour pressure, good ablation resistance and chemical inertness [68,69].

Some recent publications reveal a new experience by introducing HfC compounds towards carbon–carbon composites. Wang et al. described the possibility of obtaining a hafnium carbide coating for carbon–carbon composite substrate by using the chemical vapor deposition method [70], and another coating for carbon–carbon composites by co-deposition of hafnium (tantalum) carbon using the same chemical vapor deposition technique [71]. A different method was reported by Li et al. where the deposition of hafnium carbide on the carbon–carbon composites was possible by immersing the carbon materials in a hafnium oxychloride aqueous solution [72]. To offer protection for carbon–carbon composites, hafnium and silicon carbide multilayers were deposited under low-pressure chemical vapour deposition as coatings [73].

The high environmental temperature of aero applications has significant action upon the materials. Some tests were performed to evaluate the strengths of HfC ceramics at different temperature. In this sense, from room temperature to up to ~ 2200 °C a strength of approximatively 350 MPa was recorded, which declines with the increase of the temperature. At 2200 °C plastic deformation appeared, as a result of grain-boundary sliding. This test highlights the essential role of grain boundaries, because in HfC ceramics with smaller grain size, the decline was more considerable [14,74].

5.3. Carbon/Silicon Carbide (C/SiC) Composites

Among the ceramic materials, silicon carbide (SiC) is placed as a first choice when a high-temperature environment is present. This material is used especially for structural components of aerospace vehicles such as transportation and nuclear areas, due to the fact that SiC possesses significant thermal conductivity, remarkable specific strength and superior tribology behavior at raised temperatures. Like any other material, it also has properties that do not meet the necessary conditions, such as low fracture resistance which limits in some cases the utilization of it in applications of interest. In this sense, given the subject discussed above (see the 5.1 carbon–carbon composites subsection), the carbon fiber-reinforced silicon carbine ceramic matrix composite materials are seeming to fulfill the requirements for high temperature applications. The fracture resistance is upgraded, and also the strength is increased with the supplement of high strength fibers [14,75].

The addition of carbon fiber in the silicon carbide ceramic matrix, increase the final composite material characteristics, highlighting noticeable material properties such as high strength, superior thermal shock resistance surrounded by good oxidation resistance, low density and a specific feature of managing and maintaining the mechanical properties even if the applications are under elevated temperatures. All of these properties determine the material to use in extreme conditions including oxidizing atmosphere, as manufacturing materials for components of aero vehicles. It is well known that by obtaining a composite material the final structure will be improved together with the characteristics. A better oxidation resistance is manifested in carbon–carbon SiC composites, compared to individual materials. Due to the fact that on the surface of the substrate, the silica offers a protective layer, the behavior under oxidizing atmosphere of the composite is improved. Moreover, light weight is a property that is more accentuated in the composite material compared to the individual one, and also the economic part is ameliorated because the carbon matrix is easier to develop than silicon carbide matrix [76,77].

Another way in maintaining a suitable activity of the materials is to incorporate silicon carbide fiber in the silicon carbide matrix. The components of the aero vehicles like gas turbine engines offer the best options when its manufacture includes the utilization of silicon carbide fiber reinforced silicon carbide. To evaluate the stress rupture properties, the high temperature composite materials which consist of, basically, SiC, were investigated under 100 MPa as a moderate stress level. The results of SiC–SiC composite showed it to be able to operate at temperature beyond 1315°C. Carbon–carbon composite and carbon fiber-silicon carbide exhibited advantageous and preferred stress rupture properties at elevated temperature. Moreover, SiC–SiC composites, results with an advanced in the durability of the resistance at oxidation atmosphere, compared to carbon–carbon and carbon fiber–silicon carbide [2,77,78].

5.4. Zirconium Carbide/Silicon Carbide (ZrC/SiC) Composites

Ultra-high temperature applications include the utilization of zirconium carbide (ZrC), as one of the best options due to the fact that the exceptional properties performed with suitable activity of the ZrC under harsh conditions. At high temperatures, the ZrC composite generate a refractory oxide scale which is another advantage when it comes to oxidation [14].

Transition metal carbides have considerable properties, being in the focus of the researchers for manufacturing aero vehicles components with required properties such as high melting point, high hardness and chemical stability, which are characteristics for zirconium carbide. Moreover, ZrC possesses features like impressive hardness which is mandatory for many cutting tools or/and abrasive industries. Numerous papers, place the zirconium carbide as a suitable material for elevated temperature applications due to high corrosion resistance [79].

Rocket engine nozzles and hypersonic vehicles components during their applications, are in direct contact with aggressive environment. For this reason, the materials used in manufacturing these components have to present firstly a high melting temperature. Zirconium carbide ceramic is a promising material in this way. However, there are in this case some limits of the materials such as poor sinterability because of the fact that ZrC possesses a reduced self-diffusion coefficient and strong covalent bonding. By a poor sinterability is understood the fact that it is more complicated to reach a completely dense composites without a support from sintering additives. Because of the fact that ZrC ceramic composites may have limits in terms of their full activity under special conditions having poor thermal shock resistance and low fracture toughness, by adding SiC into ZrC the properties may be improved. The mechanical properties and oxidation resistance of ZrC are clearly enhanced after the incorporation of SiC, leading to the generation of a melted SiO_2 layer at high temperature and also to the discrepancy of thermal expansion coefficient among ZrC and SiC [80,81].

5.5. Zirconium Diboride/Silicon Carbide (ZrB₂/SiC) Composites

Another excellent candidate for applications at high temperatures is zirconium diboride (ZrB₂). This diboride is similar to zirconium carbide having attractive properties such as low density, high melting point, remarkable chemical inertness, and it is used as thermal protection barrier on the substrate of aerospace vehicles. However, in this case too the individual zirconium diboride did not reach all the required conditions because has some inconvenience such as low fracture toughness and low oxidation resistance. Moreover, it is a similarity between zirconium diboride and zirconium carbine when it comes to manufacturing completely dense samples. This process is limited by the undesirable characteristics of ZrB₂ such as strong covalent bond and reduced self-diffusion coefficient, and because of the impurities on the surface of substrate materials. Also in this case, the addition of SiC brings a benefic difference, changing the properties and increasing the mechanical properties, the thermal and oxidation resistance. In the same time, the exaggerated grain growth of zirconium diboride is avoided with the addition on silicon carbide [82].

5.6. Aluminum Oxide/Zirconium Dioxide (Al₂O₃/ZrO₂) Composites and Zirconium Dioxide/Silicon Dioxide (ZrO₂/SiO₂) Aerogels

An attractive oxide ceramic candidate for aerospace application is ZrO₂. The characteristics of this ceramic are represented by a very high melting point at a temperature of ~2700 °C, promising mechanical properties and stability in oxidative conditions [83]. The use of ZrO₂ as thermal barrier coatings has been a favorable choice for several years and even in the present time is still recommended. A difference between the traditional ZrO₂ coatings and nanostructured ZrO₂ coatings may have a large influence on the properties of the final material. The research data reveal that the nano structure of ZrO₂ has improved the properties of the material with higher toughness, lower thermal conductivity, higher bonding strength, and higher wear resistance [84]. Over the years, there has been interest regarding the use of zirconia as fully stabilized zirconia, partially stabilized zirconia and tetragonal zirconia polygonal [83].

To obtain a better performance in a special high temperature environment, the involvement of the scientific community has focused on the use of the Al₂O₃/ZrO₂ eutectic ceramic as a thermal barrier coating. This composite is obtained as a melt growth composite material, meaning of a eutectic reaction between the matrix phase and the second phase which occurs when oxide melt is solidifying. The final composite is made by Al₂O₃ and ZrO₂ being developed at the same time and together, which can possess a micro or nano structure. Due to the fact that, Al₂O₃/ZrO₂ eutectic ceramics possess excellent behavior at high temperatures such as oxidation resistance and high temperature strength became a new generation of materials used as thermal barrier coatings and may even overcome the properties of SiC at high temperatures [85–87].

Another new generation of composite materials is based on ZrO₂/SiO₂ aerogel. Silica aerogel is produced using nanoparticles as aggregate and form a three-dimensional structure by interconnecting each nanoparticle between them. However, the silica aerogels can resist only at a temperature below 600 °C if there is a need for a long working conditions [88]. In this sense, the composite of ZrO₂/SiO₂ aerogel have improved results under high temperature due to ultra-low thermal conductivity. Additionally, this composite presents more outstanding properties such as low density and a better heat insulation leading to a thermal stability at a temperature of 1000 °C [89,90]. However, there are some issues speaking about the mechanical strength and fragility of aerogels. Some augmented methods have been used in order to obtain a better result, such as including ceramic fibers or functional polymers (epoxy, polyurethane and polyethylene) by cross-linking them with the aerogels. As a comparison between organic or inorganic reinforcement, it has been demonstrated that the inorganic reinforcement is obvious and confirmed as having potential due to its supportive stability behavior at high temperature [91].

Due to the fact that the materials used in aerospace applications must have a suitable behavior at high temperatures, below are some properties that are taken into account when choosing these materials, based on those discussed in this review (Table 1).

Table 1. Melting temperature and mechanical properties of various materials used as thermal barrier coatings.

Material	Melting Temperature (°C)	Hardness (GPa)	Young's Modulus (GPa)
TaC	3427 [92]	20.6 ± 1.2 [92]	579 ± 20 [92]
HfC	3890 [93]	31.5 ± 1.3 [92]	552 ± 15 [92]
SiC	2730 [93]	25.5 [94]	450 [95]
ZrC	3530 [96]	31.3 ± 1.4 [97]	507 ± 16 [97]
ZrB_2	3245 [98]	21 [99]	490 [99]
ZrO_2	2699 [100]	11.77 [101]	171 [102]
Al_2O_3	2071 [100]	21.58 [101]	380 [102]

6. Conclusions

Thermal-barrier coatings obtained by using the electron beam physical vapour deposition technique represent a way to improve the behavior of aero vehicles in high-temperature applications. The coatings have a significant role in assuring a barrier which acts in high-temperature environments. The performance of the thermal barrier is enhanced thanks to various ceramic coats deposited on the substrate.

In order to choose suitable materials for aerospace applications, it has been proven that ceramic materials have properties that are mandatory for such applications. Ceramic materials possess low thermal conductivities and, for this reason, it is desirable for the manufacturing of components for aero vehicles to contain a large proportion of ceramic composites. The performance of the engine is increased, the temperature of the metal substrate is reduced and managed, and the lifetimes of the engines, hot sections and turbines are prolonged only by covering with thermal-barrier coatings.

The selection of materials for acting as a thermal barrier is based on the evaluation of the materials. Basic requirements are mandatory such as high melting point, low thermal conductivity, chemical inertness, good adherence to the metallic substrate, high-temperature resistance, high strength and resistance to oxidation at high temperatures. However, until now, no single material can achieve all of these mandatory conditions.

Author Contributions: Conceptualization, B.S.V.; Investigation, A.C.B., I.A.N., A.I.N.; Data curation, V.A.S.; Writing–review and editing, B.S.V., A.C.B., I.A.N., A.I.N., V.A.S.; All authors have read and agreed to the published version of the manuscript.

Funding: This research was funded by Romanian Ministry for Research and Innovation, RDI Program for Space Technology and Advanced Research - STAR, grant number 528.

Conflicts of Interest: The authors declare no conflict of interest.

References

1. Rana, S.; Fangueiro, R. *Advanced Composite Materials for Aerospace Engineering: Processing, Properties and Applications*; Woodhead Publishing: Cambridge, UK, 2016.
2. Misra, A. Composite Materials for Aerospace Propulsion Related to Air and Space Transportation. In *Lightweight Composite Structures in Transport*; Elsevier: Amsterdam, The Netherlands, 2016; pp. 305–327.
3. Zhang, X.; Chen, Y.; Hu, J. Recent advances in the development of aerospace materials. *Prog. Aerosp. Sci.* **2018**, *97*, 22–34. [CrossRef]
4. Zhu, J.-H.; Zhang, W.-H.; Xia, L. Topology optimization in aircraft and aerospace structures design. *Arch. Comput. Methods Eng.* **2016**, *23*, 595–622. [CrossRef]
5. Huang, R.; Riddle, M.; Graziano, D.; Warren, J.; Das, S.; Nimbalkar, S.; Cresko, J.; Masanet, E. Energy and emissions saving potential of additive manufacturing: The case of lightweight aircraft components. *J. Clean. Prod.* **2016**, *135*, 1559–1570. [CrossRef]

6. Rana, S.; Fangueiro, R. Advanced Composites in Aerospace Engineering. In *Advanced Composite Materials for Aerospace Engineering*; Elsevier: Amsterdam, The Netherlands, 2016; pp. 1–15.

7. McIlhagger, A.; Archer, E.; McIlhagger, R. Manufacturing Processes for Composite Materials and Components for Aerospace Applications. In *Polymer Composites in the Aerospace Industry*; Elsevier: Amsterdam, The Netherlands, 2015; pp. 53–75.

8. Toozandehjani, M.; Kamarudin, N.; Dashtizadeh, Z.; Lim, E.Y.; Gomes, A.; Gomes, C. Conventional and advanced composites in aerospace industry: Technologies revisited. *Am. J. Aerosp. Eng.* **2018**, *5*, 9–15. [CrossRef]

9. Park, S.-J.; Seo, M.-K. (Eds.) Chapter 7—Types of Composites. In *Interface Science and Technology*; Elsevier: Amsterdam, The Netherlands, 2011; Volume 18, pp. 501–629.

10. Gopal, K. Product Design for Advanced Composite Materials in Aerospace Engineering. In *Advanced Composite Materials for Aerospace Engineering*; Elsevier: Amsterdam, The Netherlands, 2016; pp. 413–428.

11. Falaschetti, M.P.; Rans, C.; Troiani, E. On the application of metal foils for improving the impact damage tolerance of composite materials. *Compos. Part B Eng.* **2017**, *112*, 224–234.

12. Jawaid, M.; Thariq, M. *Sustainable Composites for Aerospace Applications*; Woodhead Publishing: Cambridge, UK, 2018.

13. Jin, X.; Fan, X.; Lu, C.; Wang, T. Advances in oxidation and ablation resistance of high and ultra-high temperature ceramics modified or coated carbon/carbon composites. *J. Eur. Ceram. Soc.* **2018**, *38*, 1–28. [CrossRef]

14. Tang, S.; Hu, C. Design, preparation and properties of carbon fiber reinforced ultra-high temperature ceramic composites for aerospace applications: A review. *J. Mater. Sci. Technol.* **2017**, *33*, 117–130. [CrossRef]

15. Wang, Y.; Chen, Z.; Yu, S. Ablation behavior and mechanism analysis of C/SiC composites. *J. Mater. Res. Technol.* **2016**, *5*, 170–182. [CrossRef]

16. Marsh, G. Composites in commercial jets. *Reinf. Plast.* **2015**, *59*, 190–193. [CrossRef]

17. Asl, M.S.; Nayebi, B.; Ahmadi, Z.; Zamharir, M.J.; Shokouhimehr, M. Effects of carbon additives on the properties of ZrB_2–based composites: A review. *Ceram. Int.* **2018**, *44*, 7334–7348. [CrossRef]

18. Lino Alves, F.J.; Baptista, A.M.; Marques, A.T. 3—Metal and Ceramic Matrix Composites in Aerospace Engineering. In *Advanced Composite Materials for Aerospace Engineering*; Rana, S., Fangueiro, R., Eds.; Woodhead Publishing: Cambridge, UK, 2016; pp. 59–99. [CrossRef]

19. Miracle, D. Metal matrix composites—From science to technological significance. *Compos. Sci. Technol.* **2005**, *65*, 2526–2540. [CrossRef]

20. Doorbar, P.J.; Kyle-Henney, S. 4.19 Development of Continuously-Reinforced Metal Matrix Composites for Aerospace Applications. *Compr. Compos. Mater.* **2018**, *4*, 439–463.

21. Paul, A.; Jayaseelan, D.D.; Venugopal, S.; Zapata-Solvas, E.; Binner, J.; Vaidhyanathan, B.; Heaton, A.; Brown, P.M.; Lee, W. *UHTC Composites for Hypersonic Applications*; The American Ceramic Society Bulletin: Westerville, OH, USA, 2012.

22. Neuman, E.W.; Hilmas, G.E.; Fahrenholtz, W.G. Ultra-high temperature mechanical properties of a zirconium diboride–zirconium carbide ceramic. *J. Am. Ceram. Soc.* **2016**, *99*, 597–603. [CrossRef]

23. Krishnarao, R.; Alam, M.Z.; Das, D. In-situ formation of SiC, ZrB_2-SiC and ZrB_2-SiC-B4C-YAG coatings for high temperature oxidation protection of C/C composites. *Corros. Sci.* **2018**, *141*, 72–80. [CrossRef]

24. Binner, J.; Porter, M.; Baker, B.; Zou, J.; Venkatachalam, V.; Diaz, V.R.; D'Angio, A.; Ramanujam, P.; Zhang, T.; Murthy, T. Selection, processing, properties and applications of ultra-high temperature ceramic matrix composites, UHTCMCs—A review. *Int. Mater. Rev.* **2019**. [CrossRef]

25. Sadowski, T.; Golewski, P. *Loadings in Thermal Barrier Coatings of Jet Engine Turbine Blades: An Experimental Research and Numerical Modeling*; Springer: New York, NY, USA, 2016.

26. Sahith, M.S.; Giridhara, G.; Kumar, R.S. Development and analysis of thermal barrier coatings on gas turbine blades—A Review. *Mater. Today Proc.* **2018**, *5*, 2746–2751. [CrossRef]

27. Northam, M.; Rossmann, L.; Sarley, B.; Harder, B.; Park, J.-S.; Kenesei, P.; Almer, J.; Viswanathan, V.; Raghavan, S. Comparison of Electron-Beam Physical Vapor Deposition and Plasma-Spray Physical Vapor Deposition Thermal Barrier Coating Properties Using Synchrotron X-ray Diffraction. In Proceedings of the ASME Turbo Expo 2019: Turbomachinery Technical Conference and Exposition, Phoenix, AZ, USA, 17–21 June 2019.

28. Karaoglanli, A.C.; Doleker, K.M.; Ozgurluk, Y. State of the Art Thermal Barrier Coating (TBC) Materials and TBC Failure Mechanisms. In *Properties and Characterization of Modern Materials*; Springer: New York, NY, USA, 2017; pp. 441–452.

29. Xu, H.; Guo, H. *Thermal Barrier Coatings*; Elsevier: Amsterdam, The Netherlands, 2011.

30. Schmitt, M.P. Advanced Thermal Barrier Coating Materials and Design Architectures for Improved Durability. Ph.D. Dissertation, The Pennsylvania State University, University Park, PA, USA, 2016.

31. Ghosh, S. Thermal barrier ceramic coatings—A review. *Adv. Ceram. Process.* **2015**, 111–138.

32. Kumar, V.; Balasubramanian, K. Progress update on failure mechanisms of advanced thermal barrier coatings: A review. *Prog. Org. Coat.* **2016**, *90*, 54–82. [CrossRef]

33. Nandi, A.; Ghosh, S. Advanced Multi-layered Thermal Barrier Coatings—An Overview. *J. Mater. Sci. Res. Rev.* **2019**, *30*, 1–17.

34. Kumar, V.; Kandasubramanian, B. Processing and design methodologies for advanced and novel thermal barrier coatings for engineering applications. *Particuology* **2016**, *27*, 1–28. [CrossRef]

35. Zhang, B.; Song, W.; Wei, L.; Xiu, Y.; Xu, H.; Dingwell, D.B.; Guo, H. Novel thermal barrier coatings repel and resist molten silicate deposits. *Scr. Mater.* **2019**, *163*, 71–76. [CrossRef]

36. Bernard, B.; Quet, A.; Bianchi, L.; Joulia, A.; Malié, A.; Schick, V.; Rémy, B. Thermal insulation properties of YSZ coatings: Suspension plasma spraying (SPS) versus electron beam physical vapor deposition (EB-PVD) and atmospheric plasma spraying (APS). *Surf. Coat. Technol.* **2017**, *318*, 122–128. [CrossRef]

37. Karaoglanli, A.C.; Doleker, K.M.; Demirel, B.; Turk, A.; Varol, R. Effect of shot peening on the oxidation behavior of thermal barrier coatings. *Appl. Surf. Sci.* **2015**, *354*, 314–322. [CrossRef]

38. Bose, S. *High Temperature Coatings*; Elsevier: Oxford, UK, 2017.

39. Fan, W.; Bai, Y. Review of suspension and solution precursor plasma sprayed thermal barrier coatings. *Ceram. Int.* **2016**, *42*, 14299–14312. [CrossRef]

40. Soboyejo, W.O.; Obayemi, J.; Annan, E.; Ampaw, E.; Daniels, L.; Rahbar, N. Review of High Temperature Ceramics for Aerospace Applications. *Adv. Mater. Res.* **2016**, *1132*, 385. [CrossRef]

41. Bose, S. (Ed.) Chapter 7—Thermal Barrier Coatings (TBCs). In *High Temperature Coatings*, 2nd ed.; Butterworth-Heinemann: Oxford, UK, 2018; pp. 199–299. [CrossRef]

42. Singh, J.; Wolfe, D.E. Review Nano and macro-structured component fabrication by electron beam-physical vapor deposition (EB-PVD). *J. Mater. Sci.* **2005**, *40*, 1–26. [CrossRef]

43. Ali, N.; Teixeira, J.A.; Addali, A.; Saeed, M.; Al-Zubi, F.; Sedaghat, A.; Bahzad, H. Deposition of Stainless Steel Thin Films: An Electron Beam Physical Vapour Deposition Approach. *Materials* **2019**, *12*, 571. [CrossRef]

44. Zhang, G.-J.; Ni, D.-W.; Zou, J.; Liu, H.-T.; Wu, W.-W.; Liu, J.-X.; Suzuki, T.S.; Sakka, Y. Inherent anisotropy in transition metal diborides and microstructure/property tailoring in ultra-high temperature ceramics—A review. *J. Eur. Ceram. Soc.* **2018**, *38*, 371–389. [CrossRef]

45. Savino, R.; Criscuolo, L.; Di Martino, G.D.; Mungiguerra, S. Aero-thermo-chemical characterization of ultra-high-temperature ceramics for aerospace applications. *J. Eur. Ceram. Soc.* **2018**, *38*, 2937–2953. [CrossRef]

46. Behera, M.P.; Dougherty, T.; Singamneni, S. Conventional and Additive Manufacturing with Metal Matrix Composites: A Perspective. *Procedia Manuf.* **2019**, *30*, 159–166. [CrossRef]

47. Huang, Z.; Wu, L. Ultrahigh-Temperature Ceramics (UHTCs) Systems. In *Phase Equilibria Diagrams of High Temperature Non-Oxide Ceramics*; Springer: New York, NY, USA, 2018; pp. 103–162.

48. Ghasali, E.; Yazdani-Rad, R.; Asadian, K.; Ebadzadeh, T. Production of Al-SiC-TiC hybrid composites using pure and 1056 aluminum powders prepared through microwave and conventional heating methods. *J. Alloys Compd.* **2017**, *690*, 512–518. [CrossRef]

49. Koli, D.K.; Agnihotri, G.; Purohit, R. A review on properties, behaviour and processing methods for Al-nano Al_2O_3 composites. *Procedia Mater. Sci.* **2014**, *6*, 567–589. [CrossRef]

50. Balani, K.; Harimkar, S.P.; Keshri, A.; Chen, Y.; Dahotre, N.B.; Agarwal, A. Multiscale wear of plasma-sprayed carbon-nanotube-reinforced aluminum oxide nanocomposite coating. *Acta Mater.* **2008**, *56*, 5984–5994. [CrossRef]

51. Hackett, K.; Verhoef, S.; Cutler, R.A.; Shetty, D.K. Phase constitution and mechanical properties of carbides in the Ta–C system. *J. Am. Ceram. Soc.* **2009**, *92*, 2404–2407. [CrossRef]

52. Shirvanimoghaddam, K.; Khayyam, H.; Abdizadeh, H.; Akbari, M.K.; Pakseresht, A.; Ghasali, E.; Naebe, M. Boron carbide reinforced aluminium matrix composite: Physical, mechanical characterization and mathematical modelling. *Mater. Sci. Eng. A* **2016**, *658*, 135–149. [CrossRef]

53. Ghasali, E.; Pakseresht, A.H.; Alizadeh, M.; Shirvanimoghaddam, K.; Ebadzadeh, T. Vanadium carbide reinforced aluminum matrix composite prepared by conventional, microwave and spark plasma sintering. *J. Alloys Compd.* **2016**, *688*, 527–533. [CrossRef]

54. Ghasali, E.; Alizadeh, M.; Ebadzadeh, T.; hossein Pakseresht, A.; Rahbari, A. Investigation on microstructural and mechanical properties of B4C–aluminum matrix composites prepared by microwave sintering. *J. Mater. Res. Technol.* **2015**, *4*, 411–415. [CrossRef]

55. Ghasali, E.; Pakseresht, A.; Rahbari, A.; Eslami-Shahed, H.; Alizadeh, M.; Ebadzadeh, T. Mechanical properties and microstructure characterization of spark plasma and conventional sintering of Al–SiC–TiC composites. *J. Alloys Compd.* **2016**, *666*, 366–371. [CrossRef]

56. Ghasali, E.; Shirvanimoghaddam, K.; Pakseresht, A.H.; Alizadeh, M.; Ebadzadeh, T. Evaluation of microstructure and mechanical properties of Al-TaC composites prepared by spark plasma sintering process. *J. Alloys Compd.* **2017**, *705*, 283–289. [CrossRef]

57. Fahrenholtz, W.G.; Hilmas, G.E. Ultra-high temperature ceramics: Materials for extreme environments. *Scr. Mater.* **2017**, *129*, 94–99. [CrossRef]

58. Ancona, E.; Kezerashvili, R.Y. Temperature restrictions for materials used in aerospace industry for the near-Sun orbits. *Acta Astronaut.* **2017**, *140*, 565–569. [CrossRef]

59. Scarponi, C. Carbon–Carbon Composites in Aerospace Engineering. In *Advanced Composite Materials for Aerospace Engineering*; Elsevier: Amsterdam, The Netherlands, 2016; pp. 385–412.

60. Zhu, L.; Li, N.; Childs, P. Light-weighting in aerospace component and system design. *Propuls. Power Res.* **2018**, *7*, 103–119. [CrossRef]

61. Hendler, M.; Extra, S.; Lockan, M.; Bestle, D.; Flassig, P. Compressor Design in the Context of Holistic Aero Engine Design. In Proceedings of the 18th AIAA/ISSMO Multidisciplinary Analysis and Optimization Conference, Atlanta, GA, USA, 25–29 June 2018; p. 3334.

62. Jahan, A.; Edwards, K.L.; Bahraminasab, M. *Multi-Criteria Decision Analysis for Supporting the Selection of Engineering Materials in Product Design*; Butterworth-Heinemann: Oxford, UK, 2016.

63. Zhang, Y.; Wang, H.; Li, T.; Fu, Y.; Ren, J. Ultra-high temperature ceramic coating for carbon/carbon composites against ablation above 2000 K. *Ceram. Int.* **2018**, *44*, 3056–3063. [CrossRef]

64. Albano, M.; Delfini, A.; Pastore, R.; Micheli, D.; Marchetti, M. A new technology for production of high thickness carbon/carbon composites for launchers application. *Acta Astronaut.* **2016**, *128*, 277–285. [CrossRef]

65. Prasad, N.E.; Wanhill, R.J. *Aerospace Materials and Material Technologies*; Springer: New York, NY, USA, 2017; Volume 3.

66. Kumar, S.; Shekar, K.C.; Jana, B.; Manocha, L.; Prasad, N.E. C/C and C/SiC Composites for Aerospace Applications. In *Aerospace Materials and Material Technologies*; Springer: New York, NY, USA, 2017; pp. 343–369.

67. Wang, J.; Yang, G.; Zhang, F.; Xiong, Y.; Xiong, Q. The preparation and mechanical properties of carbon/carbon (C/C) composite and carbon fiber reinforced silicon carbide (Cf/SiC) composite joint by partial transient liquid phase (PTLP) diffusion bonding process. *Vacuum* **2018**, *158*, 113–116. [CrossRef]

68. Patra, N.; Al Nasiri, N.; Jayaseelan, D.D.; Lee, W.E. Thermal properties of Cf/HfC and Cf/HfC-SiC composites prepared by precursor infiltration and pyrolysis. *J. Eur. Ceram. Soc.* **2018**, *38*, 2297–2303. [CrossRef]

69. Patra, N.; Al Nasiri, N.; Jayaseelan, D.D.; Lee, W.E. Low-temperature solution synthesis of nanosized hafnium carbide using pectin. *Ceram. Int.* **2016**, *42*, 1959–1963. [CrossRef]

70. Wang, Y.-L.; Xiong, X.; Li, G.-D.; Liu, H.-F.; Chen, Z.-K.; Sun, W.; Zhao, X.-J. Ablation behavior of HfC protective coatings for carbon/carbon composites in an oxyacetylene combustion flame. *Corros. Sci.* **2012**, *65*, 549–555. [CrossRef]

71. Wang, Y.-L.; Xiong, X.; Li, G.-D.; Liu, H.-F.; Chen, Z.-K.; Sun, W.; Zhao, X.-J. Preparation and ablation properties of Hf (Ta) C co-deposition coating for carbon/carbon composites. *Corros. Sci.* **2013**, *66*, 177–182. [CrossRef]

72. Li, C.; Li, K.; Li, H.; Ouyang, H.; Zhang, Y.; Guo, L. Mechanical and thermophysical properties of carbon/carbon composites with hafnium carbide. *Ceram. Int.* **2013**, *39*, 6769–6776. [CrossRef]

73. Xue, L.; Su, Z.-A.; Yang, X.; Huang, D.; Yin, T.; Liu, C.; Huang, Q. Microstructure and ablation behavior of C/C–HfC composites prepared by precursor infiltration and pyrolysis. *Corros. Sci.* **2015**, *94*, 165–170. [CrossRef]

74. Vinci, A.; Zoli, L.; Sciti, D.; Watts, J.; Hilmas, G.E.; Fahrenholtz, W.G. Influence of fibre content on the strength of carbon fibre reinforced HfC/SiC composites up to 2100 °C. *J. Eur. Ceram. Soc.* **2019**, *39*, 3594–3603. [CrossRef]

75. Zhu, W.; Fu, H.; Xu, Z.; Liu, R.; Jiang, P.; Shao, X.; Shi, Y.; Yan, C. Fabrication and characterization of carbon fiber reinforced SiC ceramic matrix composites based on 3D printing technology. *J. Eur. Ceram. Soc.* **2018**, *38*, 4604–4613. [CrossRef]

76. Hu, C.; Hong, W.; Xu, X.; Tang, S.; Du, S.; Cheng, H.-M. Sandwich-structured C/C-SiC composites fabricated by electromagnetic-coupling chemical vapor infiltration. *Sci. Rep.* **2017**, *7*, 13120. [CrossRef]

77. Hu, C.; Tang, S.; Pang, S.; Cheng, H.-M. Long-term oxidation behaviors of C/SiC composites with a SiC/UHTC/SiC three-layer coating in a wide temperature range. *Corros. Sci.* **2019**, *147*, 1–8. [CrossRef]

78. Padture, N.P. Advanced structural ceramics in aerospace propulsion. *Nat. Mater.* **2016**, *15*, 804. [CrossRef]

79. Wang, L.; Si, L.; Zhu, Y.; Qian, Y. Solid-state reaction synthesis of ZrC from zirconium oxide at low temperature. *Int. J. Refract. Met. Hard Mater.* **2013**, *38*, 134–136. [CrossRef]

80. Cheng, Y.; Hu, P.; Zhou, S.; Zhang, X.; Han, W. Using macroporous graphene networks to toughen ZrC–SiC ceramic. *J. Eur. Ceram. Soc.* **2018**, *38*, 3752–3758. [CrossRef]

81. Wang, X.F.; Liu, J.C.; Hou, F.; Hu, J.D.; Sun, X.; Zhou, Y.C. Synthesis of ZrC–SiC powders from hybrid liquid precursors with improved oxidation resistance. *J. Am. Ceram. Soc.* **2015**, *98*, 197–204. [CrossRef]

82. Vinci, A.; Zoli, L.; Landi, E.; Sciti, D. Oxidation behaviour of a continuous carbon fibre reinforced ZrB2–SiC composite. *Corros. Sci.* **2017**, *123*, 129–138. [CrossRef]

83. Sengupta, P.; Manna, I. Advanced High-Temperature Structural Materials for Aerospace and Power Sectors: A Critical Review. *Trans. Indian Inst. Met.* **2019**, *72*, 2043–2059. [CrossRef]

84. Cui, Y.-H.; Hu, Z.-C.; Ma, Y.-D.; Yang, Y.; Zhao, C.-C.; Ran, Y.-T.; Gao, P.-Y.; Wang, L.; Dong, Y.-C.; Yan, D.-R. Porous nanostructured ZrO2 coatings prepared by plasma spraying. *Surf. Coat. Technol.* **2019**, *363*, 112–119. [CrossRef]

85. Yan, S.; Wu, D.; Niu, F.; Huang, Y.; Liu, N.; Ma, G. Effect of ultrasonic power on forming quality of nano-sized Al2O3-ZrO2 eutectic ceramic via laser engineered net shaping (LENS). *Ceram. Int.* **2018**, *44*, 1120–1126. [CrossRef]

86. Fu, L.-S.; Wang, Z.; Fu, X.-S.; Chen, G.-Q.; Zhou, W.-L. Microstructure and mechanical properties of Y2O3-doped melt-grown Al2O3-ZrO2 eutectic ceramic. *Mater. Sci. Eng. A* **2017**, *703*, 372–379. [CrossRef]

87. Chen, Y.-D.; Yang, Y.; Chu, Z.-H.; Chen, X.-G.; Wang, L.; Liu, Z.; Dong, Y.-C.; Yan, D.-R.; Zhang, J.-X.; Kang, Z.-L. Microstructure and properties of Al2O3-ZrO2 composite coatings prepared by air plasma spraying. *Appl. Surf. Sci.* **2018**, *431*, 93–100. [CrossRef]

88. Hou, X.; Zhang, R.; Fang, D. An ultralight silica-modified ZrO2–SiO2 aerogel composite with ultra-low thermal conductivity and enhanced mechanical strength. *Scr. Mater.* **2018**, *143*, 113–116. [CrossRef]

89. Hou, X.; Zhang, R.; Wang, B. Novel self-reinforcing ZrO2–SiO2 aerogels with high mechanical strength and ultralow thermal conductivity. *Ceram. Int.* **2018**, *44*, 15440–15445. [CrossRef]

90. Liu, B.; Gao, M. Highly Mixed ZrO2/SiO2 Hybrid Aerogel Deriving from Freely Tangled Weakly Branched Primary Clusters Enables Improved Thermal Stability and Excellent Thermal Insulating Performance. *ACS Appl. Nano Mater.* **2019**. [CrossRef]

91. He, J.; Zhao, H.; Li, X.; Su, D.; Ji, H.; Yu, H.; Hu, Z. Large-scale and ultra-low thermal conductivity of ZrO2 fibrofelt/ZrO2-SiO2 aerogels composites for thermal insulation. *Ceram. Int.* **2018**, *44*, 8742–8748. [CrossRef]

92. Castle, E.; Csanádi, T.; Grasso, S.; Dusza, J.; Reece, M. Processing and properties of high-entropy ultra-high temperature carbides. *Sci. Rep.* **2018**, *8*, 8609. [CrossRef]

93. Yang, X.; Zhao, H.C.; Feng, C. High-temperature protective coatings for C/SiC composites. *J. Asian Ceram. Soc.* **2014**, *2*, 305–309. [CrossRef]

94. Moema, J.S.; Papo, M.J.; Stumpf, W.E.; Slabbert, D. The role of retained austenite on performance of grinding media. *Wear* **2010**, *4*, 5.

95. Auciello, O.; Birrell, J.; Carlisle, J.A.; Gerbi, J.E.; Xiao, X.; Peng, B.; Espinosa, H.D. Materials science and fabrication processes for a new MEMS technology based on ultrananocrystalline diamond thin films. *J. Phys. Condens. Matter* **2004**, *16*, R539. [CrossRef]

96. Gale, T.; Totemeir, W.C. *Smithells Metals Reference Book*; Butterworth-Heinemann: Boston, MA, USA, 2004.

97. Balko, J.; Csanádi, T.; Sedlák, R.; Vojtko, M.; KovalLíková, A.; Koval, K.; Wyzga, P.; Naughton-Duszová, A. Nanoindentation and tribology of VC, NbC and ZrC refractory carbides. *J. Eur. Ceram. Soc.* **2017**, *37*, 4371–4377. [CrossRef]

98. Sonber, J.; Murthy, T.; Sairam, K.; Chakravartty, J. Effect of NdB6 addition on densification and properties of ZrB$_2$. *Ceram. Silikáty* **2016**, *60*, 41–47. [CrossRef]

99. Mitra, R.; Upender, S.; Mallik, M.; Chakraborty, S.; Ray, K.K. Mechanical, Thermal and Oxidation Behaviour of Zirconium Diboride Based Ultra-High Temperature Ceramic Composites. In Proceedings of the 9th ICKEM Key Engineering Materials, Oxford, UK, 31 March–1 April 2019; pp. 55–68.

100. Vinodkumar, T.; Reddy, B.M. *Chapter 3—Catalytic Combustion over Cheaper Metal Oxides in Catalytic Combustion*; Nova Science Publisher: Hauppauge, NY, USA, 2011; pp. 105–140.

101. Alsebaie, A.M. Characterisation of Alumina-Zirconia Composites Produced by Micron-Sized Powders. Master's Thesis, Dublin City University, Dublin, Ireland, 2006.

102. El-Mahallawi, I.S.; Shash, A.Y.; Amer, A.E. Nanoreinforced cast Al-Si alloys with Al$_2$O$_3$, TiO$_2$ and ZrO$_2$ nanoparticles. *Metals* **2015**, *5*, 802–821. [CrossRef]

MDPI

Review

Recent Advances in Magnetite Nanoparticle Functionalization for Nanomedicine

Roxana Cristina Popescu [1,2], Ecaterina Andronescu [1] and Bogdan Stefan Vasile [1,*]

[1] National Research Center for Micro and Nanomaterials, Department of Science and Oxide Materials and Nanomaterials, Politehnica University of Bucharest, 060042 Bucharest, Romania; roxpopescu@yahoo.co.uk (R.C.P.); ecaterina.andronescu@upb.ro (E.A.)

[2] Department of Life and Environmental Physics, "Horia Hulubei" National Institute for Physics and Nuclear Engineering, 077125 Magurele, Romania

* Correspondence: bogdan.vasile@upb.ro; Tel.: +40-727-589-960

Received: 31 October 2019; Accepted: 11 December 2019; Published: 16 December 2019

Abstract: Functionalization of nanomaterials can enhance and modulate their properties and behaviour, enabling characteristics suitable for medical applications. Magnetite (Fe_3O_4) nanoparticles are one of the most popular types of nanomaterials used in this field, and many technologies being already translated in clinical practice. This article makes a summary of the surface modification and functionalization approaches presented lately in the scientific literature for improving or modulating magnetite nanoparticles for their applications in nanomedicine.

Keywords: magnetite nanoparticles; Fe_3O_4; functionalization; surface modification; conjugation; nanomedicine; biocompatibility; clinical translation

1. Introduction

As a preponderance of biological processes begin and take place at molecular level, it is understandable why diagnosis and therapeutic solutions have been sought at the nanoscale. The use of nanoparticles in medicine is determined by the processes occurring at the bio-interface. In this context, manipulation of surface properties is highly important as it can determine the fate and functionality of the nano-system and can be achieved through the application of different surface functionalization.

During the last few years, magnetite (Fe_3O_4) nanoparticles have been attracting interest, especially in the area of clinical-oriented medical applications, many of which have already been approved by Food and Drug Administration (FDA), such as diagnosis [1,2], hyperthermia cancer treatment [3] or combating iron deficiencies [4]. This was possible due to their properties like biocompatibility [5–8], biodegradability [9–11], magnetic behaviour [12,13] and the possibility of easy functionalization [14,15]. Other possible uses of these nanoparticles might be in fields like catalysis [16,17], environmental remediation [18–20], electronics [21–23].

The route of synthesis enables controlling not only the chemical composition, but also the size, shape, surface properties and magnetic properties. The chemical methods for synthesis offer the advantage that the resulting nanoparticles can be functionalized at the end of the process, which ensures improved stability compared to non-functionalized materials and conservation of magnetic properties.

One of the most common and easiest chemical methods for magnetite nanoparticles synthesis is the co-precipitation developed by Massart in 1981 [24]. The method resides in the reaction between the ferric and ferrous ions in a basic medium. Different ferric and ferrous salts can be used as precursors (like chlorides, sulfates) and different bases, such as sodium hydroxide [25,26], ammonia [27,28]. The molar ratio of the precursor ions is usually 2:1 (Fe(III): Fe(II)), however, smaller ratios can be employed (such as 1.5:1), as the oxidation of Fe^{2+} can occur [29] and the pH of the precipitation solution should be kept between pH = 9–14 [26,30]. Also, a low concentration of O_2 is favorable, in order to prevent the oxidation of the nanoparticles and loss of magnetic properties [27]. A non-oxidant medium can be assured by the addition of nitrogen, in gas form or dissolved (such as in ammonia solution). Typically, the synthesis is undertaken in low-heat conditions (about 80 °C [31]), however, room temperature reactions can take place [32]. Moreover, the introduction of surfactants or other organic molecules in the reaction medium (the precipitation base) or in the precursor mixture, can influence the size, shape and surface properties of the resulting nanoparticles [33,34] through the formation of small micelles which limit the space of nucleation and growth available for the nanoparticle. Interactions between the torganic phase and the terminal groups of the nanoparticles might be facilitated and in situ conjugations of the magnetite nanoparticles can take place [35].

The advantages of the co-precipitation method are rapidity, ease, reproducibility and high-yield synthesis, however, the main disadvantage is given by the fact that, in order to obtain a narrow size distribution of the resulting nanoparticles, some reaction parameters must be strictly assured [36]. Table 1 summarized how reaction parameters influence the properties of the resulting nanoparticles in the co-precipitation of the ferric and ferrous ions.

Table 1. Influence of reaction parameters on the properties of magnetite nanoparticles resulting from the co-precipitation method.

No.	Reaction Parameter	Property	Measure	Reference
1	Fe^{3+}/Fe^{2+} ratio	Iron oxide phase	Directly proportional	[37]
		Magnetism	Inversely proportional	[38,39]
		Dimension	Directly proportional	[39,40]
2	pH value	Iron oxide phase	Inversely proportional	[41]
		Magnetism	Inversely proportional	[38,42]
		Dimension	Insignificant	[42]
3	Type of base	Iron oxide phase	Depending on the type of base	[26]
		Magnetism	Depending on the type of base	[26]
		Dimension	Depending on the type of base	[26]
4	Temperature	Iron oxide phase	Directly proportional	[43]
		Magnetism	Inversely proportional	[44]
		Dimension	Inversely proportional	[40,45]
5	Concentration of precursors	Dimension	Directly proportional	[40]
6	pH of the precursor solution	Iron oxide phase		[40]
		Magnetism		[40]
		Dimension	Directly proportional	[40]
7	Addition of surfactants	Dimension	Directly proportional	[38,46,47]
		Surface charge	Dependent on the surfactant	[47]
		Composition	Dependent on the surfactant	[47]
		Shape	Dependent on the surfactant	[33]
		Magnetisation	Dependent of the surfactant	[47]

The solvothermal method is the second most popular method for the obtaining of magnetite nanoparticles and is performed in the presence of solvents, using temperatures that are higher than the boiling points of the solvents. The reaction is performed inside an enclosed system, like the autoclave, at high pressures. The composition of solvents influences the shape and size of the nanoparticles [48] however, the size is significantly determined by the temperature and duration of reactions. Different mixtures of agents such as tri ethylene glycol [49], oleylamine and ethylene glycol [50], or benzyl ether [51]. can be added in the solvent mixture in order to act as reducing agents for the precursor(s), leading to the synthesis of highly stable functionalized magnetite nanoparticles.

The hydrothermal method is based on the use of high temperatures and pressures to obtain single Fe_3O_4 crystals [52]. Saturation of the precursors is required to initiate crystallization and this is enabled by a temperature difference between the precursors (crystallization area) and an aqueous area in the autoclave.

The microemulsion method uses micelles as nanoreactors for the nucleation and growth of magnetite nanoparticles in a limited space [53]. Thus, one main advantage of this method would be low polydispersity indices of the resulting nanoparticles and controlled morphology of these. Moreover, the nanoparticles are in situ functionalized through encapsulation [54,55].

Lately, a lot on non-conventional methods have been used in order to obtain magnetite nanoparticles. For example, the gas flame synthesis leads to highly dispersed nanoparticles with low polydispersity indices being obtained [56,57]; moreover in situ functionalization can be applied [58].

A rigorous control of the parameters of the synthesis method leads to crystalline nanoparticles with unique mineralogical phase composition being obtained. Magnetite nanoparticles have inverse spinel structure, with a face centred cubic lattice, where the iron ions are placed in the interstitial sites.

Moreover, a controlled synthesis assures and conserves the native properties of magnetite nanoparticles, such as the property of superparamagnetism, with high magnetic susceptibility, which in the absence of magnetic field shows null magnetization [59,60]. Temperature can randomly change the orientation of the magnetic spins, but this effect can also occur after a certain time (Neel relaxation time), due to the magnetic anisotropy of the nanoparticle. Placing Fe_3O_4 nanoparticles in an exterior magnetic field causes the orientation of the nanoparticles magnetic moments with the magnetic field, while alternated magnetic fields repeatedly change the orientation of the magnetic moments, with an energy loss, converted to thermal energy. In order to preserve the magnetic property of Fe_3O_4 nanoparticles, different functionalization approaches are employed.

The fate of magnetite nanoparticles in the human body is highly dependent on size, surface properties and terminal functional groups. It has been proved that the physical characteristics of the nanoparticles, such as size [61–63] and shape [64–67], influence their relationship with living cells. Additionally, surface properties [68,69], not only dictate the interaction with the biological barriers (membranes, vascular lumens), but can also modulate the way in which the nano-complex is perceived by the cells and tissues. In nanomedicine, this can dictate the effectiveness towards clinical translation. A rigorous control of the physical and chemical properties of magnetite nanoparticles can, most of the time, decide the fate of the nano-system and its ability to fulfil the requirements for which it has been designed and developed [70]. The route of administration can also determine the outcome of the nanoparticles, as they can encounter more or less biological barriers in their way to the targeted area.

Ma et al. [71] made a study on Kumming mice that were daily injected intraperitoneally during 1 week with different concentrations of Fe_3O_4 nanoparticles (0, 5, 10, 20, 40 mg/kg), the subjects presenting lesions and the impairment of the hepatic and renal tissues, by means of oxidative mechanisms; the maximum recommended dose was 5 mg/kg. Wang et al. [72] determined the presence of Fe_3O_4 nanoparticles in the brain after the intraperitoneal injection. Following intragastric administration of 600 mg/kg magnetite nanoparticles to mice [73], a maximum of concentration was determined in lungs and kidneys after 6 h of administration, in liver, brain, stomach and small intestine after 24 h, in heart and spleen after 3 days, respectively in peripheral blood after 5 days. Intravenous injection (15 mg/kg, 5 times) in C57BL/6 mice determined an accumulation of magnetite nanoparticles in liver, lungs and spleen, which were degraded to non-magnetic iron oxide species [74].

Due to the high surface-to-volume ratio, as a result of the nanometric dimension many hydroxyl terminal groups are available for conjugation with other molecules (Figure 1). It is this property that enables a lot of practicable approaches for surface modification, in order to alter and modulate the physical and chemical behaviour of magnetite nanoparticles. This review article discusses different approaches of functionalization for magnetite nanoparticles applications in medicine.

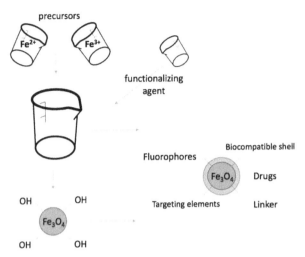

Figure 1. Schematic representation of the two main types of magnetite nanoparticle functionalization processes for medical applications: in situ, respectively, post-synthesis functionalization.

2. Functionalization of Magnetite Nanoparticles

Besides their advantages, magnetite nanoparticles have some major flaws, like rapid agglomeration, chemical reactivity, high surface energy, oxidation, which might alter their biocompatibility, properties and performance. In order to prevent these unwanted events, different surface functionalization is applied.

The functionalization refers to the conjugation of different molecules. In case of nanoparticles, this process determines a modification of the surface chemistry, which leads to changes in the physical, chemical and biological properties.

There are different types of functionalization. Depending on the time when it is done, the functionalization process can be in situ [75–77], in case the conjugation takes place simultaneously with the nucleation process of the nanoparticle, during the synthesis or post-synthesis [78,79], when the functionalization reaction(s) is (are) done after the synthesis of the nanoparticles (Figure 1).

By taking into consideration the chemistry of functionalization, either non-covalent or covalent bindings can take place between the surface modifying molecule and the magnetite nanoparticles. The non-covalent conjugation [80–82] mainly takes place through interactions that are based on the receptor-ligand affinity principle. Some examples are the electrostatic interactions, entrapment into secondary elements (like polymeric films) or π-π stacking. In this case, mostly ionic bonds appear, following the transfer of one electron from a metallic to a non-metallic atom and the electrostatic interaction between the resulting ions.

In the case of covalent binding [83,84], different chemical reactions can take place during the functionalization process, such as substitution (nucleophile or electrophile), addition (nucleophile or electrophile), elimination, oxidation, reduction, polymerization, or esterification, in presence of different catalysts. In order to conjugate the desired molecule on the surface of the magnetite nanoparticles, intermediary linkers can be used, such as oleic acid [85], aminoproliltriethoxy silane [86], 3-(trimethoxysilyl) propyl methacrylte [87].

Sometimes, the preferred approach is to have a non-specific physical sorption, which would give a less stable conjugation (in case of delivery application or to facilitate the degradation of the nano-system), but chemical sorption can also be employed. In this case, covalent bonds can appear between identical atoms or different atoms which share electrons, each atom participating with one electron. This appears for non-metal elements. These are classified as non-polar covalent bonds (between the same type of atoms), covalent bonds between different atoms, coordinative bonds (when two electrons are shared).

Metallic bonds are chemical bonds that form between metal elements. It is very rare that this interaction takes place between the Fe atoms in the oxide structure of magnetite and other metals, when developing core-shell metallic nanoparticles. For example, Han C.W. et al. [88] has obtained Fe_3O_4-Au core-shell nanoparticles by in situ vacuum annealing of dumbbell-like Au-Fe_3O_4 nanoparticles obtained by epitaxial growth of magnetite on Au nanoparticles. The process was undertaken using a transmission electron microscope and recorded. During the annealing, the gold nanoparticles transformed into a gold nano-film, which was melting the surface of the magnetite nanoparticle, simultaneously with the reduction of the Fe_3O_4 nanoparticle, taking place a strong metal-support bonding between the two components.

Different approaches of magnetite nanoparticles functionalization (Figure 1) will be discussed in the following sections, depending on the type of conjugation agent (inorganic or organic) and related to their biomedical applications.

3. Inorganic Functionalization of Magnetite Nanoparticles

3.1. Oxides

Among oxides, SiO_2 (silica) coating is one of the most commonly used approaches for nanoparticle surface modification, especially in the case of iron oxides like magnetite. This is mainly determined by the properties induced by silica coating of Fe_3O_4 nanoparticles, such as reducing the aggregation phenomena and thus improving the stability of the resulting functionalized nanoparticles [89], but also enhancing their biocompatibility [65,90].

There are several methods that can be used for SiO_2 conjugation on magnetite nanoparticles. The most frequently encountered approach is the sol-gel (Stoeber) method, which is based on the hydrolysis of tetraethoxysilane (TEOS) in an alcoholic medium, using ammonia as catalyst [91,92]. The method is popular due to its ease, but also due to the ability to obtain monodisperse-coated nanoparticles, with controlled dimension and shape. By using this approach, the chemical composition and structure, as well as magnetic properties of the Fe_3O_4 nanoparticles, are preserved.

Another precise, but more elaborate method for the obtaining of $Fe_3O_4@SiO_2$ nanoparticles is the microemulsion method, which can be either water-in-oil (W/O) or oil-in-water (O/W). Such methods are usually employed for obtaining of Fe_3O_4 nanoparticles and the in situ functionalization [93]. This method can also be microwave assisted [93,94].

Mesoporous silica, such as MCM-41 or SBA-15 have grown in interest due to their biocompatibility [95–97] and highly controlled porosity [98–100], which enable their use as controlled drug delivery platforms [101,102]. In order to obtain mesoporous silica-coated magnetite nanoparticles, a similar approach as in $Fe_3O_4@$ amorphous SiO_2 can be employed, but additionally an organic agent is used as template for the pore structure [103–105]. Such agents can be cetyltrimethylammonium bromide (CTAB), cetyltrimethylammonium chloride, n-octylamine, tetrapropylammonium bromide (TPABr) [106], triblock polymers like (EO)x-(PO)y-(EO)x (Pluronic L101, P103, P104, P105, F108) [107].

Due to their high porosity, the mesoporous silica-coated magnetite nanoparticles can absorb high quantities of therapeutic agents. Moreover, SiO_2 is dissolved in acidic environment, such as in the tumor microenvironment, inflammation, bacterial biofilm, or the endo-lysosomal compartments of the cells, making silica-functionalized Fe_3O_4 great stimuli-responsive materials for the controlled delivery of therapeutic agents [108–110].

Other Si-based molecules have been used as functionalization agents for magnetite nanoparticles, in order to increase their stability or be used as linkers for further surface conjugation. Some examples are (3-aminopropyl)triethoxysilane (APTES) [111–113], 3-Aminopropyltrimethoxysilane (APTS) [114], (3-Mercaptopropyl)trimethoxysilane (MPTS) [115], triethoxy vinyl silane (VTES) [116], aminosilane [117,118]. Table 2 summarizes some recent examples of Fe3O4@SiO2 nano-systems and their applications in biomedicine.

Table 2. Recent approaches in Fe_3O_4-SiO_2 based nanostructures conjugates.

No.	System Description	Application	Type of Conjugation	Evaluation	Reference
1	$Fe_3O_4@SiO_2$	Magnetic resonance imaging contrast substance as in vivo stem cell tracker	Negatively charged Fe_3O_4@citrate act as seeds for Si precursor; encapsulation using sol gel method;	Determination of distribution and chemical changes dynamics of $Fe_3O_4@SiO_2$; high chemical stability; distribution in cytoplasm;	[119]
2	$Fe_3O_4@SiO_2$/anti-rHBsAg (Hepatitis B surface antigen)	Purification of recombinant Hepatitis B for vaccine production;	In situ functionalization; encapsulation using sol gel method;	In vitro isolation of rHBsAg antigen from Pichia pastoris yeast	[120]
3	$Fe_3O_4@SiO_2$	Plasmid DNA purification	$SiCl_4$ cross-linker between $Fe_3O_4@NH_3$ and (3-aminopropyl)triethoxysilane (APTES); encapsulation using sol gel method;	Efficient in vitro plasmid DNA purification from *E. Coli* DH5a cells	[121]
4	Fe_3O_4@boronic acid/mesoporous (m) SiO_2	Magnetic and pH triggered drug release;	–	Biocompatibility and high uptake in MC3T3-E1 cells; Controlled drug release and good magnetic properties;	[122]
5	$Fe_3O_4@mSiO_2$/catalase (CAT)	Enzyme protection in catalysis;	Encapsulation in SiO_2 using TMOS (tetramethoxysilane) functionalization with APTES for CAT conjugation and growth of $mSiO_2$ using CTAB as template and TMOS;	Good stability and catalytic activity	[123]
6	Fe_3O_4@oleic acid@$mSiO_2$/5-Fluorouracil	Drug delivery for cancer therapy;	In situ Fe_3O_4@oleic acid were functionalized with CTAB through weak interaction (Van der Waals); hydrolisation of tetraethoxysilane (TEOS) on Fe_3O_4/CTAB; encapsulation in $mSiO_2$ using the inversed microemulsion method;	In vitro biocompatibility for MCF-7 cells; efficient drug loading;	[124]

Numerous metal oxides have been used as functionalizing agents to modify the surface of magnetite nanoparticles, in order to obtain composites with improved functions. ZnO-conjugated Fe_3O_4 nanoparticles have been developed in order to implement photocatalytic properties to the developed nano-systems. This phenomenon appears due to high oxygen vacancies on the surface of the nanoparticles and due to the fact that the electron-hole pairs induced by photon-triggering are inhibited by Fe^{3+} ions [125]. Similar photocatalytic effects are given by $Fe_3O_4@TiO_2$ nanoparticles [102]. Shi L et al. [31] obtained $Fe_3O_4@TiO_2$ core-shell nanoparticles using post-synthesis functionalizing based on a hydrothermal approach. Similarly, Zhang L et al. [126] and Choi K-H et al. [127] used the solvothermal synthesis for microsphere preparation.

3.2. Metals

The surface conjugation of Fe_3O_4 with different metals has been employed in order to improve the biocompatibility of magnetite nanoparticles and to induce an inert character to the final nano-structure. The metal coating of Fe_3O_4 nanoparticle surface can be done either directly or through an intermediate functionalizing layer.

Core-shell magnetite@gold nanoparticles are interesting for their multifunctionality. The direct route to obtain this type of nano-composites is by directly reducing Au^{3+} ions on the surface of the Fe_3O_4 nanoparticles, using reducing agents such as sodium citrate [60,128], sodium borohydride [129], and hydroxylamine hydrochloride [130]. Through this method mostly result dumbbell-like, core-satellite, or sometimes star-shaped structures, but core-shell nanoparticles can only result after multiple repetitions of the coating procedure. The main disadvantage of this method is the low yield synthesis, as many gold nanoparticles result [131]. Moreover, the reduction of Au^{3+} into Au^0 takes place at the boiling point of the watery solution (80–90 °C), which might lead to an oxidation of Fe_3O_4 and loss of magnetic properties.

Also, a more efficient direct method of conjugation might be in situ functionalization, through the organic synthesis approach [132]. Usually, these routes employ different agents to reduce $Fe(acac)_3$ [132] or $FeO(OH)$ [133] in presence of $HAuCl_4$ which is simultaneously reduced, forming core-shell structures. Other organic molecules, such as oleic acid [134] are used to act as reducing and stabilization agents at the same time.

The use of an intermediary layer between the previously-synthesized magnetite nanoparticles and the gold layer acts as a "glue" between the two components. In situ or post-synthesis functionalization of iron oxide nanoparticles is undertaken, in order to obtain a functional layer that can either attract the Au^{3+} ions, which are afterwards reduced to Au^0 using a third substance [135,136], or the conjugated molecules act as a reducing agent themselves [134].

Fe_3O_4-Au conjugated nanoparticles have applications in medical imaging. Due to the presence and properties of both magnetite and gold phases, such nanoparticles can be used as a contrast substance in both magnetic resonance imaging (MRI), computer tomography (CT) and photoacoustic imaging (PA). For attempt, Hu Y et al. [137] developed Fe_3O_4@Au nano-systems starting from Fe_3O_4@Ag@citric acid as seeds for Au^{3+}. The resulting star-shaped nanoparticles were functionalized with polyethyleneimine (PEI) to improve stability and folic acid to induce the targeting ability (Figure 2). Ge Y et al. [138] used antibody (McAb) cetuximab (C225) conjugation of Fe_3O_4@Au to induce targeting ability for glioblastoma. The functionality of Fe_3O_4@Au nano-composites for targeted tumor imaging has been proved in vivo [137–139].

Other possible application of magnetite-gold nano-conjugates refers to their use in cancer radiotherapy, following their activation with different types of radiation: ionizing radiation (IR) [140,141], near-infrared (NIR) radiation [134,142] and radiofrequency (RF) [143] radiation. Radiotherapy mediated by nanoparticles has been considered as an approach that overcomes the resistance of tumor cells to radiotherapy and/or chemotherapy [144–147].

Generally, the use of metal elements to radiosensitize tumor cells is based on increasing the photoelectrical absorption, after their accumulation inside the malignant tissue. The high atomic number elements absorb most of the radiation compared to the surrounding healthy tissues and, due to the photoelectric and Compton effects, lower energy photons, Auger secondary electrons and low-energy secondary electrons are released [148–151]. Also, an enhanced production of reactive oxygen species occurs, due to the formation of secondary electrons and photons, but also due to the high surface reactivity of the nanoparticles [152,153]. This affects directly the DNA of the tumor cells. Moreover, nanoparticles can directly interact with the DNA, forming bonds or intercalating intro the DNA chain [154,155]. The biological outcome is oxidative stress, cell-cycle disruption and DNA repair inhibition [148] in the tumor cells.

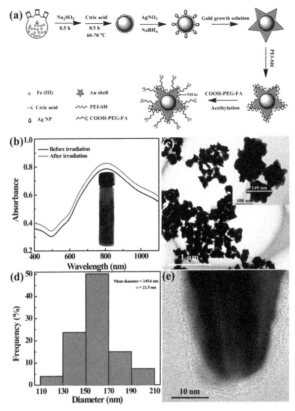

Figure 2. Star-shaped gold-conjugated Fe_3O_4 nanoparticles; functionalization with organic molecules (polyethyleneimine, PEI): (**a**) schematic representation of the synthesis and conjugation processes; (**b**) ultraviolet–visible (UV–VIS) spectra for (non-) irradiated the nano-constructs; (**c**) transmission electron micrograph (TEM) of the resulted nanoconstructs; (**d**) histogram distribution of size; (**e**) high-resolution TEM (HR-TEM) of the resulted nanoconstructs; reprinted from [137].

The radiofrequency ablation (RFA) as a new method for cancer treatment has recently attracted more interest due to the fact that it does not harm normal tissues, when using frequencies from 10 kHz–900 MHz; the radiation has high penetration capability and non-ionizing effects on the tissues. The mechanism of toxicity upon cancer cells is produced by the induced thermal disruption determined due to the friction appearing in the ionic collisions of the biomolecules, when aligning in the alternating current flow [156]. RF-responsive nanomaterials have been proposed as probes for the treatment procedure, because of their ability to produce heat due to the resistance heating (in conductive materials, such as gold [153]), respectively magnetic heating (in magnetic materials, such as magnetite [157]). Gold-conjugated magnetite nanoparticles are excellent candidates for RFA treatment of cancer [142,158].

Another possible application of gold-conjugated magnetite nanoparticles is biosensing, due to the surface plasmon resonance property of gold [159–162]. Moreover, further functionalization of Fe_3O_4@gold with different antibodies gives the ability of specific targeting of cells, which together with the magnetic properties of the nano-systems enable their applicability in cell sorting or cell separation [163,164].

Platinum-conjugated magnetite nanoparticles also have possible applications in radiotherapy enhancement. Also an inert noble metal, Pt has an atomic number higher than Au, being able to induce higher radiosensitizing effects [165,166]. Ma M. et al. [167] used a "glue" layer, DMSA (meso-2, 3-dimercaptosuccinic acid), for Pt ions that were reduced using $NaBH_4$ on the surface of previously-synthesized magnetite nanoparticles, in order to obtain dumbbell-like structures. A similar approach was employed by Wu D et al. [168] who used MnO_2 as intermediary layer for Pt ions absorption followed by reduction on the surface of the $Fe_3O_4@MnO_2$ nano-conjugate.

Silver coated magnetite nanoparticles can be obtained using the same approaches as gold-magnetite conjugates. Their applications in the medical field vary from catalysis [169], contrast substance in medical imaging [170,171], radiation therapy [172], the most frequent application being given by their anti-microbial properties [173]. Chang M et al. [174] obtained $Fe_3O_4@Ag$ nanoparticles using in situ functionalization and proved their effect against *E. coli* strains. Brollo M. E et al. [175] synthesized brick-like nano-composites using a thermal decomposition method and in situ conjugation.

4. Carbon-Based Functionalization of Magnetite Nanoparticles

The carbon-based functionalization of magnetite nanoparticles is treated separately from the (in)organic sections, as both inorganic (such as SiC [176]), as well as organic (graphene, carbon nanotubes) and $Fe_3O_4@C$ composites are approached.

The majority of $Fe_3O_4@C$ composites applications are in electronics (used as supercapacitors [177], anode materials in lithium-ion batteries [178], absorbents [177]). These materials can be obtained by in situ or post-synthesis functionalization, using the hydrothermal approach [179–181].

For applications in the biomedical field, the conjugation of magnetite nanoparticles and carbon-based nanostructures, such as graphene, carbon nanotubes or fullerenes are more often encountered. Amide bonding is a very frequent approach in conjugation of Fe_3O_4 and carbon-based nanoparticles [158], alongside with click chemistry. These types of reactions are modular reactions like cycloadditions, nucleophilic ring-openings, carbon multiple bond additions and non-aldol carbonyl reactions [182]. The most common type in functionalizing carbon-based nanomaterials is Cu(I)-catalysed azide-alkyne 3+2 cycloaddition (CuAAC) [183]. Table 3 presents recent exampled of Fe_3O_4-carbon nanoparticles conjugates.

Table 3. Recent approaches in Fe_3O_4-carbon-based nanostructures conjugates.

No.	System Description	Application	Type of Conjugation	Evaluation	Reference
1.	Fe_3O_4 @APS–graphene/ 5-Fluorouracil	Drug-delivery systems for cancer treatment;	Amide bonding using 1-ethyl-3-(3-dimethylaminopropyl) carbodiimide	In vitro drug release at acidic pH; efficient in vitro internalizing in hepatocarcinoma HepG2 cells; biocompatibility of the carrier nanoparticles;	[184]
2.	$Fe_3O_4@$ APTES/graphene oxide (GO)/doxorubicin	Drug-delivery systems and imaging diagnosis in cancer management;	Amide bonding using N-(3-Dimethylaminopropyl)-N′-ethylcarbodiimide hydrochloride (EDC)	In vitro low cytotoxicity compared to GO; superparamegnetic properties and 10.7 r2/r1 relaxivity; fluorescence in VIS; high doxorubicin loading and 2.5 fold higher efficiency; (Figure 3)	[185]
3.	$Fe_3O_4@$azide-sodium ascorbate-GO@ alkyne	Efficient absorbent and removal of dyes;	Click chemistry approach between the azide functional groups on the Fe_3O_4, sodium L-ascorbate and alkyne functional groups on GO;	Superparamagnetic properties; efficient absorbent and removal of dyes;	[186]
4.	$Fe_3O_4@GO$	Magnetic fluids;	Absorption;	Improvement of friction and wear performances with magnetic field;	[187]
5.	Polyvinyl alcohol (PVA)/ $Fe_3O_4@$ carbon nanotubes (CNTs)	Absorbent and dye removal; Anti-bacterial effects;	–	Optimal dye removal and anti-bacterial properties;	[188]

Table 3. *Cont.*

No.	System Description	Application	Type of Conjugation	Evaluation	Reference
6.	Fe$_3$O$_4$/multi walled CNTs/laser scribed graphene/chitosan/glassy carbon electrode	Detection of heavy metals	–	Electrode for the determination of Cd^{2+} and Pb^{2+} using square wave anodic stripping voltammetry; wide linear range; ultralow detection limit; excellent repeatability, reproducibility, stability;	[189]
7.	Single-walled CNTs-PEG-Fe$_3$O$_4$@ carbon quantum dots (CQD)/doxorubicin/sgc8c aptamer	Targeted photodynamic and photothermal ablation of tumor cells; controlled drug delivery; targeted imaging using fluorescence and magnetic resonance imaging (MRI)	Through polyethylene glycol (PEG) linker using amide bonding;	Near infrared triggered production of reactive oxygen species and heat; good imaging properties; good biocompatibility of the carrier and cellular internalization; high drug loading ability; selective accumulation at tumor site in human adenocarcinoma (HeLa) tumor-bearing mice intravenously injected with the system;	[190]
8.	GO-Chitosan/Fe$_3$O$_4$/ glucose oxidase	Glucose biosensor and magnetic resonance imaging;	–	Good glucose biosensing ability;	[191]

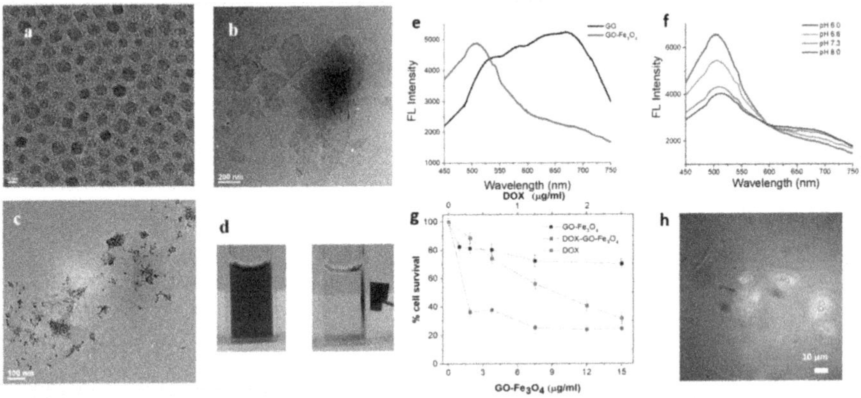

Figure 3. Fe$_3$O$_4$@(3-aminopropyl)triethoxysilane (APTES)-graphene oxide nano-system for drug delivery and diagnosis in cancer: (**a**) TEM of Fe$_3$O$_4$ nanoparticles; (**b**) TEM of graphene oxide; (**c**) TEM of Fe$_3$O$_4$-graphene oxide conjugates; (**d**) magnetic manipulation of Fe$_3$O$_4$-graphene oxide conjugates in aqueous solution; (**e**) fluorescence specra of graphene oxide and Fe$_3$O$_4$-graphene oxide conjugates; (**f**) fluorescence specra of Fe$_3$O$_4$-graphene oxide conjugates at different pH; (**g**) HeLa cell survival (%) after incubation with equivalent concentrations of Fe$_3$O$_4$-graphene oxide conjugates, Fe$_3$O$_4$-graphene oxide conjugates loaded with doxorubicin, respectively doxorubicin; (**h**) fluorescence image of internalized Fe$_3$O$_4$-graphene oxide conjugates in HeLa cells; adapted from [185].

5. Organic Functionalization of Magnetite Nanoparticles

The functionalization of magnetite nanoparticles with organic compounds is mostly done in order to improve their stability [192] and biocompatibility [193]. Another reason would be to improve their interaction with biological barriers (cellular membranes, vascular endothelium, blood-brain barrier) and facilitate the nanoparticles' passage through these [194,195].

Furthermore, magnetite nanoparticles have a hydrophobic character which favours the adsorption of serum proteins, causing not only blood clogging, but also leading to the opsonisation phenomenon. Through this, the nanoparticles are immediately collected by the cells of the mononuclear phagocyte system and eliminated from systemic circulation. In order to improve the pharmacological kinetics of the magnetite nanoparticles, functionalization with hydrophilic polymers, such as polyethylene glycol (PEG) [196] is applied.

In case of controlled delivery of therapeutic substances, organic materials and especially polymers are the best stimuli-responsive materials (responsive to changes in temperature, pH, light). Fe_3O_4 nanoparticles functionalized with biocompatible responsive polymers are ideal for such applications, as the magnetite core enables magnetic targeting properties of the system, while the soft shell encapsulates large quantities of drug molecules.

Also, polymers enable many available functional groups for the conjugation of other molecules. Thus, specific molecules can be conjugated for targeting certain type of cells or area of the body (like folic acid [197,198], L-3,4- dihydroxyphenylalanine (L-DOPA) [199], riboflavin [200], arginine-glycine-aspartate (RGD) [201] for cancer targeting) and/or light-responsive molecules for detection and imaging (such as fluorescein isotiocianate-FITC [202]). Moreover, Fe_3O_4 can be used as contrast substance in MRI because of its ability to alter the spin-spin relaxation time T2 of the surrounding water protons [203]. Given all these properties, functionalized magnetite nanoparticles can be used as multifunctional platforms for cancer detection and therapy.

Organic materials for magnetite nanoparticles functionalization will be discussed in separate sections as follows: small molecules and surfactants, lipids, polymers, phytochemicals, respectively drug molecules.

5.1. Small Molecules and Surfactants

Functionalization of magnetite nanoparticles with amphiphilic molecules (surfactants) has been proved as a good solution to improve the stability of the suspensions [204,205]. However, surfactants can rather have a toxic behaviour and are not recommended for biological applications [206–208].

Instead, functionalization with small molecules was proposed. Oleic acid is the most common small lipophilic molecules used for the functionalization of magnetite nanoparticles. Fe_3O_4@oleic acid has good stability [209], biocompatibility [210] and can be used for further functionalization: oleic acid can act either as a "glue" layer to conjugate other compounds [211] or as a starting point in ligand exchange approach [212,213].

Functionalization of magnetite nanoparticles with small molecules or surfactants is mostly done in situ using solvothermal [51,214,215] or microemulsion [53,216] approaches, however, post-synthesis conjugation can also be done [217,218].

Figure 4 [219] illustrates an approach for oleic acid capping of magnetite nanoparticles and the morphological and hydrodynamic properties of the resulting functionalized nanoparticles, in comparison with bare Fe_3O_4.

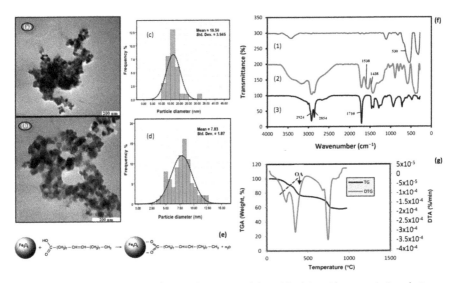

Figure 4. Surface conjugation of magnetite nanoparticles with oleic acid: transmission electron microscopy (TEM) image for (**a**) bare Fe$_3$O$_4$, respectively (**b**) oleic acid conjugated Fe$_3$O$_4$; particle diameter distribution for (**c**) bare Fe$_3$O$_4$, respectively (**d**) oleic acid conjugated Fe$_3$O$_4$; (**e**) schematic representation of the capping principle; (**f**) Fourier transform infrared (FTIR) spectra of Fe$_3$O$_4$ (**1**) Fe$_3$O$_4$/oleic acid (**2**), respectively oleic acid (**3**); (**g**) thermogravimetric analysis (TGA) and differential thermogravimetric analysis (DTA) curves for oleic acid conjugated Fe$_3$O$_4$; adapted from [219].

5.2. Lipids

Lipids are the main component of cellular membranes, thus conjugation with magnetite nanoparticles would be ideal for biomedical applications. Lipid-coated nanoparticles favour the interaction with and passage through biological membranes [220,221], enhancing the biocompatibility of Fe$_3$O$_4$ nanoparticles [197,222] and preventing the opsonisation phenomenon [223]. The obtaining of lipid-conjugated magnetite nanoparticles is most of the time done through encapsulation [224,225].

5.3. Polymers

The functionalization of magnetite nanoparticles with polymers can be undertaken using both in situ and post-synthesis functionalizing. It is very common in case of co-precipitation method for Fe$_3$O$_4$ synthesis to introduce polymer molecules in the precipitation solution, in order to determine the simultaneous functionalization, nucleation and growth of the nanoparticles [226,227]. In this case, mostly non-covalent bonds (electrostatic forces) appear between the polymers and magnetite nanoparticles.

The latter method starts from previously synthesized magnetite nanoparticles that can be conjugated with different polymers through the available hydroxyl groups on their surface. These are mostly condensation reactions. One approach is through the ester bond formation. Also, intermediate linkers can be used, such as APTES, which enable amine terminal groups on the surface of the magnetite nanoparticles. These can be then coupled with different polymers through an amide bond formation.

The main reason for polymer surface functionalization of magnetite nanoparticles is the increase of stability, as the polymeric molecules act as splicing agents between the magnetic nanoparticles, preventing their aggregation. The longer the polymeric chain, the higher the stability of the nanoparticles. However, this can produce an inverse effect, as a reduced magnetic response can occur when stimulating the functionalized nanoparticles with an exterior magnetic field.

Polyethylene glycol (PEG) is the most widely used polymer for magnetite nanoparticles functionalization. PEG with different molecular weights are employed, in order to modulate the hydrodynamic properties of the resulting nano-composites and to improve their stability [228,229]. Other frequently used polymers for Fe_3O_4 nanoparticles functionalization are polyethyleneimine (PEI) [230,231], glucose [232–234], dextran [235,236], and chitosan [237–239]. Table 4 summarizes some examples of polymer-functionalized magnetite nanoparticles and their applications.

Table 4. Recent approaches in Fe_3O_4-polymer-based nanostructures conjugates.

No.	System Description	Application	Type of Conjugation	Evaluation	Reference
1.	Fe_3O_4@ poly(polyethylene glycol methacrylate-co-acrylic acid) (P(PEGMA-AA))	Hyperthermia and MRI contrast substance;	Electrostatic interactions between the acrylic acid and positively-charged Fe3O4;	Improved stability and salt tolerance; excellent blood compatibility; formation of blood protein corona; resistance to cell internalization; improvement of contrast in MRI;	[240,241]
2.	Fe_3O_4/methyl methacrylate/ethylene glycol dimethacrylate/hydroxyl ethyl methacrylate/gemcitabine	Hyperthermia and drug delivery for cancer therapy	–	Good incorporation of drug; temperature triggered release; (Figure 5)	[242]
3.	Fe_3O_4@PEG/Doxorubicin	Drug delivery and hyperthermia in cancer treatment;	In situ conjugation	pH responsive release of drug; no cytotoxicity of Fe_3O_4@PEG for human fibroblasts; Fe_3O_4@PEG/Doxorubicin showed good internalization and cytotoxicity for mouse skin fibrosarcoma; good magnetic properties;	[243]
4.	Fe_3O_4@ poly(lactic-co-glycolicacid) (PLGA)-PEG@ folic acid/curcumin	Targeted drug delivery for cancer treatment;	Encapsulation;	High drug loading and delivery; high in vitro targeting efficiency for cervical carcinoma; in vitro induction of apoptosis and reduction of tumor cell proliferation;	[244]
5.	Fe_3O_4@ C/carboxymethyl cellulose/chitosan/diclofenac sodium	Controlled drug delivery;	In situ conjugation and subsequent electrostatic conjugation;	High drug-loading efficiency; pH sensitive drug delivery;	[245]
6.	Fe_3O_4@ dextran	–	Covalent binding via electron pairing;	–	[246]
7.	Fe_3O_4@dextran	Near-infrared (NIR) photothermal ablation of tumor cells;	In situ encapsulation;	In vitro biocompatibility; in vitro and in vivo tumor growth inhibition after NIR activation;	[247]
8.	Fe_3O_4@ poly ε acrylic acid-gelatin/ hydroxyapatite/ polycaprolactone	Bone tissue engineering scaffolds for hyperthermia cancer treatment;	Electrostatic interactions between the acrylic acid and positively-charged Fe3O4;	Characterisation of the magnetic behaviour for hyperthermia applications;	[248]

Table 4. *Cont.*

No.	System Description	Application	Type of Conjugation	Evaluation	Reference
9.	Fe$_3$O$_4$/poly-L-lactide (PLLA) nanofibers	Bone tissue engineering;	–	In vivo evaluation on tibia defect rabbit model; computer tomography and histological investigations revealed higher bone-healing potential than conventional PLLA	[249]

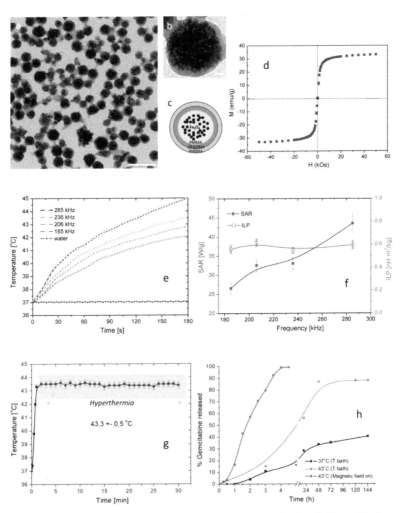

Figure 5. MagP-OH particles: (**a**) TEM image, scale 200 nm, (**b**) TEM detail, scale 20 nm, (**c**) schematic representation of MagP, (**d**) magnetisation curve of MagP, (**e**) time evolution of temperature for various frequencies, (**f**) Specific Absorption Rate (SAR) and Intrinsic Loss Power (ILP) for Ha = 16.2 kA/m, (**g**) hyperthermia measurement, (**h**) drug release measurement; adapted from [242].

Maier-Hauff K. group has studied the effects of soft polymer coated Fe_3O_4 nanoparticle-mediated hyperthermia combined with external beam radiotherapy on glioblastoma multiforme patients [250–252]. Nowadays, this treatment plan has been clinically approved and used by MagForce [3].

Hyperthermia is a therapeutic procedure for cancer which rises the temperature of the tissue to about 41–45 °C for a certain period of time [253]. Tumor cells are sensitive to these temperatures, while normal healthy cells endure temperatures up to 46–47 °C. Nanoparticle-mediated magnetic hyperthermia uses the magnetic property of Fe_3O_4 nanoparticles to produce thermal energy [254]. The nanoparticles are exposed to external alternated magnetic fields which cause successive (de) magnetization, the supplementary energy to reach the relaxation state being converted to thermal energy [255].

5.4. Phytochemicals

Phytochemicals are chemical products derived from plants, which might have beneficial effects on human health. Conjugation of magnetite nanoparticles with different phytochemicals was done in order to improve their biocompatibility [256,257] and induce certain therapeutic properties (antibacterial [32,258–260], anticancer [11,261]). Mostly, these plant-originated chemicals are used as reducing agents for the iron precursors [262,263] during the synthesis of the nanoparticles. This process enables an in situ functionalization of the resulting materials with molecules in the plant extracts, which are mostly rich in hydroxyl groups. However, post-synthesis functionalization can also be employed [256].

In traditional medicine, phytochemicals have been used extensively due to their potential therapeutic activity, continuing to be the basis of alternative therapeutic approaches even today, in cancer therapy [264,265], anti-microbial applications [258,266], anti-inflammatory approaches [267,268], anti-viral and immune system enhancement [269]. Moreover, folic acid has been used extensively as targeting agent for tumour cells [270,271], as these cells exhibit a higher density of folic acid receptors on the membrane, compared to healthy cells.

In the case of anti-bacterial applications, one important branch refers to combating the medical devices associated infections and biofilm formation, one approach for preventing antibiotic resistant bacteria contamination being the use of alternative medicine. Figure 6 illustrates the compositional structure and biological characterisation of matrix-assisted pulsed laser evaporation (MAPLE) deposited Fe_3O_4@*Cinnamomum verum* thin films. These have been developed in the idea of implant surface modification with anti-bacterial potential. Such substrates are biocompatible for eukaryote cells (in the surrounding tissues) and exhibit a toxic effect against prokaryote (bacterial) cells.

5.5. Drug Molecules

Magnetite-based nano-systems have been broadly used as drug-delivery systems [272–275]. A direct conjugation of the drug with the functional groups of magnetite is mostly undertaken in order to assure a targeted transport of the therapeutic molecules at the site of action through magnetic directing. Weak bonding (such as non-covalent interactions) between the two components is preferred, in order to allow facile delivery of the drug. Strong interactions may affect the chemical structure of the drug molecule and determine therapeutic properties loss.

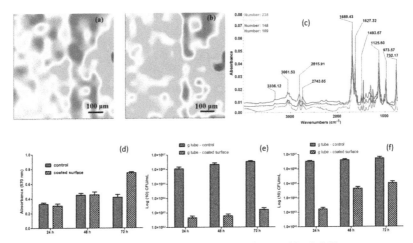

Figure 6. Matrix-assisted pulsed laser evaporation (MAPLE)-deposited $Fe_3O_4@Cinnamomum\ verum$ at fluence F = 400 mJ/cm^2: Infrared microscopy-distribution of intensity of (**a**) 2815 cm^{-1}, (**b**) 1689 cm^{-1}, (**c**) IR spectra; (**d**) biocompatibility evaluation for endothelial cells; antibacterial evaluation—*S. aureus* biofilm formation (**e**), respectively, *E. coli* biofilm formation (**f**) [32].

6. Conclusions

In the context of the advancement of magnetite nanoparticles implications in nanomedicine, a high control of their hydrodynamic and biocompatibility properties should be guaranteed, besides the fulfilment of their main biomedical function. This can be assured through the conjugation of secondary components. This review summarizes the latest advances in various approaches for Fe_3O_4 nanoparticles functionalization for nanomedicine applications:

- Multifunctionality of Fe_3O_4 nanoparticles is given by its properties (magnetism, biocompatibility);
- They have many applications in the medical field, among which a few have been approved by the FDA for clinical use (MRI contrast substance, magnetic hyperthermia, iron deficiency supplement);
- The route of synthesis also determines the surface functionality among other properties;
- Surface functionalization determines an alteration of the surface chemistry, leading to changes in the physical, chemical and biological properties;
- Classification of functionalization processes. Depending on: time of functionalization (in situ, respectively post synthesis), chemistry of functionalization (non-covalent and covalent), chemistry of the functionalizing agent (inorganic and organic);
- Non-specific physical sorption is preferred in applications such as drug delivery systems;
- Among the oxides, SiO_2 coating of magnetite nanoparticles is the most common because it enhances the biocompatibility and stability of the nanoparticles; some common approaches to obtain this conjugation are the sol-gel method, respectively, microemulsion;
- The mesoporous silica coating is biocompatible and offers high controlled porosity; is good for drug delivery applications;
- Metal oxide (ZnO, TiO_2) functionalization has photocatalytic applications;
- Surface functionalization of magnetite nanoparticles with metals induces an inert character; the most popular approach in this category is the conjugation of Fe_3O_4 with gold because of its biocompatibility and multifunctionality; approaches to obtain this type of nanoparticles are: reduction of gold ions on the surface of magnetite nanoparticles, respectively, the organic synthesis approach; the final applications are numerous: medical imaging (MRI, CT, PA), radiosensitiation, radiofrequency ablation, biosensing, cell sorting;

- Carbon-Fe_3O_4 nano-composites mostly have applications in electronics, but also in biosensing and drug delivery systems; in order to obtain these materials, the direct precipitation of magnetite nanoparticles on the surface of the carbon nanomaterial can be applied or a hydrothermal approach for in situ functionalization;
- The conjugation of magnetite nanoparticles with organic molecules has the advantage of improving the stability, biocompatibility and interaction with biological membranes of the Fe_3O_4; mostly has applications in the development of drug delivery systems;
- Surfactants have been used to improve the stability of the magnetite nano-constructs, but can have toxic effects;
- Lipid-encapsulated nanoparticles enhance the biocompatibility of the magnetite nanoparticles and improve their interaction with biological membranes, while preventing opsonisation;
- The functionalization of Fe_3O_4 with polymers is the type of surface modification most encountered for these nanoparticles and can be undertaken both in situ (through electrostatic interactions) or post-synthesis (through condensation); it increases the stability and biocompatibility of magnetite nanoparticles, leading to applications in medical imaging, hyperthermia treatment of cancer, drug delivery systems, tissue engineering;
- A polymer-coated Fe_3O_4 nanoparticle (MagForce) has been approved by the FDA for use in hyperthermia treatment of cancer;
- Drug-delivery systems based on magnetite nanoparticles can be developed for commercial medicines or phytochemicals; the therapeutic molecule can be directly conjugated on the Fe_3O_4 surface or can be attached through an intermediate layer;
- Phytochemicals-Fe_3O_4 are popular alternative medicines with antimicrobial, antitumor, anti-inflammatory or antiviral applications; conjugation with magnetite nanoparticles can be undertaken through both weak and strong interactions;
- Conventional drugs are mostly attached through strong interactions from the magnetite nanoparticles.

Author Contributions: The authors contributions are as follows: writing—original draft preparation, R.C.P.; writing—review and editing, B.S.V., R.C.P.; visualization, B.S.V., E.A.; supervision, E.A.

Funding: This research was funded by Operational Programme Human Capital of the Ministry of European Funds through the Financial Agreement 51668/09.07.2019, SMIS code 124705.

References

1. Amag Pharmaceuticals. Available online: http://www.amagpharma.com/our-products/ (accessed on 19 November 2019).
2. Wáng, Y.X.J.; Idée, J.M. A comprehensive literatures update of clinical researches of superparamagnetic resonance iron oxide nanoparticles for magnetic resonance imaging. *Quant. Imaging Med. Surg.* **2017**, *7*, 88–122. [CrossRef] [PubMed]
3. MagForce. Fighting Cancer with Nanomedicine. Available online: http://www.magforce.de/en/home.html (accessed on 7 October 2019).
4. Feraheme Ferumoxytol Injection. Available online: https://www.feraheme.com (accessed on 19 November 2019).
5. Zhang, D.; Du, Y. The Biocompatibility Study of Fe3O4 Magnetic Nanoparticles Used in Tumor Hyperthermia. In Proceedings of the 2006 1st IEEE International Conference on Nano/Micro Engineered and Molecular Systems, Zhuhai, China, 18–21 January 2006; pp. 339–342. [CrossRef]
6. Chen, D.; Tang, Q.; Li, X.; Zhou, X.; Zhang, J.; Xue, W.-Q.; Xiang, J.-Y.; Guo, C.-Q. Biocompatibility of magnetic Fe3O4 nanoparticles and their cytotoxic effect on MCF-7 cells. *Int. J. Nanomed.* **2012**, *7*, 4973–4982. [CrossRef] [PubMed]

7. Sun, J.; Zhou, S.; Hou, P.; Yang, Y.; Weng, J.; Li, X.; Li, M. Synthesis and characterization of biocompatible Fe$_3$O$_4$ nanoparticles. *J. Biomed. Mater. Res.* **2007**, *80A*, 333–341. [CrossRef] [PubMed]

8. Tian, Q.; Ning, W.; Wang, W.; Yuan, X.; Bai, Z. Synthesis of size-controllable Fe$_3$O$_4$ magnetic submicroparticles and its biocompatible evaluation in vitro. *J. Cent. South Univ.* **2016**, *23*, 2784–2791. [CrossRef]

9. Tseng, W.-K.; Chieh, J.-J.; Yang, Y.-F.; Chiang, C.-K.; Chen, Y.-L.; Yang, S.Y.; Horng, H.-E.; Yang, H.-C.; Wu, C.-C. A Noninvasive Method to Determine the Fate of Fe$_3$O$_4$ Nanoparticles following Intravenous Injection Using Scanning SQUID Biosusceptometry. *PLoS ONE* **2012**, *7*, e48510. [CrossRef]

10. Gu, L.; Fang, R.H.; Sailor, M.J.; Park, J.H. In vivo clearance and toxicity of monodisperse iron oxide nanocrystals. *ACS Nano* **2012**, *6*, 4947–4954. [CrossRef]

11. Yew, Y.P.; Shameli, K.; Miyake, M.; Khairudin, N.B.B.A.; Mohamad, S.E.B.; Naiki, T.; Lee, K.X. Green biosynthesis of superparamagnetic magnetite Fe$_3$O$_4$ nanoparticles and biomedical applications in targeted anticancer drug delivery system: A review. *Arab. J. Chem.* **2018**, 1–22. [CrossRef]

12. Patsula, V.; Moskvin, M.; Dutz, S.; Horák, D. Size-dependent magnetic properties of iron oxide nanoparticles. *J. Phys. Chem. Solids* **2016**, *88*, 24–30. [CrossRef]

13. Li, Q.; Kartikowati, C.W.; Horie, S.; Ogi, T.; Iwaki, T.; Okuyama, K. Correlation between particle size/domain structure and magnetic properties of highly crystalline Fe$_3$O$_4$ nanoparticles. *Sci. Rep.* **2017**, *7*, 9894. [CrossRef]

14. Darwish, M.S.A.; Nguyen, N.H.A.; Ševců, A.; Stibor, I. Functionalized Magnetic Nanoparticles and Their Effect on Escherichia coli and Staphylococcus aureus. *J. Nanomater.* **2015**, *2015*, 416012–416022. [CrossRef]

15. Wu, W.; He, Q.; Jiang, C. Magnetic Iron Oxide Nanoparticles: Synthesis and Surface Functionalization Strategies. *Nanoscale Res. Lett.* **2008**, *3*, 397–415. [CrossRef] [PubMed]

16. Li, Z.-X.; Luo, D.; Li, M.-M.; Xing, X.-F.; Ma, Z.-Z.; Xu, H. Recyclable Fe$_3$O$_4$ Nanoparticles Catalysts for Aza-Michael Addition of Acryl Amides by Magnetic Field. *Catalysts* **2017**, *7*, 219. [CrossRef]

17. Alishiri, T.; Oskooei, H.A.; Heravi, M.M. Fe$_3$O$_4$ Nanoparticles as an Efficient and Magnetically Recoverable Catalyst for the Synthesis of α,β-Unsaturated Heterocyclic and Cyclic Ketones under Solvent-Free Conditions. *Synth. Commun.* **2013**, *43*, 3357–3362. [CrossRef]

18. Araújo, R.; Castro, A.C.M.; Fiúza, A. The Use of Nanoparticles in Soil and Water Remediation Processes. *Mater. Today Proc.* **2015**, *2*, 315–320. [CrossRef]

19. Jiang, B.; Lian, L.; Xing, Y.; Zhang, N.; Chen, Y.; Lu, P.; Zhang, D. Advances of magnetic nanoparticles in environmental application: Environmental remediation and (bio)sensors as case studies. *Environ. Sci. Pollut. Res.* **2018**, *25*, 30863–30879. [CrossRef]

20. Gutierrez, A.M.; Dziubla, T.D.; Hilt, J.Z. Recent advances on iron oxide magnetic nanoparticles as sorbents of organic pollutants in water and wastewater treatment. *Rev. Environ. Health* **2017**, *32*, 111–117. [CrossRef]

21. De Teresa, J.M.; Fernández-Pacheco, A.; Morellon, L.; Orna, J.; Pardo, J.A.; Serrate, D.; Algarabel, P.A.; Ibarra, M.R. Magnetotransport properties of Fe$_3$O$_4$ thin films for applications in spin electronics. *Microelectron. Eng.* **2007**, *84*, 1660–1664. [CrossRef]

22. Guo, L.; Sun, H.; Qin, C.; Li, W.; Wang, F.; Song, W.; Du, J.; Zhong, F.; Ding, Y. Flexible Fe$_3$O$_4$ nanoparticles/N-doped carbon nanofibers hybrid film as binder-free anode materials for lithium-ion batteries. *Appl. Sur. Sci.* **2018**, *459*, 263–270. [CrossRef]

23. Salimi, P.; Norouzi, O.; Pourhosseini, S.E.M. Two-step synthesis of nanohusk Fe$_3$O$_4$ embedded in 3D network pyrolytic marine biochar for a new generation of anode materials for Lithium-Ion batteries. *J. Alloys Compd.* **2019**, *786*, 930–937. [CrossRef]

24. Massart, R. Preparation of aqueous magnetic liquids in alkaline and acidic media. *IEEE Trans. Magn.* **1981**, *17*, 1247–1248. [CrossRef]

25. Kalantari, K.; Ahmad, M.B.; Shameli, K.; Bin Hussein, M.Z.; Khandanlou, R.; Khanehzaei, H. Size-Controlled Synthesis of Fe$_3$O$_4$ Magnetic Nanoparticles in the Layers of Montmorillonite. *J. Nanomater.* **2014**, *2014*, 739485–739494. [CrossRef]

26. Mascolo, M.C.; Pei, Y.; Ring, T.A. Room Temperature Co-Precipitation Synthesis of Magnetite Nanoparticles in a Large pH Window with Different Bases. *Materials* **2013**, *6*, 5549–5567. [CrossRef] [PubMed]

27. Mo, Z.; Zhang, C.; Guo, R.; Meng, S.; Zhang, J. Synthesis of Fe$_3$O$_4$ Nanoparticles Using Controlled Ammonia Vapor Diffusion under Ultrasonic Irradiation. *Ind. Eng. Chem. Res.* **2011**, *50*, 63534–63539. [CrossRef]

28. Grumezescu, A.M.; Gestal, M.C.; Holban, A.M.; Grumezescu, V.; Vasile, B.S.; Mogoanta, L.; Iordache, F.; Bleotu, C.; Mogosanu, G.D. Biocompatible Fe_3O_4 increases the efficacy of amoxicillin delivery against Gram-positive and Gram-negative bacteria. *Molecules* **2014**, *19*, 5013–5027. [CrossRef] [PubMed]

29. Jiang, W.; Lai, K.L.; Hu, H.; Zeng, X.-B.; Lan, F.; Liu, F.; Liu, K.-X.; Wu, Y.; Gu, Z.-W. The effect of $[Fe^{3+}]/[Fe^{2+}]$ molar ratio and iron salts concentration on the properties of superparamagnetic iron oxide nanoparticles in the water/ethanol/toluene system. *J. Nanoparticle Res.* **2011**, *13*, 5135–5145. [CrossRef]

30. Ali, A.; Zafar, H.; Zia, M.; Ul Haq, I.; Phull, A.R.; Ali, J.S.; Hussain, A. Synthesis, characterization, applications, and challenges of iron oxide nanoparticles. *Nanotechnol. Sci. Appl.* **2016**, *9*, 49–67. [CrossRef]

31. Shi, L.; Huang, J.; He, Y. Recyclable purification-evaporation systems based on $Fe_3O_4@TiO_2$ nanoparticles. *Energy Procedia* **2017**, *142*, 356–361. [CrossRef]

32. Anghel, A.G.; Grumezescu, A.M.; Chirea, M.; Grumezescu, V.; Socol, G.; Iordache, F.; Oprea, A.E.; Anghel, I.; Holban, A.M. MAPLE Fabricated $Fe_3O_4@$Cinnamomum verum Antimicrobial Surfaces for Improved Gastrostomy Tubes. *Molecules* **2014**, *19*, 8981–8994. [CrossRef]

33. Shen, L.; Qiao, Y.; Guo, Y.; Meng, S.; Yang, G.; Wu, M.; Zhao, J. Facile co-precipitation synthesis of shape-controlled magnetite nanoparticles. *Ceram. Int.* **2014**, *40*, 1519–1524. [CrossRef]

34. Singh, A.K.; Srivastava, O.N.; Singh, K. Shape and Size-Dependent Magnetic Properties of Fe_3O_4 Nanoparticles Synthesized Using Piperidine. *Nanoscale Res. Lett.* **2017**, *12*, 298–305. [CrossRef]

35. Shah, S.T.; Yehya, W.A.; Saad, O.; Simarani, K.; Chowdhury, Z.; Alhadi, A.A.; Al-Ani, L.A. Surface Functionalization of Iron Oxide Nanoparticles with Gallic Acid as Potential Antioxidant and Antimicrobial Agents. *Nanomaterials* **2017**, *7*, 306. [CrossRef]

36. Zhu, N.; Ji, H.; Yu, P.; Niu, J.; Farooq, M.U.; Akram, M.W.; Udego, I.O.; Li, H.; Niu, X. Surface Modification of Magnetic Iron Oxide Nanoparticles. *Nanomaterials* **2018**, *8*, 810. [CrossRef] [PubMed]

37. Lassoued, A.; Dkhil, B.; Gadri, A.; Ammar, S. Control of the shape and size of iron oxide (α-Fe_2O_3) nanoparticles synthesized through the chemical precipitation method. *Res. Phys.* **2017**, *7*, 3007–3015. [CrossRef]

38. Li, J.L.; Li, D.C.; Zhang, S.L.; Cui, H.C.; Wang, C. Analysis of the factors affecting the magnetic characteristics of nano-Fe_3O_4 particles. *Chin. Sci. Bull.* **2011**, *8*, 803–810. [CrossRef]

39. Barbosa Salviano, L.; da Silva Cardoso, T.M.; Cordeiro Silva, G.; Silva Dantas, S.; de Mello Ferreira, A. Microstructural Assessment of Magnetite Nanoparticles (Fe_3O_4) Obtained by Chemical Precipitation Under Different Synthesis Conditions. *Mater. Res.* **2018**, *21*, e20170764. [CrossRef]

40. Meng, H.; Zhang, Z.; Zhao, F.; Qiu, T.; Yang, J. Orthogonal optimization design for preparation of Fe_3O_4 nanoparticles via chemical coprecipitation. *Appl. Surf. Sci.* **2013**, *280*, 679–685. [CrossRef]

41. Andrade, A.I.; Souza, D.M.; Pereira, M.C.; Fabris, J.D.; Domingues, R.Z. pH effect on the synthesis of magnetite nanoparticles by the chemical reduction-precipitation method. *Quim. Nova* **2010**, *33*, 524–527. [CrossRef]

42. Ramadan, W.; Karim, M.; Hannoyer, B.; Saha, S. Effect of pH on the Structural and Magnetic Properties of Magnetite Nanoparticles Synthesized by Co-Precipitation. *Adv. Mater. Res.* **2012**, *324*, 129–132. [CrossRef]

43. Kalska-Szostko, B.; Wykowska, U.; Satula, D.; Nordblad, P. Thermal treatment of magnetite nanoparticles. *Beilstein J. Nanotechnol.* **2015**, *6*, 1385–1396. [CrossRef]

44. Niu, J.M.; Zheng, Z.G. Effect of Temperature on Fe_3O_4 Nanoparticles prepared by Coprecipitation Method. *Adv. Mater. Res.* **2014**, *900*, 172–176. [CrossRef]

45. Saragi, T.; Depi, B.L.; Butarbutar, S.; Permana, B. The impact of synthesis temperature on magnetite nanoparticles size synthesized by co-precipitation method. *J. Phys. Conf. Ser.* **2018**, *1013*, 012190. [CrossRef]

46. Fayas, A.P.A.; Vinod, E.M.; Joseph, J.; Ganesan, R.; Pandey, R.K. Dependence of pH and surfactant effect in the synthesis of magneite (Fe_3O_4) nanopaticles and its properties. *J. Magn. Magn. Mater.* **2010**, *322*, 400–404. [CrossRef]

47. Filippousi, M.; Angelakeris, M.; Katsikini, M.; Paloura, E.; Esthimiopoulos, I.; Wang, Y.; Zamboulis, D.; Van Tendeloo, G. Surfactant Effects on the Structural and Magnetic Properties of Iron Oxide Nanoparticles. *J. Phys. Chem. C* **2014**, *118*, 16209–16217. [CrossRef]

48. Fatima, H.; Lee, D.-W.; Yun, H.J.; Kim, K.-S. Shape-controlled synthesis of magnetic Fe_3O_4 nanoparticles with different iron precursors and capping agents. *RSC Adv.* **2018**, *8*, 22917–22923. [CrossRef]

49. Fotukian, S.M.; Barati, A.; Soleymani, M.; Alizadeh, A.M. Solvothermal synthesis of $CuFe_2O_4$ and Fe_3O_4 nanoparticles with high heating efficiency for magnetic hyperthermia application. *J. Alloys Compd.* **2019**, 152548–152556. [CrossRef]

50. Zhang, W.; Shen, F.; Hong, R. Solvothermal synthesis of magnetic Fe_3O_4 microparticles via self-assembly of Fe_3O_4 nanoparticles. *Particuology* **2011**, *9*, 179–186. [CrossRef]

51. Qi, M.; Zhang, K.; Li, S.; Wu, J.; Pham-Hui, C.; Diao, X.; Xiao, D.; He, H. Superparamagnetic Fe_3O_4 nanoparticles: Synthesis by a solvothermal process and functionalization for a magnetic targeted curcumin delivery system. *New J. Chem.* **2016**, *40*, 4480–4491. [CrossRef]

52. Yan, J.; Mo, S.; Nie, J.; Chen, W.; Shen, X.; Hu, J.; Hao, G.; Tong, H. Hydrothermal synthesis of monodisperse Fe_3O_4 nanoparticles based on modulation of tartaric acid. *Colloids Surf. A Physicochem. Eng. Asp.* **2009**, *340*, 109–114. [CrossRef]

53. Lu, T.; Wang, J.; Yin, J.; Wang, A.; Wang, X.; Zhang, T. Surfactant effects on the microstructures of Fe_3O_4 nanoparticles synthesized by microemulsion method. *Colloids Surf. A Physicochem. Eng. Asp.* **2013**, *436*, 675–683. [CrossRef]

54. Su, H.; Han, X.; He, L.; Deng, L.; Yu, K.; Jiang, H.; Wu, C.; Jia, Q.; Shan, S. Synthesis and characterization of magnetic dextran nanogel doped with iron oxide nanoparticles as magnetic resonance imaging probe. *Int. J. Biol. Macromol.* **2019**, *128*, 768–774. [CrossRef]

55. Pham, X.N.; Nguyen, T.P.; Pham, T.N.; Tran, N.T.T.; Tran, T.V.T. Synthesis and characterization of chitosan-coated magnetite nanoparticles and their application in curcumin drug delivery. *Adv. Nat. Sci. Nanosci. Nanotechnol.* **2016**, *7*, 045010–045019. [CrossRef]

56. Unni, M.; Uhl, A.M.; Savliwala, S.; Savitzky, B.H.; Dhavalikar, R.; Garraud, N.; Arnold, D.P.; Kourkoutis, L.F.; Andrew, J.S.; Rinaldi, C. Thermal Decomposition Synthesis of Iron Oxide Nanoparticles with Diminished Magnetic Dead Layer by Controlled Addition of Oxygen. *ACS Nano* **2017**, *11*, 2284–2303. [CrossRef] [PubMed]

57. Kumfer, B.M.; Shinoda, K.; Jeyadevan, B.; Kennedy, I.M. Gas-phase flame synthesis and properties of magnetic iron oxide nanoparticles with reduced oxidation state. *J. Aerosol Sci.* **2010**, *41*, 257–265. [CrossRef] [PubMed]

58. Lassenberger, A.; Gruenewald, T.A.; van Oostrum, P.D.J.; Rennhofer, H.; Amenitsch, H.; Zirbs, R.; Lichtenegger, H.C.; Reimhult, E. Monodisperse Iron Oxide Nanoparticles by Thermal Decomposition: Elucidating Particle Formation by Second-Resolved in Situ Small-Angle X-ray Scattering. *Chem. Mater.* **2017**, *29*, 4511–4522. [CrossRef] [PubMed]

59. Wei, Y.; Han, B.; Hu, X.; Lin, Y.; Wang, X.; Deng, X. Synthesis of Fe_3O_4 Nanoparticles and their Magnetic Properties. *Procedia Eng.* **2012**, *27*, 632–637. [CrossRef]

60. Xu, C.; Lu, X.; Dai, H. The Synthesis of Size-Adjustable Superparamagnetism Fe_3O_4 Hollow Microspheres. *Nanoscale Res. Lett.* **2017**, *12*, 234–244. [CrossRef] [PubMed]

61. Zhang, S.; Li, J.; Lykotrafitis, G.; Bao, G.; Suresh, S. Size-Dependent Endocytosis of Nanoparticles. *Adv. Mater.* **2009**, *21*, 419–424. [CrossRef]

62. Bannunah, A.M.; Vllasaliu, D.; Lord, J.; Stolnik, S. Mechanisms of Nanoparticle Internalization and Transport Across an Intestinal Epithelial Cell Model: Effect of Size and Surface Charge. *Mol. Pharm.* **2014**, *11*, 4363–4373. [CrossRef]

63. Shang, L.; Nienhaus, K.; Nienhaus, G.U. Engineered nanoparticles interacting with cells: Size matters. *J. Nanobiotechnol.* **2014**, *12*, 5–16. [CrossRef]

64. Gratton, S.E.A.; Ropp, P.A.; Pohlhaus, P.D.; Luft, J.C.; Madden, V.J.; Napier, M.E.; DeSimone, J.M. The effect of particle design on cellular internalization pathways. *Proc. Natl. Acad. Sci. USA* **2008**, *105*, 11613–11618. [CrossRef]

65. Xie, X.; Liao, J.; Shao, X.; Li, Q.; Lin, Y. The Effect of shape on Cellular Uptake of Gold Nanoparticles in the forms of Stars, Rods, and Triangles. *Sci. Rep.* **2017**, *7*, 3827–3836. [CrossRef] [PubMed]

66. Toy, R.; Peiris, P.M.; Ghaghada, K.B.; Karathanasis, E. Shaping cancer nanomedicine: The effect of particle shape on the in vivo journey of nanoparticles. *Nanomedicine* **2014**, *9*, 121–134. [CrossRef] [PubMed]

67. Chen, L.; Xiao, S.; Zhu, H.; Liang, H. Shape-dependent internalization kinetics of nanoparticles by membranes. *Soft Matter* **2016**, *12*, 2632–2641. [CrossRef] [PubMed]

68. Cui, Y.N.; Xu, Q.X.; Davoodi, P.; Wang, D.P.; Wang, C.H. Enhanced intracellular delivery and controlled drug release of magnetic PLGA nanoparticles modified with transferrin. *Acta Pharmacol. Sin.* **2017**, *38*, 943–953. [CrossRef] [PubMed]

69. Georgieva, J.V.; Kalicharan, D.; Couraud, P.-O.; Romero, I.A.; Weksler, B.; Hoekstra, D.; Zuhorn, I.S. Surface Characteristics of Nanoparticles Determine Their Intracellular Fate in and Processing by Human Blood–Brain Barrier Endothelial Cells In Vitro. *Mol. Ther.* **2011**, *19*, 318–325. [CrossRef] [PubMed]

70. Arias, L.S.; Pessan, J.P.; Vieira, A.P.M.; Lima, T.M.T.; Delbem, A.C.B.; Monteiro, D.R. Iron Oxide Nanoparticles for Biomedical Applications: A Perspective on Synthesis, Drugs, Antimicrobial Activity, and Toxicity. *Antibiotics* **2018**, *7*, 46. [CrossRef]

71. Ma, P.; Luo, Q.; Chen, J.; Gan, Y.; Du, J.; Ding, S.; Xi, Z.; Yang, X. Intraperitoneal injection of magnetic Fe_3O_4-nanoparticle induces hepatic and renal tissue injury via oxidative stress in mice. *Int. J. Nanomed.* **2012**, *7*, 4809–4818. [CrossRef]

72. Wang, Y.; Qin, N.; Chen, S.; Zhao, J.; Yang, X. Oxidative-damage effect of Fe_3O_4 nanoparticles on mouse hepatic and brain cells in vivo. *Front. Biol.* **2013**, *8*, 549–555. [CrossRef]

73. Wang, J.; Chen, Y.; Chen, B.; Ding, J.; Xia, G.; Gao, C.; Cheng, J.; Jin, N.; Zhou, Y.; Li, X.; et al. Pharmacokinetic parameters and tissue distribution of magnetic Fe_3O_4 nanoparticles in mice. *Int. J. Nanomed.* **2010**, *5*, 861–866. [CrossRef]

74. Mejías, R.; Gutiérrez, L.; Salas, G.; Pérez-Yagüe, S.; Zotes, T.M.; Lázaro, F.J.; Morales, M.P.; Barber, D.F. Long term biotransformation and toxicity of dimercaptosuccinic acid-coated magnetic nanoparticles support their use in biomedical applications. *J. Control. Release* **2013**, *171*, 225–233. [CrossRef]

75. De Tercero, M.D.; Bruns, M.; Martínez, I.G.; Türk, M.; Fehrenbacher, U.; Jennewein, S.; Barner, L. Continuous Hydrothermal Synthesis of In Situ Functionalized Iron Oxide Nanoparticles: A General Strategy to Produce Metal Oxide Nanoparticles With Clickable Anchors. *Part. Part. Syst. Charact.* **2013**, *30*, 229–234. [CrossRef]

76. De Tercero, M.D.; Gonzáles Martínez, I.; Herrmann, M.; Bruns, M.; Kübel, C.; Jennewein, S.; Fehrenbacher, U.; Barner, L.; Türk, M. Synthesis of in situ functionalized iron oxide nanoparticles presenting alkyne groups via a continuous process using near-critical and supercritical water. *J. Supercrit. Fluids* **2013**, *82*, 83–95. [CrossRef]

77. Karimzadeh, I.; Aghazadeh, M.; Doroudi, T.; Ganjali, M.R.; Kolivand, P.H. Superparamagnetic Iron Oxide (Fe_3O_4) Nanoparticles Coated with PEG/PEI for Biomedical Applications: A Facile and Scalable Preparation Route Based on the Cathodic Electrochemical Deposition Method. *Adv. Phys. Chem.* **2017**, *2017*, 9437487–9437494. [CrossRef]

78. Bini, R.A.; Marques, R.F.C.; Santos, F.J.; Chaker, J.A.; Jafelicci, M. Synthesis and functionalization of magnetite nanoparticles with different amino-functional alkoxysilanes. *J. Magn. Magn. Mater.* **2012**, *324*, 534–539. [CrossRef]

79. Rudakovskaya, P.G.; Gerasimov, V.M.; Metelkina, O.N.; Beloglazkina, E.K.; Savchenko, A.G.; Shchetinin, I.V.; Salikhov, S.V.; Abakumov, M.A.; Klyachko, N.L.; Golovin, Y.I.; et al. Synthesis and characterization of PEG-silane functionalized iron oxide(II, III) nanoparticles for biomedical application. *Nanotechnol. Russ.* **2015**, *10*, 896–903. [CrossRef]

80. Wang, L.; Li, Y.R.; Li, J.; Zou, S.; Stach, E.A.; Takeuchi, K.J.; Takeuchi, E.S.; Marschilok, A.C.; Wong, S.S. Correlating Preparative Approaches with Electrochemical Performance of Fe_3O_4-MWNT Composites Used as Anodes in Li-Ion Batteries. *J. Solid State Sci. Technol.* **2017**, *6*, M3122–M3131. [CrossRef]

81. Jokerst, J.V.; Lobovkina, T.; Zare, R.N.; Gambhir, S.S. Nanoparticle PEGylation for imaging and therapy. *Nanomedicine* **2011**, *6*, 715–728. [CrossRef]

82. Wei, W.; Bai, F.; Fan, H. Surfactant-Assisted Cooperative Self-Assembly of Nanoparticles into Active Nanostructures. *iScience* **2019**, *11*, 272–293. [CrossRef]

83. Chen, L.; Wu, L.; Liu, F.; Qi, X.; Ge, Y.; Shen, S. Azo-functionalized Fe_3O_4 nanoparticles: A near-infrared light triggered drug delivery system for combined therapy of cancer with low toxicity. *J. Mater. Chem. B* **2016**, *4*, 3660–3669. [CrossRef]

84. Gawali, S.L.; Barick, K.C.; Shetake, N.G.; Rajan, V.; Pandey, B.N.; Kumar, N.N.; Priyadarsini, N.K.I.; Hassan, P.A. pH-Labile Magnetic Nanocarriers for Intracellular Drug Delivery to Tumor Cells. *ACS Omega* **2019**, *47*, 11728–11736. [CrossRef]

85. Sharma, K.S.; Ningthoujam, R.S.; Dubey, A.K.; Chattopadhyay, A.; Phapale, S.; Juluri, R.R.; Mukherjee, S.; Tewari, R.; Shetake, N.G.; Pandey, B.N.; et al. Synthesis and characterization of monodispersed water dispersible Fe₃O₄ nanoparticles and in vitro studies on human breast carcinoma cell line under hyperthermia condition. *Sci. Rep.* **2018**, *8*, 14766–14777. [CrossRef] [PubMed]

86. Demin, A.M.; Krasnov, V.P.; Charushin, V.N. Covalent Surface Modification of Fe₃O₄ Magnetic Nanoparticles with Alkoxy Silanes and Amino Acids. *Mendeleev Commun.* **2013**, *23*, 14–16. [CrossRef]

87. Arsalani, N.; Fattahi, H.; Nazarpoor, M. Synthesis and characterization of PVP-functionalized superparamagnetic Fe₃O₄ nanoparticles as an MRI contrast agent. *eXPRESS Polym. Lett.* **2010**, *4*, 329–338. [CrossRef]

88. Han, C.W.; Choksi, T.; Milligan, C.A.; Majumdar, P.; Manto, M.J.; Cui, Y.; Sang, X.; Unocic, R.R.; Zemlyanov, D.Y.; Wang, C.; et al. A Discovery of Strong Metal-Support Bonding in Nano-engineered Au-Fe₃O₄ Dumbbell-like Nanoparticles by In-situ Transmission Electron Microscopy. *Nano Lett.* **2017**, *17*, 4576–4582. [CrossRef] [PubMed]

89. Abbas, M.; Torati, S.R.; Iqbal, S.A.; Kim, C.G. A novel and rapid approach for the synthesis of biocompatible and highly stable Fe₃O₄/SiO₂ and Fe₃O₄/C core/shell nanocubes and nanorods. *New J. Chem.* **2017**, *41*, 2724–2734. [CrossRef]

90. Khosroshahi, M.E.; Ghazanfari, L.; Tahriri, M. Characterisation of binary (Fe₃O₄/SiO₂) biocompatible nanocomposites as magnetic fluid. *J. Exp. Nanosci.* **2011**, *6*, 580–595. [CrossRef]

91. Guo, X.; Mao, F.; Wang, W.; Yang, Y.; Bai, Z. Sulfhydryl-Modified Fe₃O₄@SiO₂ Core/Shell Nanocomposite: Synthesis and Toxicity Assessment in Vitro. *ACS Appl. Mater. Interfaces* **2015**, *7*, 14983–14991. [CrossRef]

92. Nikmah, A.; Taufiq, A.; Hidayat, A. Synthesis and Characterization of Fe₃O₄/SiO₂ nanocomposites. *IOP Conf. Ser. Earth Environ. Sci.* **2019**, *276*, 012046.

93. Ding, H.L.; Zhang, X.Y.; Wang, S.; Xu, J.M.; Xu, S.C.; Li, G.H. Fe₃O₄@SiO₂ Core/Shell Nanoparticles: The Silica Coating Regulations with a Single Core for Different Core Sizes and Shell Thicknesses. *Chem. Mater.* **2012**, *2423*, 4572–4580. [CrossRef]

94. Liu, C.Y.; Puig, T.; Obradors, X.; Ricart, S.; Ros, J. Ultra-fast microwave-assisted reverse microemulsion synthesis of Fe₃O₄@SiO₂ core–shell nanoparticles as a highly recyclable silver nanoparticle catalytic platform in the reduction of 4-nitroaniline. *RSC Adv.* **2016**, *6*, 88762–88769. [CrossRef]

95. Zhang, Y.; Yan, J.; Liu, S. Biocompatibility and biomedical applications of functionalized mesoporous silica nanoparticles. *Biointerface Res. Appl. Chem.* **2014**, *4*, 767–775.

96. Tang, F.; Li, L.; Chen, D. Mesoporous Silica Nanoparticles: Synthesis. Biocompatibility and Drug Delivery. *Adv. Mater.* **2012**, *24*, 1504–1534. [CrossRef]

97. Asefa, T.; Tao, Z. Biocompatibility of Mesoporous Silica Nanoparticles. *Chem. Res. Toxicol.* **2012**, *25*, 2265–2284. [CrossRef]

98. Narayan, R.; Nayak, U.Y.; Raichur, A.M.; Garg, S. Mesoporous Silica Nanoparticles: A Comprehensive Review on Synthesis and Recent Advances. *Pharmaceutics* **2018**, *10*, 118. [CrossRef] [PubMed]

99. Isa, E.D.M.; Ahmad, H.; Rahman, M.B.A. Optimization of Synthesis Parameters of Mesoporous Silica Nanoparticles Based on Ionic Liquid by Experimental Design and Its Application as a Drug Delivery Agent. *J. Nanomater.* **2019**, *2019*, 4982054–4982062. [CrossRef]

100. Jorge, J.; Verelst, M.; de Castro, G.R.; Martines, M.A.U. Synthesis parameters for control of mesoporous silica nanoparticles (MSNs). *Biointerface Res. Appl. Chem.* **2016**, *6*, 1520–1524.

101. Sun, J.-G.; Jiang, Q.; Zhang, X.-P.; Shan, K.; Liu, B.-H.; Zhao, C.; Yan, B. Mesoporous silica nanoparticles as a delivery system for improving antiangiogenic therapy. *Int. J. Nanomed.* **2019**, *14*, 1489–1501. [CrossRef] [PubMed]

102. Vallet-Regí, M.; Colilla, M.; Izquierdo-Barba, I.; Manzano, M. Mesoporous Silica Nanoparticles for Drug Delivery: Current Insights. *Molecules* **2017**, *23*, 47. [CrossRef] [PubMed]

103. Ye, F.; Laurent, S.; Fornara, A.; Astolfi, L.; Qin, J.; Roch, A.; Martini, A.; Toprak, M.S.; Muller, R.N.; Muhammed, M. Uniform mesoporous silica coated iron oxide nanoparticles as a highly efficient, nontoxic MRI T2 contrast agent with tunable proton relaxivities. *Contrast Media Mol. Imaging* **2012**, *7*, 460–468. [CrossRef] [PubMed]

104. Sharafi, Z.; Bakhshi, B.; Javidi, J.; Adrangi, S. Synthesis of Silica-coated Iron Oxide Nanoparticles: Preventing Aggregation without Using Additives or Seed Pretreatment. *Iran. J. Pharm. Res.* **2018**, *17*, 386–395. [PubMed]

105. Yin, N.Q.; Wu, P.; Yang, T.H.; Wang, M. Preparation and study of a mesoporous silica-coated Fe₃O₄ photothermal nanoprobe. *RSC Adv.* **2017**, *7*, 9123–9129. [CrossRef]

106. Venkatathri, N. Synthesis of mesoporous silica nanosphere using different templates. *Solid State Commun.* **2007**, *143*, 493–497. [CrossRef]

107. Kipkemboi, P.; Fogden, A.; Alfredsson, V.; Flostroem, K. Triblock Copolymers as Templates in Mesoporous Silica Formation: Structural Dependence on Polymer Chain Length and Synthesis Temperature. *Langmuir* **2001**, *17*, 5398–5402. [CrossRef]

108. Peralta, M.E.; Jadhav, S.A.; Magnacca, G.; Scalarone, D.; Mártire, D.O.; Parolo, M.E.; Carlos, L. Synthesis and in vitro testing of thermoresponsive polymer-grafted core-shell magnetic mesoporous silica nanoparticles for efficient controlled and targeted drug delivery. *J. Colloid Interface Sci.* **2019**, *544*, 198–205. [CrossRef]

109. Park, S.S.; Jung, M.H.; Lee, Y.-S.; Bae, J.-H.; Kim, S.-H.; Ha, C.-S. Functionalised mesoporous silica nanoparticles with excellent cytotoxicity against various cancer cells for pH-responsive and controlled drug delivery. *Mater. Des.* **2019**, *184*, 108187–108197. [CrossRef]

110. Li, T.; Geng, T.; Md, A.; Banerjee, P.; Wang, B. Novel scheme for rapid synthesis of hollow mesoporous silica nanoparticles (HMSNs) and their application as an efficient delivery carrier for oral bioavailability improvement of poorly water-soluble BCS type II drugs. *Colloids Surf. B Biointerfaces* **2019**, *176*, 185–193. [CrossRef] [PubMed]

111. Rowley, J.; Abu-Zahra, N.H. Synthesis and characterization of polyethersulfone membranes impregnated with (3-aminopropyltriethoxysilane) APTES-Fe_3O_4 nanoparticles for As(V) removal from water. *J. Environ. Chem. Eng.* **2019**, *7*, 102875–102885. [CrossRef]

112. Liang, X.X.; Ouyang, X.K.; Wang, S.; Yang, L.-Y.; Huang, F.; Ji, C.; Chen, X. Efficient adsorption of Pb(II) from aqueous solutions using aminopropyltriethoxysilane-modified magnetic attapulgite@chitosan (APTS-Fe_3O_4/APT@CS) composite hydrogel beads. *Int. J. Biol. Macromol.* **2019**, *137*, 741–750. [CrossRef]

113. Langeroudi, M.P.; Binaeian, E. Tannin-APTES modified Fe_3O_4 nanoparticles as a carrier of Methotrexate drug: Kinetic, isotherm and thermodynamic studies. *Mater. Chem. Phys.* **2018**, *218*, 210–217. [CrossRef]

114. Arum, Y.; Yun-Ok, O.; Kang, H.W.; Seok-Hwan, A.; Junghwan, O. Chitosan-Coated Fe_3O_4 Magnetic Nanoparticles as Carrier of Cisplatin for Drug Delivery. *Fish. Aquat. Sci.* **2015**, *18*, 89–98. [CrossRef]

115. Zhang, S.; Zhang, Y.; Liu, J.; Xu, Q.; Xiao, H.; Wang, X.; Xu, H.; Zhou, J. Thiol modified Fe_3O_4@SiO_2 as a robust, high effective, and recycling magnetic sorbent for mercury removal. *Chem. Eng. J.* **2013**, *226*, 30–38. [CrossRef]

116. Badragheh, S.; Zeeb, M.; Olyai, M.R.T.B. Silica-coated magnetic iron oxide functionalized with hydrophobic polymeric ionic liquid:a promising nanoscale sorbent for simultaneous extraction of antidiabetic drugs from human plasma prior to their quantitation by HPLC. *RSC Adv.* **2018**, *8*, 30550–30561. [CrossRef]

117. Rego, G.N.A.; Mamani, J.B.; Souza, T.K.F.; Nucci, M.P.; Silva, H.R.D.; Gamarra, L.F. Therapeutic evaluation of magnetic hyperthermia using Fe_3O_4-aminosilane-coated iron oxide nanoparticles in glioblastoma animal model. *Einstein* **2019**, *17*, 1–9. [CrossRef] [PubMed]

118. Shaleri Kardar, Z.S.; Beyki, M.H.; Shemirani, F. Bifunctional aminosilane-functionalized Fe_3O_4 nanoparticles as efficient sorbent for preconcentration of cobalt ions from food and water samples. *Res. Chem. Intermed.* **2017**, *43*, 4079–4096. [CrossRef]

119. Tian, F.; Chen, G.; Yi, P.; Zhang, J.; Li, A.; Zhang, J.; Zheng, L.; Deng, Z.; Shi, Q.; Peng, R.; et al. Fates of Fe_3O_4 and Fe_3O_4@SiO_2 nanoparticles in human mesenchymal stem cells assessed by synchrotron radiation-based techniques. *Biomaterials* **2014**, *35*, 6412–6421. [CrossRef] [PubMed]

120. Mostafaei, M.; Hosseini, S.N.; Khatami, M.; Javidanbardan, A.; Sepahy, A.A.; Asadi, E. Isolation of recombinant Hepatitis B surface antigen with antibody-conjugated superparamagnetic Fe_3O_4/SiO_2 core-shell nanoparticles. *Protein Expr. Purif.* **2018**, *145*, 1–6. [CrossRef]

121. Fan, Q.; Guan, Y.; Zhang, Z.; Xu, G.; Yang, Y.; Guo, C. A new method of synthesis well-dispersion and dense Fe_3O_4@SiO_2 magnetic nanoparticles for DNA extraction. *Chem. Phys. Lett.* **2019**, *715*, 7–13. [CrossRef]

122. Gan, Q.; Lu, X.; Yuan, Y.; Qian, J.; Zhou, H.; Lu, X.; Shi, J.; Liu, C. A magnetic, reversible pH-responsive nanogated ensemble based on Fe_3O_4 nanoparticles-capped mesoporous silica. *Biomaterials* **2011**, *32*, 1932–1942. [CrossRef]

123. Cui, J.; Sun, B.; Lin, T.; Feng, Y.; Jia, S. Enzyme shielding by mesoporous organosilica shell on Fe_3O_4@silica yolk-shell nanospheres. *Int. J. Biol. Macromol.* **2018**, *117*, 673–682. [CrossRef]

124. Asgari, M.; Soleymani, M.; Miri, T.; Barati, A. A robust method for fabrication of monodisperse magnetic mesoporous silica nanoparticles with core-shell structure as anticancer drug carriers. *J. Mol. Liquids* **2019**, *292*, 111367–111375. [CrossRef]

125. Wang, J.; Yang, J.; Li, X.; Wang, D.; Wei, B.; Song, H.; Li, X.; Fu, S. Preparation and photocatalytic properties of magnetically reusable Fe$_3$O$_4$@ZnO core/shell nanoparticles. *Phys. E* **2016**, *75*, 66–71. [CrossRef]

126. Zhang, L.; Wu, Z.; Chen, L.; Zhang, L.; Li, X.; Xu, H.; Wang, H.; Zhu, G. Preparation of magnetic Fe$_3$O$_4$/TiO$_2$/Ag composite microspheres with enhanced photocatalytic activity. *Solid State Sci.* **2016**, *52*, 42–48. [CrossRef]

127. Choi, K.H.; Min, J.; Park, S.Y.; Park, B.J.; Jung, J.S. Enhanced photocatalytic degradation of tri-chlorophenol by Fe$_3$O$_4$@TiO$_2$@Au photocatalyst under visible-light. *Ceram. Int.* **2019**, *45*, 9477–9482. [CrossRef]

128. Li, Y.; Liu, J.; Zhong, Y.; Zhang, J.; Wang, Z.; Wang, L.; An, Y.; Lin, M.; Gao, Z.; Zhang, D. Biocompatibility of Fe$_3$O$_4$@Au composite magnetic nanoparticles in vitro and in vivo. *Int. J. Nanomed.* **2011**, *6*, 2805–2819. [CrossRef]

129. Nalluri, S.R.; Nagarjuna, R.; Patra, D.; Ganesan, R.; Balaji, G. Large Scale Solid-state Synthesis of Catalytically Active Fe$_3$O$_4$@M (M = Au, Ag and Au-Ag alloy) Core-shell Nanostructures. *Sci. Rep.* **2019**, *9*, 6603–6614. [CrossRef] [PubMed]

130. Ma, L.L.; Borwankar, A.U.; Willsey, B.W.; Yoon, K.Y.; Tam, J.O.; Sokolov, K.V.; Feldman, M.D.; Milner, T.E.; Johnston, K.P. Growth of textured thin Au coatings on iron oxide nanoparticles with near infrared absorbance. *Nanotechnology* **2013**, *24*, 025606–025620. [CrossRef] [PubMed]

131. Sanchez, L.M.; Alvarez, V.A. Advances in Magnetic Noble Metal/Iron-Based Oxide Hybrid Nanoparticles as Biomedical Devices. *Bioengineering* **2019**, *6*, 75. [CrossRef] [PubMed]

132. Ma, C.; Shao, H.; Zhan, S.; Hou, P.; Zhang, X.; Chai, Y.; Liu, H. Bi-phase dispersible Fe$_3$O$_4$@Au core–shell multifunctional nanoparticles: Synthesis, characterization and properties. *Compos. Interfaces* **2019**, *26*, 537–549. [CrossRef]

133. Zhu, H.; Zhu, E.; Ou, G.; Gao, L.; Chen, J. Fe$_3$O$_4$-Au and Fe$_2$O$_3$-Au Hybrid Nanorods: Layer-by-Layer Assembly Synthesis and Their Magnetic and Optical Properties. *Nanoscale Res. Lett.* **2010**, *5*, 1755–1761. [CrossRef] [PubMed]

134. Lu, Q.; Dai, X.; Zhang, P.; Tan, X.; Zhong, Y.; Yao, C.; Song, M.; Song, G.; Zhang, G.; Peng, G.; et al. Fe$_3$O$_4$@Au composite magnetic nanoparticles modified with cetuximab for targeted magneto-photothermal therapy of glioma cells. *Int. J. Nanomed.* **2018**, *13*, 2491–2505. [CrossRef] [PubMed]

135. Wang, L.; Wang, L.; Luo, J.; Fan, Q.; Suzuki, M.; Suzuki, I.S.; Engelhard, M.H.; Lin, Y.; Kim, N.; Jian, Q.; et al. Monodispersed Core−Shell Fe$_3$O$_4$@Au Nanoparticles. *J. Phys. Chem. B* **2005**, *109*, 21593–21601. [CrossRef] [PubMed]

136. Xu, Z.; Hou, Y.; Sun, S. Magnetic Core/Shell Fe$_3$O$_4$/Au and Fe$_3$O$_4$/Au/Ag Nanoparticles with Tunable Plasmonic Properties. *J. Am. Chem. Soc.* **2007**, *129*, 8698–8699. [CrossRef] [PubMed]

137. Hu, Y.; Wang, R.; Wang, S.; Ding, L.; Li, J.; Luo, Y.; Wang, X.; Shen, M.; Shi, X. Multifunctional Fe$_3$O$_4$ @ Au core/shell nanostars: A unique platform for multimode imaging and photothermal therapy of tumors. *Sci. Rep.* **2016**, *6*, 28325–28337. [CrossRef] [PubMed]

138. Ge, Y.; Zhong, Y.; Ji, G.; Lu, Q.; Dai, X.; Guo, Z.; Zhang, P.; Peng, G.; Zhang, K.; Li, Y. Preparation and characterization of Fe$_3$O$_4$@Au-C225 composite targeted nanoparticles for MRI of human glioma. *PLoS ONE* **2018**, *13*, e0195703. [CrossRef]

139. Kang, N.; Xu, D.; Han, Y.; Lv, X.; Chen, Z.; Zhou, T.; Ren, L.; Zhou, X. Magnetic targeting core/shell Fe$_3$O$_4$/Au nanoparticles for magnetic resonance/photoacoustic dual-modal imaging. *Mater. Sci. Eng. C* **2019**, *98*, 545–549. [CrossRef]

140. Klein, S.; Hübner, J.; Menter, C.; Distel, L.V.R.; Neuhuber, W.; Kryschi, C. A Facile One-Pot Synthesis of Water-Soluble, Patchy Fe$_3$O$_4$-Au Nanoparticles for Application in Radiation Therapy. *Appl. Sci.* **2019**, *9*, 15. [CrossRef]

141. Hu, R.; Zheng, M.; Wu, J.; Li, C.; Shen, D.; Yang, D.; Li, L.; Ge, M.; Chang, Z.; Dong, W. Core-Shell Magnetic Gold Nanoparticles for Magnetic Field-Enhanced Radio-Photothermal Therapy in Cervical Cancer. *Nanomaterials* **2017**, *7*, 111. [CrossRef]

142. Li, J.; Hu, Y.; Yang, J.; Wei, P.; Sun, W.; Shen, M.; Zhang, G.; Shi, X. Hyaluronic acid-modified Fe$_3$O$_4$@Au core/shell nanostars for multimodal imaging and photothermal therapy of tumors. *Biomaterials* **2015**, *38*, 10–21. [CrossRef]

143. Park, S.I.; Chung, S.H.; Kim, H.C.; Lee, S.G.; Lee, S.J.; Kim, H.; Kim, H.; Jeong, S.W. Prolonged heating of Fe$_3$O$_4$–Au hybrid nanoparticles in a radiofrequency solenoid coil. *Colloids Surf. A Physicochem. Eng. Asp.* **2018**, *538*, 304–309. [CrossRef]

144. Her, S.; Jaffray, D.A.; Allen, C. Gold nanoparticles for applications in cancer radiotherapy: Mechanisms and recent advancements. *Adv. Drug Deliv. Rev.* **2017**, *109*, 84–101. [CrossRef]

145. Yu, B.; Liu, T.; Du, X.; Luo, Z.; Zheng, W.; Chen, T. X-ray-responsive selenium nanoparticles for enhanced cancer chemo-radiotherapy. *Colloids Surf. B Biointerfaces* **2016**, *139*, 180–189. [CrossRef] [PubMed]

146. Mattea, F.; Vedelago, J.; Malano, F.; Gomez, C.; Strumia, M.C.; Valente, M. Silver nanoparticles in X-ray biomedical applications. *Radiat. Phys. Chem.* **2017**, *130*, 442–450. [CrossRef]

147. Berbeco, R.I.; Ngwa, W.; Makrigiorgos, G.M. Localized Dose Enhancement to Tumor Blood Vessel Endothelial Cells via Megavoltage X-rays and Targeted Gold Nanoparticles: New Potential for External Beam Radiotherapy. *Int. J. Radiat. Oncol. Biol. Phys.* **2011**, *81*, 270–276. [CrossRef] [PubMed]

148. Generalov, R.; Kuan, W.B.; Chen, W.; Kristensen, S.; Juzenas, P. Radiosensitizing effect of zinc oxide and silica nanocomposites on cancer cells. *Colloids Surf. B Biointerfaces* **2015**, *129*, 79–86. [CrossRef]

149. Wolfe, T.; Chatterjee, D.; Lee, J.; Grant, J.D.; Bhattarai, S.; Tailor, R.; Goodrich, G.; Nicolucci, P.; Krishnan, S. Targeted gold nanoparticles enhance radiosensitization of prostate tumors to megavoltage radiation therapy in vivo. *Nanomedicine* **2015**, *11*, 1277–1283. [CrossRef]

150. Miladi, I.; Aloy, M.T.; Armandy, E.; Mowat, P.; Kryza, D.; Magne, N.; Tillement, O.; Lux, F.; Billotey, C.; Janier, M.; et al. Combining ultrasmall gadolinium-based nanoparticles with photon irradiation overcomes radioresistance of head and neck squamous cell carcinoma. *Nanomedicine* **2015**, *11*, 247–257. [CrossRef]

151. Liu, Y.; Liu, X.; Jin, X.; He, P.; Zheng, X.; Dai, Z.; Ye, F.; Zhao, T.; Chen, W.; Li, Q. The dependence of radiation enhancement effect on the concentration of gold nanoparticles exposed to low- and high- LET radiations. *Eur. J. Med. Phys.* **2015**, *31*, 210–218. [CrossRef]

152. Roduner, E. Size matters: Why nanomaterials are different. *Chem. Soc. Rev.* **2006**, *35*, 583–592. [CrossRef]

153. Eustis, S.; El-Sayed, M.A. Why gold nanoparticles are more precious than pretty gold: Noble metal surface plasmon resonance and its enhancement of the radiative and nonradiative properties of nanocrystals of different shapes. *Chem. Soc. Rev.* **2006**, *35*, 209–217. [CrossRef]

154. Prado-Gotor, R.; Grueso, E. A kinetic study of the interaction of DNA with gold nanoparticles: Mechanistic aspects of the interaction. *Phys. Chem. Chem. Phys.* **2011**, *13*, 1479–1489. [CrossRef]

155. Li, K.; Zhao, X.; Hammer, B.K.; Du, S.; Chen, Y. Nanoparticles Inhibit DNA Replication by Binding to DNA: Modeling and Experimental Validation. *ACS Nano* **2013**, *7*, 9664–9674. [CrossRef] [PubMed]

156. Glazer, E.S.; Curley, S.A. Non-invasive radiofrequency ablation of malignancies mediated by quantum dots, gold nanoparticles and carbon nanotubes. *Ther. Deliv.* **2011**, *2*, 1325–1330. [CrossRef] [PubMed]

157. Li, J.; Liang, Z.; Zhong, X.; Zhao, Z.; Li, J. Study on the Thermal Characteristics of Fe_3O_4 Nanoparticles and Gelatin Compound for Magnetic Fluid Hyperthermia in Radiofrequency Magnetic Field. *IEEE Trans. Magn.* **2014**, *50*, 1–4. [CrossRef]

158. Liu, B.; Zhou, J.; Zhang, B.; Qu, J. Synthesis of $Ag@Fe_3O_4$ Nanoparticles for Photothermal Treatment of Ovarian Cancer. *J. Nanomater.* **2019**, *2019*, 6457968–6457976. [CrossRef]

159. Wang, J.; Sun, Y.; Wang, L.; Zhu, X.; Zhang, H.; Song, D. Surface plasmon resonance biosensor based on Fe_3O_4/Au nanocomposites. *Colloids Surf. B Biointerfaces* **2010**, *81*, 600–606. [CrossRef]

160. Zhao, Y.; Zhang, W.; Lin, Y.; Du, D. The vital function of Fe_3O_4@Au nanocomposite for hydrolase biosensor design and its application in detection of methyl parathion. *Nanoscale* **2013**, *5*, 1121–1126. [CrossRef]

161. Liu, F.M.; Nie, J.; Qin, Y.N.; Yin, W.; Hou, C.J.; Huo, D.Q.; He, B.; Xia, T.C. A biomimetic sensor based on specific receptor ETBD and Fe_3O_4@Au/MoS_2/GN for signal enhancement shows highly selective electrochemical response to ultra-trace lead (II). *J. Solid State Electrochem.* **2017**, *21*, 3257–3268. [CrossRef]

162. Li, S.; Liang, J.; Zhou, Z.; Li, G. An electrochemical immunosensor for AFP measurement based on the magnetic Fe_3O_4@Au@CS nanomaterials. *IOP Conf. Ser. Mater. Sci. Eng.* **2018**, *382*, 022017. [CrossRef]

163. Cui, Y.R.; Hong, C.; Zhou, Y.L.; Li, Y.; Gao, X.M.; Zhang, X.X. Synthesis of orientedly bioconjugated core/shell Fe_3O_4@Au magnetic nanoparticles for cell separation. *Talanta* **2011**, *85*, 1246–1252. [CrossRef]

164. Li, J.; Zou, S.; Gao, J.; Liang, J.; Zhou, H.; Liang, L.; Wu, W. Block copolymer conjugated Au-coated Fe_3O_4 nanoparticles as vectors for enhancing colloidal stability and cellular uptake. *J. Nanobiotechnol.* **2017**, *15*, 56–67. [CrossRef]

165. Li, Y.; Yun, K.H.; Lee, H.; Goh, S.H.; Suh, Y.G.; Choi, Y. Porous platinum nanoparticles as a high-Z and oxygen generating nanozyme for enhanced radiotherapy in vivo. *Biomaterials* **2019**, *197*, 12–19. [CrossRef] [PubMed]

166. Samadi, A.; Klingberg, H.; Jauffred, L.; Kjaer, A.; Bendix, P.M.; Oddershede, L.B. Platinum nanoparticles: A non-toxic, effective and thermally stable alternative plasmonic material for cancer therapy and bioengineering. *Nanoscale* **2018**, *10*, 9097–9107. [CrossRef] [PubMed]

167. Ma, M.; Xie, J.; Zhang, Y.; Chen, Z.; Gu, N. Fe$_3$O$_4$@Pt nanoparticles with enhanced peroxidase-like catalytic activity. *Mater. Lett.* **2013**, *105*, 36–39. [CrossRef]

168. Wu, D.; Ma, H.; Zhang, Y.; Jia, H.; Yan, T.; Wei, Q. Corallite-like Magnetic Fe$_3$O$_4$@MnO$_2$@Pt Nanocomposited as Multiple Signal Amplifiers for the Detection of Carcinoembryonic Antigen. *ACS Appl. Mater. Interfaces* **2015**, *7*, 18786–18793. [CrossRef] [PubMed]

169. Khaghani, S.; Ghanbari, D.J. Magnetic and photo-catalyst Fe$_3$O$_4$–Ag nanocomposite: Green preparation of silver and magnetite nanoparticles by garlic extract. *Mater. Sci. Mater. Electron.* **2017**, *28*, 2877–2886. [CrossRef]

170. Gao, G.; Wang, K.; Huang, P.; Zhang, Y.; Bao, C.; Cui, D. Superparamagnetic Fe$_3$O$_4$–Ag hybrid nanocrystals as a potential contrast agent for CT imaging. *Cryst. Eng. Commun.* **2012**, *14*, 7556–7559. [CrossRef]

171. Sadat, M.E.; Baghbador, M.K.; Dunn, A.W.; Wagner, H.P.; Ewing, R.C.; Zhang, J.; Xu, H.; Pauletti, G.M.; Mast, D.B.; Shi, D. Photoluminescence and photothermal effect of Fe$_3$O$_4$ nanoparticles for medical imaging and therapy. *Appl. Phys. Lett.* **2014**, *105*, 091903. [CrossRef]

172. Zhang, X.; Liu, Z.; Lou, Z.; Chen, F.; Chang, S.; Miao, Y.; Zhou, Z.; Hu, X.; Feng, J.; Ding, Q.; et al. Radiosensitivity enhancement of Fe$_3$O$_4$@Ag nanoparticles on human glioblastoma cells. *Artif. Cells Nanomed. Biotechnol.* **2018**, *46* (Suppl. S1), 975–984. [CrossRef]

173. Nguyen, T.N.L.; Do, T.V.; Nguyen, T.V.; Dao, P.H.; Trinh, V.T.; Mac, V.P.; Nguyen, A.H.; Dinh, D.A.; Nguyen, T.A.; Vo, T.K.A.; et al. Antimicrobial activity of acrylic polyurethane/Fe$_3$O$_4$-Ag nanocomposite coating. *Prog. Org. Coat.* **2019**, *132*, 15–20. [CrossRef]

174. Chang, M.; Lin, W.S.; Xiao, W.; Chen, Y.N. Antibacterial Effects of Magnetically-Controlled Ag/Fe$_3$O$_4$ Nanoparticles. *Materials* **2018**, *11*, 659. [CrossRef]

175. Brollo, M.E.F.; Lopez-Ruiz, R.; Muraca, D.; Figueroa, S.J.A.; Pirota, K.R.; Knobel, M. Compact Ag@Fe$_3$O$_4$ Core-shell Nanoparticles by Means of Single-step Thermal Decomposition Reaction. *Sci. Rep.* **2014**, *4*, 6839–6845. [CrossRef] [PubMed]

176. Kim, M.; Kim, J. Synergistic interaction between pseudocapacitive Fe3O4 nanoparticles and highly porous silicon carbide for high-performance electrodes as electrochemical supercapacitors. *Nanotechnology* **2017**, *28*, 195401–195414.

177. Fan, H.; Niu, R.; Duan, J.; Liu, W.; Shen, W. Fe$_3$O$_4$@Carbon Nanosheets for All-Solid-State Supercapacitor Electrodes. *ACS Appl. Mater. Interfaces* **2016**, *8*, 19475–19483. [CrossRef] [PubMed]

178. Zeng, Z.; Zhao, H.; Wang, J.; Lv, P.; Zhang, T.; Xia, Q. Nanostructured Fe$_3$O$_4$@C as anode material for lithium-ion batteries. *J. Power Sources* **2014**, *248*, 15–21. [CrossRef]

179. Zhang, Z.; Kong, J. Novel magnetic Fe$_3$O$_4$@C nanoparticles as adsorbents for removal of organic dyes from aqueous solution. *J. Hazard. Mater.* **2011**, *193*, 325–329. [CrossRef] [PubMed]

180. Mao, G.Y.; Yang, W.J.; Bu, F.X.; Jiang, D.M.; Zhao, Z.J.; Zhang, Q.H.; Fang, Q.C.; Jiang, J.S. One-step hydrothermal synthesis of Fe$_3$O$_4$@C nanoparticles with great performance in biomedicine. *J. Mater. Chem. B* **2014**, *2*, 4481–4488. [CrossRef]

181. Da Costa, T.R.; Baldi, E.; Figueiró, A.; Colpani, G.L.; Silva, L.L.; Zanetti, M.; de Mello, J.M.M.; Fiori, M.A. Fe$_3$O$_4$@C core-shell nanoparticles as adsorbent of ionic zinc: Evaluating of the adsorptive capacity. *Mater. Res.* **2019**, *22* (Suppl. S1), e20180847. [CrossRef]

182. Hein, C.D.; Liu, X.M.; Wang, D. Click chemistry, a powerful tool for pharmaceutical sciences. *Pharm. Res.* **2008**, *25*, 2216–2230. [CrossRef]

183. Campidelli, S. Click Chemistry for Carbon Nanotubes Functionalization. *Curr. Org. Chem.* **2011**, *15*, 1151–1159. [CrossRef]

184. Fan, X.; Jiao, G.; Zhao, W.; Jin, P.; Li, X. Magnetic Fe$_3$O$_4$-graphene composites as targeted drug nanocarriers for pH-activated release. *Nanoscale* **2013**, *5*, 1143–1152. [CrossRef]

185. Gonzalez-Rodriguez, R.; Campbell, E.; Naumov, A. Multifunctional graphene oxide/iron oxide nanoparticles for magnetic targeted drug delivery dual magnetic resonance/fluorescence imaging and cancer sensing. *PLoS ONE* **2019**, *14*, e0217072. [CrossRef]

186. Namvari, M.; Namazi, H. Clicking graphene oxide and Fe$_3$O$_4$ nanoparticles together: An efficient adsorbent to remove dyes from aqueous solutions. *Int. J. Environ. Sci. Technol.* **2014**, *11*, 1527–1536. [CrossRef]

187. Arukali Sammaiah, A.; Huang, W.; Wang, X. Synthesis of magnetic Fe_3O_4/graphene oxide nanocomposites and their tribological properties under magnetic field. *Mater. Res. Express* **2018**, *5*, 105006–105015. [CrossRef]

188. Sadeghfar, F.; Ghaedi, M.; Asfaram, A.; Jannesar, R.; Javadian, H.; Pezeshkpour, V. Polyvinyl alcohol/Fe_3O_4@carbon nanotubes nanocomposite: Electrochemical-assisted synthesis, physicochemical characterization, optical properties, cytotoxicity effects and ultrasound-assisted treatment of aqueous based organic compound. *J. Ind. Eng. Chem.* **2018**, *65*, 349–362. [CrossRef]

189. Xu, Z.; Fan, X.; Ma, Q.; Tang, B.; Lu, Z.; Zhang, J.; Mo, G.; Ye, J.; Ye, J. A sensitive electrochemical sensor for simultaneous voltammetric sensing of cadmium and lead based on Fe_3O_4/multiwalled carbon nanotube/laser scribed graphene composites functionalized with chitosan modified electrode. *Mater. Chem. Phys.* **2019**, *238*, 121876–121877. [CrossRef]

190. Zhang, M.; Wang, W.; Cui, Y.; Chu, X.; Sun, B.; Zhou, N.; Shen, J. Magnetofluorescent Fe_3O_4/carbon quantum dots coated single-walled carbon nanotubes as dual-modal targeted imaging and chemo/photodynamic/photothermal triple-modal therapeutic agents. *Chem. Eng. J.* **2018**, *338*, 526–538. [CrossRef]

191. Zhang, W.; Li, X.; Zhou, R.; Wu, H.; Shi, H.; Yu, S.; Liu, Y. Multifunctional glucose biosensors from Fe_3O_4 nanoparticles modified chitosan/ graphene nanocomposites. *Sci. Rep.* **2015**, *5*, 11129–11138. [CrossRef]

192. Patsula, V.; Horák, D.; Kučka, J.; Mackova, H.; Lobaz, V.; Francova, P.; Herynek, V.; Heizer, T.; Paral, P.; Sefc, L. Synthesis and modification of uniform PEG-neridronate-modified magnetic nanoparticles determines prolonged blood circulation and biodistribution in a mouse preclinical model. *Sci. Rep.* **2019**, *9*, 10765–10777. [CrossRef]

193. Yuan, G.; Yuan, Y.; Xu, K.; Luo, Q. Biocompatible PEGylated Fe_3O_4 nanoparticles as photothermal agents for near-infrared light modulated cancer therapy. *Int. J. Mol. Sci.* **2014**, *15*, 18776–18788. [CrossRef]

194. Blanco, E.; Shen, H.; Ferrari, M. Principles of nanoparticle design for overcoming biological barriers to drug delivery. *Nat. Biotechnol.* **2015**, *33*, 941–951. [CrossRef]

195. Zhou, Y.; Peng, Z.; Seven, E.S.; Leblanc, R.M. Crossing the blood-brain barrier with nanoparticles. *J. Control. Release* **2018**, *270*, 290–303. [CrossRef]

196. Suk, J.S.; Xu, Q.; Kim, N.; Hanes, J.; Ensign, L.M. PEGylation as a strategy for improving nanoparticle-based drug and gene delivery. *Adv. Drug Deliv. Rev.* **2016**, *99 Pt A*, 28–51. [CrossRef]

197. Mostaghasi, E.; Zarepour, A.; Zarrabi, A. Folic acid armed Fe_3O_4-HPG nanoparticles as a safe nano vehicle for biomedical theranostics. *J. Taiwan Inst. Chem. Eng.* **2018**, *82*, 33–41. [CrossRef]

198. Avedian, N.; Zaaeri, F.; Daryasari, M.P.; Javar, H.A.; Khoobi, M. pH-sensitive biocompatible mesoporous magnetic nanoparticles labeled with folic acid as an efficient carrier for controlled anticancer drug delivery. *J. Drug Deliv. Sci. Technol.* **2018**, *44*, 323–332. [CrossRef]

199. Hashemi-Moghaddam, H.; Zavareh, S.; Gazi, E.M.; Jamili, M. Assessment of novel core–shell Fe_3O_4@poly l-DOPA nanoparticles for targeted Taxol® delivery to breast tumor in a mouse model. *Mater. Sci. Eng. C* **2018**, *93*, 1036–1043. [CrossRef] [PubMed]

200. Wu, C.Y.; Chen, Y.C. Riboflavin immobilized Fe_3O_4 magnetic nanoparticles carried with n-butylidenephthalide as targeting-based anticancer agents, Artificial Cells. *Nanomed. Biotechnol.* **2019**, *47*, 210–220. [CrossRef]

201. Arriortua, O.K.; Insausti, M.; Lezama, L.; de Muro, I.G.; Garaio, E.; de la Fuente, J.M.; Fratila, R.M.; Morales, M.P.; Costa, R.; Eceiza, M.; et al. RGD-Functionalized Fe_3O_4 nanoparticles for magnetic hyperthermia. *Colloids Surf. B Biointerfaces* **2018**, *165*, 315–324. [CrossRef] [PubMed]

202. Lu, X.; Jiang, R.; Fan, Q.; Zhang, L.; Zhang, H.; Yang, M.; Ma, Y.; Wang, L.; Huang, W. Fluorescent-magnetic poly(poly(ethyleneglycol)monomethacrylate)-grafted Fe_3O_4 nanoparticles from post-atom-transfer-radical-polymerization modification: Synthesis, characterization, cellular uptake and imaging. *J. Mater. Chem.* **2012**, *22*, 6965–6973. [CrossRef]

203. Stephen, Z.R.; Kievit, F.M.; Zhang, M. Magnetite Nanoparticles for Medical MR Imaging. *Mater. Today* **2011**, *14*, 330–338. [CrossRef]

204. Wang, Z.; Lam, A.; Acosta, E. Suspensions of Iron Oxide Nanoparticles Stabilized by Anionic Surfactants. *J. Surfactants Deterg.* **2013**, *16*, 397–407. [CrossRef]

205. Choi, Y.W.; Lee, H.; Song, Y.; Sohn, D. Colloidal stability of iron oxide nanoparticles with multivalent polymer surfactants. *J. Colloid Interface Sci.* **2015**, *443*, 8–12. [CrossRef]

206. Zhang, Y.; Newton, B.; Lewis, E.; Fu, P.P.; Kafoury, R.; Ray, P.C.; Yu, H. Cytotoxicity of organic surface coating agents used for nanoparticles synthesis and stability. *Toxicol. Vitr.* **2015**, *29*, 762–768. [CrossRef] [PubMed]

207. Soares, P.I.P.; Lochte, F.; Echeverria, C.; Pereira, L.C.J.; Coutinho, J.T.; Ferreira, I.M.M.; Novo, C.M.M.; Borges, J.P.M.R. Thermal and magnetic properties of iron oxide colloids: Influence of surfactants. *Nanotechnology* **2015**, *26*, 425704–425715. [CrossRef] [PubMed]

208. Gonzales, M.; Mitsumori, L.M.; Kushleika, J.V.; Rosenfeld, M.E.; Krishnan, K.M. Cytotoxicity of iron oxide nanoparticles made from the thermal decomposition of organometallics and aqueous phase transfer with Pluronic F127. *Contrast Media Mol. Imaging* **2010**, *5*, 286–293. [CrossRef] [PubMed]

209. Chen, M.; Shen, H.; Li, X.; Ruan, J.; Yan, W.Q. Magnetic fluids' stability improved by oleic acid bilayer-coated structure via one-pot synthesis. *Chem. Pap.* **2016**, *70*, 1642–1648. [CrossRef]

210. Coricovac, D.E.; Moacă, E.A.; Pinzaru, I.; Citu, C.; Soica, C.; Mihail, C.V.; Pacurariu, C.; Tutelyan, V.A.; Tsatsakis, A.; Dehelean, C.A. Biocompatible Colloidal Suspensions Based on Magnetic Iron Oxide Nanoparticles: Synthesis, Characterization and Toxicological Profile. *Front. Pharmacol.* **2017**, *8*, 154–172. [CrossRef] [PubMed]

211. Mulder, W.J.M.; Strijkers, G.J.; van Tilborg, G.A.F.; Cormode, D.P.; Fayad, Z.A.; Nicolay, K. Nanoparticulate assemblies of amphiphiles and diagnostically active materials for multimodality imaging. *Acc. Chem. Res.* **2009**, *42*, 904–914. [CrossRef]

212. Yang, J.; Pinar, A. Understanding the role of grafted polystyrene chain conformation in assembly of magnetic nanoparticles. *Phys. Rev. E* **2014**, *90*, 042601.

213. Nagesha, D.K.; Plouffe, B.D.; Phan, M.; Lewis, L.H.; Sridhar, S.; Murthy, S.K. Functionalization-induced improvement in magnetic properties of Fe_3O_4 nanoparticles for biomedical applications. *J. Appl. Phys.* **2009**, *105*, 07B317. [CrossRef]

214. Zhang, G.; Qie, F.; Hou, J.; Luo, S.; Luo, L.; Sun, X.; Tan, T. One-pot solvothermal method to prepare functionalized Fe_3O_4 nanoparticles for bioseparation. *J. Mater. Res.* **2012**, *27*, 1006–1013. [CrossRef]

215. Ooi, F.; DuChene, J.S.; Qiu, J.; Graham, J.O.; Engelhard, M.H.; Cao, G.; Gai, Z.; Wei, W.D. A Facile Solvothermal Synthesis of Octahedral Fe_3O_4 Nanoparticles. *Small* **2015**, *11*, 2649–2653. [CrossRef]

216. Kekalo, K.; Koo, K.; Zeitchick, E.; Baker, I. Microemulsion Synthesis of Iron Core/Iron Oxide Shell Magnetic Nanoparticles and Their Physicochemical Properties. *Mater. Res. Soc. Symp. Proc.* **2012**, *1416*, 1–11. [CrossRef] [PubMed]

217. Baharuddin, A.A.; Ang, B.C.; Hussein, N.A.A.; Andriyana, A.; Wong, Y.H. Mechanisms of highly stabilized ex-situ oleic acid-modified iron oxide nanoparticles functionalized with 4-pentynoic acid. *Mater. Chem. Phys.* **2018**, *203*, 212–222. [CrossRef]

218. Justin, C.; Samrot, A.V.; Sruthi, D.P.; Sahithya, C.S.; Bhavya, K.S.; Saipriya, C. Preparation, characterization and utilization of coreshell super paramagnetic iron oxide nanoparticles for curcumin delivery. *PLoS ONE* **2018**, *13*, e0200440. [CrossRef] [PubMed]

219. Mahdavi, M.; Ahmad, M.B.; Haron, M.J.; Namvar, F.; Nadi, B.; Rahman, M.Z.A.; Amin, J. Synthesis, Surface Modification and Characterisation of Biocompatible Magnetic Iron Oxide Nanoparticles for Biomedical Applications. *Molecules* **2013**, *18*, 7533–7548. [CrossRef] [PubMed]

220. Luchini, A.; Vitiello, G. Understanding the Nano-bio Interfaces: Lipid-Coatings for Inorganic Nanoparticles as Promising Strategy for Biomedical Applications. *Front. Chem.* **2019**, *7*, 343–359. [CrossRef]

221. Moghimi, S.M.; Szebeni, J. Stealth liposomes and long circulating nanoparticles: Critical issues in pharmacokinetics, opsonization and protein-binding properties. *Prog. Lipid Res.* **2003**, *42*, 463–478. [CrossRef]

222. Gogoi, M.; Jaiswal, M.K.; Dev Sarma, H.; Bahadur, D.; Banerjee, R. Biocompatibility and therapeutic evaluation of magnetic liposomes designed for self-controlled cancer hyperthermia and chemotherapy. *Integr. Biol.* **2017**, *9*, 555–565. [CrossRef]

223. Ramishetti, S.; Huang, L. Intelligent design of multifunctional lipid-coated nanoparticle platforms for cancer therapy. *Ther. Deliv.* **2012**, *3*, 1429–1445. [CrossRef]

224. Wijaya, A.; Hamad-Schifferli, K. High-Density Encapsulation of Fe_3O_4 Nanoparticles in Lipid Vesicles. *Langmuir* **2007**, *23*, 9546–9550. [CrossRef]

225. Liang, J.; Zhang, X.; Miao, Y.; Li, J.; Gan, Y. Lipid-coated iron oxide nanoparticles for dual-modal imaging of hepatocellular carcinoma. *Int. J. Nanomed.* **2017**, *12*, 2033–2044. [CrossRef]

226. Radoń, A.; Drygała, A.; Hawełek, L.; Łukowiec, D. Structure and optical properties of Fe₃O₄ nanoparticles synthesized by co-precipitation method with different organic modifiers. *Mater. Charact.* **2017**, *131*, 148–156. [CrossRef]

227. Anbarasu, M.; Anandan, M.; Chinnasamy, E.; Gopinath, V.; Balamurugan, K. Synthesis and characterization of polyethylene glycol (PEG) coated Fe₃O₄ nanoparticles by chemical co-precipitation method for biomedical applications. *Spectrochim. Acta Mol. Biomol. Spectrosc.* **2015**, *135*, 536–539. [CrossRef] [PubMed]

228. Yang, J.; Zou, P.; Yang, L.; Cao, J.; Sun, Y.; Han, D.; Yang, S.; Wang, Z.; Chen, G.; Wang, B.; et al. A comprehensive study on the synthesis and paramagnetic properties of PEG-coated Fe₃O₄ nanoparticles. *Appl. Surf. Sci.* **2014**, *303*, 425–432. [CrossRef]

229. Gao, G.; Qiu, P.; Qian, Q.; Zhou, N.; Wang, K.; Song, H.; Fu, H.; Cui, D. PEG-200-assisted hydrothermal method for the controlled-synthesis of highly dispersed hollow Fe₃O₄ nanoparticles. *J. Alloys Compd.* **2013**, *574*, 340–344. [CrossRef]

230. Wang, R.; Degirmenci, V.; Xin, H.; Li, Y.; Wang, L.; Chen, J.; Hu, X.; Zhang, D. PEI-Coated Fe₃O₄ Nanoparticles Enable Efficient Delivery of Therapeutic siRNA Targeting REST into Glioblastoma Cells. *Int. J. Mol. Sci.* **2018**, *19*, 2230. [CrossRef]

231. Ping, T.; Wang, Q.; Zhou, Y.; Nie, J. Reducing oxygen inhibition by Fe₃O₄@PEI nanoparticles co-initiator. *J. Photochem. Photobiol. Chem.* **2019**, *373*, 171–175. [CrossRef]

232. Sun, X.; Zheng, C.; Zhang, F.; Yang, Y.; Wu, G.; Yu, A.; Guan, N. Size-controlled synthesis of magnetite (Fe₃O₄) nanoparticles coated with Glucose and Gluconic Acid from a Single Fe(III) Precursor by a Sucrose Bifunctional Hydrothermal Method. *J. Phys. Chem. C* **2009**, *113*, 16002–16008. [CrossRef]

233. Sari, A.Y.; Eko, A.S.; Candra, K.; Hasibuan, D.P.; Ginting, M.; Sebayang, P.; Simamora, P. Synthesis, Properties and Application of Glucose Coated Fe₃O₄ Nanoparticles Prepared by Co-precipitation Method. *IOP Conf. Ser. Mater. Sci. Eng.* **2017**, *214*, 012021. [CrossRef]

234. Barbaro, D.; Di Bari, L.; Gandin, V.; Evangelisti, C.; Vitulli, G.; Schiavi, E.; Marzano, C.; Ferretti, A.M.; Salvadori, P. Glucose-coated superparamagnetic iron oxide nanoparticles prepared by metal vapour synthesis are electively internalized in a pancreatic adenocarcinoma cell line expressing GLUT1 transporter. *PLoS ONE* **2015**, *10*, e0123159. [CrossRef]

235. Predescu, A.M.; Matei, E.; Berbecaru, A.C.; Pantilimon, C.; Dragan, C.; Vidu, R.; Predescu, C.; Kuncser, V. Synthesis and characterization of dextran-coated iron oxide nanoparticles. *R. Soc. Open Sci.* **2018**, *5*, 171525–171536. [CrossRef]

236. Zhang, Q.; Liu, Q.; Du, M.; Vermorken, A.; Cui, Y.; Zhang, L.; Guo, L.; Ma, L.; Chen, M. Cetuximab and Doxorubicin loaded dextran-coated Fe₃O₄ magnetic nanoparticles as novel targeted nanocarriers for non-small cell lung cancer. *J. Magn. Magn. Mater.* **2019**, *481*, 122–128. [CrossRef]

237. Shen, M.; Yu, Y.; Fan, G.; Jin, Y.M.; Tang, W.; Jia, W. The synthesis and characterization of monodispersed chitosan-coated Fe₃O₄ nanoparticles via a facile one-step solvothermal process for adsorption of bovine serum albumin. *Nanoscale Res. Lett.* **2014**, *9*, 296–304. [CrossRef] [PubMed]

238. Veisi, H.; Sajjadifar, S.; Biabri, P.M.; Hemmati, S. Oxo-vanadium complex immobilized on chitosan coated-magnetic nanoparticles (Fe₃O₄): A heterogeneous and recyclable nanocatalyst for the chemoselective oxidation of sulfides to sulfoxides with H₂O₂. *Polyhedron* **2018**, *153*, 240–247. [CrossRef]

239. Lotfi, S.; Bahari, A.; Mahjoub, S. In vitro biological evaluations of Fe₃O₄compared with core–shell structures of chitosan-coated Fe₃O₄ and polyacrylic acid-coated Fe₃O₄ nanoparticles. *Res. Chem. Intermed.* **2019**, *45*, 3497–3512. [CrossRef]

240. Illés, E.; Szekeres, M.; Tóth, I.Y.; Farkas, K.; Foeldesi, I.; Szabo, A.; Ivan, B.; Tombacz, E. PEGylation of Superparamagnetic Iron Oxide Nanoparticles with Self-Organizing Polyacrylate-PEG Brushes for Contrast Enhancement in MRI Diagnosis. *Nanomaterials* **2018**, *8*, 776. [CrossRef]

241. Illés, E.; Szekeres, M.; Tóth, I.Y.; Szabó, Á.; Iván, B.; Turcu, R.; Vékás, L.; Zupkó, I.; Jaics, G.; Tombácz, E. Multifunctional PEG-carboxylate copolymer coated superparamagnetic iron oxide nanoparticles for biomedical application. *J. Magn. Magn. Mater.* **2018**, *451*, 710–720. [CrossRef]

242. Iglesias, G.R.; Reyes-Ortega, F.; Checa Fernandez, B.L.; Delgado, Á.V. Hyperthermia-Triggered Gemcitabine Release from Polymer-Coated Magnetite Nanoparticles. *Polymers* **2018**, *10*, 269. [CrossRef]

243. Dutta, B.; Shetake, N.G.; Gawali, S.L.; Barick, B.K.; Barick, K.C.; Babu, P.D.; Pandey, B.N.; Priyadarsini, K.I.; Hassan, P.A. PEG mediated shape-selective synthesis of cubic Fe₃O₄ nanoparticles for cancer therapeutics. *J. Alloys Compd.* **2018**, *737*, 347–355. [CrossRef]

244. You, L.; Liu, X.; Fang, Z.; Xu, Q.; Zhang, Q. Synthesis of multifunctional Fe$_3$O$_4$@PLGA-PEG nano-niosomes as a targeting carrier for treatment of cervical cancer. *Mater. Sci. Eng.* **2019**, *94*, 291–302. [CrossRef]

245. Sun, X.; Shen, J.; Yu, D.; Ouyang, X. Preparation of pH-sensitive Fe$_3$O$_4$@C/carboxymethyl cellulose/chitosan composite beads for diclofenac sodium delivery. *Int. J. Biol. Macromol.* **2019**, *127*, 594–605. [CrossRef]

246. Sakaguchi, M.; Makino, M.; Ohura, T.; Yamamoto, K.; Enomoto, Y.; Takase, H. Surface modification of Fe$_3$O$_4$ nanoparticles with dextran via a coupling reaction between naked Fe$_3$O$_4$ mechano-cation and naked dextran mechano-anion: A new mechanism of covalent bond formation. *Adv. Powder Technol.* **2019**, *30*, 795–806. [CrossRef]

247. Wang, F.; Li, X.; Li, W.; Bai, H.; Gao, Y.; Ma, J.; Liu, W.; Xi, G. Dextran coated Fe$_3$O$_4$ nanoparticles as a near-infrared laser-driven photothermal agent for efficient ablation of cancer cells in vitro and in vivo. *Mater. Sci. Eng.* **2018**, *90*, 46–56. [CrossRef] [PubMed]

248. Banobre-Lopez, M.; Pineiro-Redondo, Y.; Sandri, M.; Tampieri, A.; De Santis, R.; Dediu, V.A.; Rivas, J. Hyperthermia Induced in Magnetic Scaffolds for Bone Tissue Engineering. *IEEE Trans. Magn.* **2014**, *50*, 1–7. [CrossRef]

249. Lai, W.Y.; Feng, S.W.; Chan, Y.H.; Chang, W.J.; Wang, H.T.; Huang, H.M. In Vivo Investigation into Effectiveness of Fe$_3$O$_4$/PLLA Nanofibers for Bone Tissue Engineering Applications. *Polymers* **2018**, *10*, 804. [CrossRef]

250. Maier-Hauff, K.; Rothe, R.; Scholz, R.; Gneveckow, U.; Wust, P.; Thiesen, B.; Feussner, A.; von Deimling, A.; Waldoefner, N.; Felix, R.; et al. Intracranial thermotherapy using magnetic nanoparticles combined with external beam radiotherapy: Results of a feasibility study on patients with glioblastoma multiforme. *J. Neurooncol.* **2007**, *81*, 53–60. [CrossRef]

251. Maier-Hauff, K.; Ulrich, F.; Nestler, D.; Niehoff, H.; Wust, P.; Thiesen, B.; Orawa, H.; Budach, V.; Jordan, A. Efficacy and safety of intratumoral thermotherapy using magnetic iron-oxide nanoparticles combined with external beam radiotherapy on patients with recurrent glioblastoma multiforme. *J. Neurooncol.* **2011**, *103*, 317–324. [CrossRef]

252. Grauer, O.; Jaber, M.; Hess, K.; Weckesser, M.; Schwindt, W.; Maring, S.; Woelfer, J.; Stummer, W. Combined intracavitary thermotherapy with iron oxide nanoparticles and radiotherapy as local treatment modality in recurrent glioblastoma patients. *J. Neurooncol.* **2019**, *141*, 83–84. [CrossRef]

253. Chatterjee, D.K.; Diagaradjane, P.; Krishnan, S. Nanoparticle-mediated hyperthermia in cancer therapy. *Ther. Deliv.* **2011**, *2*, 1001–1014. [CrossRef]

254. Goya, G.F.; Lima, E., Jr.; Arelaro, A.D.; Torres, T.E.; Rechenberg, H.R.; Rossi, L.; Marquina, C.; Ibarra, M.R. Magnetic Hyperthermia with Fe$_3$O$_4$ nanoparticles: The Influence of Particle Size on Energy Absorption. *IEEE Trans. Magn.* **2008**, *44*, 4444–4447. [CrossRef]

255. Kaur, P.; Aliru, M.L.; Chadha, A.S.; Asea, A.; Krishnan, S. Hyperthermia using nanoparticles—Promises and pitfalls. *Int. J. Hyperth.* **2016**, *32*, 76–88. [CrossRef]

256. Temelie, M.; Popescu, R.C.; Cocioaba, D.; Vasile, B.S.; Savu, D. Biocompatibility study of magnetic nanoparticles synthesized using a green method. *Rom. J. Phys.* **2018**, *63*, 703–716.

257. Rasouli, E.; Basirun, W.J.; Rezayi, M.; Shameli, K.; Nourmohammadi, E.; Khandanlou, R.; Izadiyan, Z.; Sarkarizi, H.K. Ultrasmall superparamagnetic Fe$_3$O$_4$ nanoparticles: Honey-based green and facile synthesis and in vitro viability assay. *Int. J. Nanomed.* **2018**, *2018*, 6903–6911. [CrossRef] [PubMed]

258. Sathishkumar, G.; Logeshwaran, V.; Sarathbabu, S.; Jha, P.K.; Jeyaraj, M.; Rajkuberan, C.; Senthilkumar, N.; Sivaramakrishnan, S. Green synthesis of magnetic Fe$_3$O$_4$ nanoparticles using Couroupitaguianensis Aubl. fruit extract for their antibacterial and cytotoxicity activities. *Artif. Cells Nanomed. Biotechnol.* **2018**, *46*, 589–598. [CrossRef] [PubMed]

259. Chifiriuc, C.; Grumezescu, V.; Grumezescu, A.M.; Saviuc, C.; Lazăr, V.; Andronescu, E. Hybrid magnetite nanoparticles/Rosmarinus officinalis essential oil nanobiosystem with antibiofilm activity. *Nanoscale Res. Lett.* **2012**, *7*, 209–216. [CrossRef]

260. Rădulescu, M.; Andronescu, E.; Holban, A.M.; Vasile, B.S.; Iordache, F.; Mogoantă, L.; Mogoșanu, G.D.; Grumezescu, A.M.; Georgescu, M.; Chifiriuc, M.C. Antimicrobial Nanostructured Bioactive Coating Based on Fe$_3$O$_4$ and Patchouli Oil for Wound Dressing. *Metals* **2016**, *6*, 103. [CrossRef]

261. Sadeghzadeh, H.; Pilehvar-Soltanahmadi, Y.; Akbarzadeh, A.; Dariushnejad, H.; Sanjarian, F.; Zarghami, N. The Effects of Nanoencapsulated Curcumin-Fe$_3$O$_4$ on Proliferation and hTERT Gene Expression in Lung Cancer Cells. *Anti-Cancer Agents Med. Chem.* **2017**, *17*, 1363–1373. [CrossRef]

262. Ruíz-Baltazar, A.J.; Reyes-López, S.Y.; Mondragón-Sánchez, M.L.; Robles-Cortés, A.I.; Pérez, R. Eco-friendly synthesis of Fe_3O_4 nanoparticles: Evaluation of their catalytic activity in methylene blue degradation by kinetic adsorption models. *Res. Phys.* **2019**, *12*, 989–995. [CrossRef]

263. Yew, Y.P.; Shameli, K.; Miyake, M.; Kuwano, N.; Khairudin, N.B.B.A.; Mohamad, S.E.B.; Lee, K.X. Green Synthesis of Magnetite (Fe_3O_4) Nanoparticles Using Seaweed (Kappaphycus alvarezii) Extract. *Nanoscale Res. Lett.* **2016**, *11*, 276–283. [CrossRef]

264. Hosseini, A.; Ghorbani, A. Cancer therapy with phytochemicals: Evidence from clinical studies. *Avicenna J. Phytomed.* **2015**, *5*, 84–97.

265. Shivani Thoidingjam, S.; Tiku, A.B. Therapeutic efficacy of Phyllanthus emblica-coated iron oxide nanoparticles in A549 lung cancer cell line. *Nanomedicine* **2019**, *14*, 2355–2371. [CrossRef]

266. Ramirez-Nuñez, A.L.; Jimenez-Garcia, L.F.; Goya, G.F.; Sanz, B.; Santoyo-Salazar, J. In vitro magnetic hyperthermia using polyphenol-coated $Fe_3O_4@\gamma Fe_2O_3$ nanoparticles from Cinnamomun verumand Vanilla planifolia: The concert of green synthesis and therapeutic possibilities. *Nanotechnology* **2018**, *29*, 074001. [CrossRef] [PubMed]

267. Lewinska, A.; Adamczyk-Grochala, J.; Bloniarz, D.; Olszowka, J.; Kulpa-Greszta, M.; Litwinienko, G.; Tomaszewska, A.; Wnuk, M.; Pazik, R. AMPK-mediated senolytic and senostatic activity of quercetin surface functionalized Fe_3O_4 nanoparticles during oxidant-induced senescence in human fibroblasts. *Redox Biol.* **2020**, *28*, 101337. [CrossRef] [PubMed]

268. Ebrahimpour, S.; Esmaeili, A.; Beheshti, S. Effect of quercetin-conjugated superparamagnetic iron oxide nanoparticles on diabetes-induced learning and memory impairment in rats. *Int. J. Nanomed.* **2018**, *2018*, 6311–6324. [CrossRef] [PubMed]

269. Dorniani, D.; Hussein, M.Z.; Kura, A.U.; Fakurazi, S.; Shaari, A.H.; Ahmad, Z. Preparation of Fe_3O_4 magnetic nanoparticles coated with gallic acid for drug delivery. *Int. J. Nanomed.* **2012**, *7*, 5745–5756. [CrossRef] [PubMed]

270. Li, L.; Gao, F.; Jiang, W.; Wu, X.; Cai, Y.; Tang, J.; Gao, X.; Gao, F. Folic acid-conjugated superparamagnetic iron oxide nanoparticles for tumor-targeting MR imaging. *Drug Deliv.* **2016**, *23*, 1726–1733. [CrossRef] [PubMed]

271. Choi, K.H.; Nam, K.C.; Cho, G.; Jung, J.S.; Park, B.J. Enhanced Photodynamic Anticancer Activities of Multifunctional Magnetic Nanoparticles (Fe_3O_4) Conjugated with Chlorin e6 and Folic Acid in Prostate and Breast Cancer Cells. *Nanomaterials* **2018**, *8*, 722. [CrossRef]

272. Popescu, R.C.; Andronescu, E.; Vasile, B.S.; Trusca, R.; Boldeiu, A.; Mogoanta, L.; Mogosanu, G.D.; Temelie, M.; Radu, M.; Grumezescu, A.M.; et al. Fabrication and Cytotoxicity of Gemcitabine-Functionalized Magnetite Nanoparticles. *Molecules* **2017**, *22*, 1080. [CrossRef]

273. Zhang, Q.; Liu, J.; Yuan, K.; Zhang, Z.; Zhang, X.; Fang, X. A multi-controlled drug delivery system based on magnetic mesoporous Fe_3O_4 nanopaticles and a phase change material for cancer thermo-chemotherapy. *Nanotechnology* **2017**, *28*, 405101–405131. [CrossRef]

274. Xia, K.; Lyu, Y.; Yuan, W.; Wang, G.; Stratton, H.; Zhang, S.; Wu, J. Nanocarriers of Fe_3O_4as a Novel Method for Delivery of the Antineoplastic Agent Doxorubicin into HeLa Cells in vitro. *Front. Oncol.* **2019**, *9*, 250–257. [CrossRef]

275. Hu, X.; Wang, Y.; Zhang, L.; Xu, M.; Zhang, J.; Dong, W. Dual-pH/Magnetic-Field-Controlled Drug Delivery Systems Based on $Fe_3O_4@SiO_2$-Incorporated Salecan Graft Copolymer Composite Hydrogels. *ChemMedChem* **2017**, *12*, 1600–1609. [CrossRef]

MDPI

St. Alban-Anlage 66

4052 Basel

Switzerland

Tel. +41 61 683 77 34

Fax +41 61 302 89 18

www.mdpi.com

Nanomaterials Editorial Office

E-mail: nanomaterials@mdpi.com

www.mdpi.com/journal/nanomaterials

CPSIA information can be obtained
at www.ICGtesting.com
Printed in the USA
LVHW070202180820
663483LV00011B/812

9 783039 289677